No Longer Property of
EWU Libraries

Urban Regional Economics

Concepts, Tools, Applications

Urban Regional Economics

Concepts, Tools, Applications

■

Wilbur R. Maki and Richard W. Lichty

Iowa State University Press / Ames

WILBUR R. MAKI, PH.D., is Professor Emeritus, Department of Applied Economics, University of Minnesota. Dr. Maki taught and directed research projects on state and regional economic issues. His work focused on building U.S. county-level and sub-county database systems for assessing industry employment, transportation, and related infrastructure requirements of local economies engaged in domestic and foreign trade. He was Minnesota State Economist in the mid-eighties and a member of various economics advisory groups in state and local government.

RICHARD W. LICHTY, PH.D., is a Distinguished Teaching Professor of Business and Economics, University of Minnesota, Duluth, and Research Director of the Center for Economic Development. He is past president of the Mid-Continent Regional Science Association and serves on the Board of Editors for that association's journal. Throughout a teaching career of over three decades, Dr. Lichty's primary focus has been education at the undergraduate level.

Iowa State University Press
2121 South State Avenue, Ames, Iowa 50014

Orders: 1-800-862-6657
Office: 1-515-292-0140
Fax: 1-515-292-3348
Web site: www.isupress.edu

♾ Printed on acid-free paper in the United States of America

First edition, 2000

Library of Congress Cataloging-in-Publication Data
Maki, Wilbur R.
Urban regional economics : concepts, tools, applications / Wilbur R. Maki and Richard W. Lichty.
p. cm.
Includes index.
ISBN 0-8138-2679-9
1. Urban economics. 2. Regional economics. I. Lichty, Richard W. II. Title.

HT321 .M23 2000
330.9173'2—dc21 99-059767

The last digit is the print number: 9 8 7 6 5 4 3 2 1

To Karl A. Fox, who developed the early outlines and content of urban regional economics and the building of functional and viable local neighborhoods and communities linked together in the pursuit of common needs and goals.

Contents

Preface

Urban regional economics is a new area in which economists can specialize. From its inception, it has been a heavily applied part of the economics discipline, but with a good grounding in basic economics theories and methods.

Urban regional economics owes much to the regional sciences because it extends their basic principles in order to deal with our most critical and troublesome issues. Some examples of urban regional economics concerns include nurturing the new engines of economic growth in the metropolitan areas most closely linked to each other on a global scale, strengthening the trading linkages between these areas and the urban and rural areas and communities beyond their borders that together form the new economic regions, and bolstering a new civic awareness of these measures and others that make democracy work within each of our many functional economic communities. A great strength of the regional sciences is their dedication to interdisciplinary work. We support these efforts and expand this work as we address these important issues.

We approach our immediate task from a somewhat different perspective. We include some of the traditional theories and models, but as points of departure. Many of our examples are from very practical applications of the theories and models of urban regional economics. Most of these stem from our own past research, often sponsored by local government and private entities. Topics such as the economic effects of local monopolies in air transportation, the pros and cons of major league team owners demanding even greater subsidies to keep their teams in place, and the need to make communities work by nurturing strong neighborhoods with good access to jobs and services are included in this immediate task.

We begin with what is a somewhat traditional look at regional delineation around the concept and the reality of local labor market areas. We try to establish why such a view of the geography of economic activity is preferred to others. Our main efforts along these lines include the following:

1. While paying initial attention to location and land use, we shift our focus to local markets, central places, exports, and the economic base of a locality, and to how each of these is supported by strategically located infrastructure and transfer systems.
2. Our discussion of tools is confined to the simplest and most useful for the sorts of demands that emerge in the study of urban regional economies.
3. Our presentation of applications covers the gamut of strategic local and regional issues relating to the role and importance of central cities and downtown districts as transportation and communications nodes of global transfer systems; of other urban labor market areas outside the central cities and their downtown districts that are the most numerous of functional economic communities; and of rural areas and communities, which are now facing critical problems of survival as many government programs are put to rest.

In short, we apply the concepts and tools of economics that have much to offer. We also cover the new literature, including writers such as Michael Porter on the competitive advantage of nations and regions, Robert Putnam on the decline of civility and civic awareness, and others who are influencing the future of problem-focused economics thought and practice.

Urban Regional Economics

Concepts, Tools, Applications

1

Introduction

Urban and regional economics are new branches of economics. The newness stems, in part, from a long-standing belief among many economists that area, distance, and related problems of migration and trade cancel at the national level. One region's losses are another region's gains. The nation's comparative advantage, according to this view, is an aggregate of its productive capabilities—the contributions of all remuneratively employed human, natural, and physical resources to the nation's gross domestic product. However, the realities of location, organizational scale, productivity of the employed local resources, and their consequences in providing one locality with a comparative advantage over another in some form of productive activity temper this view.

A local labor market, centered on a major metropolitan area, for example, offers a favorable environment for business start-ups and expansion, certainly more favorable than a local labor market in a distant, sparsely populated rural area. Newly established metropolitan area businesses, as well as long-established expanding enterprises, find an abundance of workers nearby. These workers are ready and able to work for these businesses, with both employers and employees secure in the knowledge that other employment opportunities exist or will emerge quickly if the current one fails to meet the expectations of either party. When these businesses—of which there are many—produce highly differentiated, high-order, high value-added products, they generate high earnings for their workers and owners.

The more distant rural area labor markets lack the diversity and scale to support a variety of business enterprises. Start-up businesses are few, and business expansions are limited by the size of the economic base, that is, the area's export-producing industries. This is changing for some areas as new systems of communication and transportation allow even the remotest businesses to become closely linked to the institutions and activities of major metropolitan centers. For some transitional rural areas—those between the major metropolitan areas and the distant rural areas—industry growth may be even greater than in the metropolitan areas. This location advantage often occurs because of lower site and labor costs.

Local labor markets serve to allocate labor resources in a commuting area to their "highest and best use." An urban real estate market serves to allocate the available space for residential, commercial, industrial, and public uses. Local product markets also serve this idyllic purpose, but with even greater differentiation than the markets for land and labor, both in product variety and in ways of bringing together buyers and sellers. Some local markets, like the ones for computers, are simply a part of the corresponding national and international product markets, with much competition from other producing areas or supply sources and other locally available products. Other local markets, like air transportation, function more as spatial monopolies than as truly competitive markets, exercising much control over the supply and pricing of services provided. Each local market operates in its unique local environment, adapting to differences in local regulations, community awareness, community organization, and inter-area competition.

A simple model of a local economy that starts with the production of two commodities, such as wheat and cloth, with equal productivity among residents, with no economies of size, and with travel only by foot, represents a frequently used textbook approach to the study of local economies. It contrasts sharply, however, with the reality of economic activity in any (except the most primitive) rural area. The coexistence of a wheat field with a manufacturing plant is perhaps more useful as a model of urban overspill into a rural periphery than of an isolated rural economy.

Overview of Urban Regional Economics

Urban regional economics addresses various specific questions in a present-day setting that are neglected by other fields of inquiry. Examples include the issue of local labor markets—their formation and organization, and access to them; or of local monopolies, such as a single business restricting access to transportation facilities and services for moving people and products from one central place to another. Even the location of a particular activity at one site rather than another becomes a level of question addressed by urban regional economic modeling. What accounts for the existing business locations? What differences in local income and wealth are attributed to these businesses, and what difference, if any, does all this make to the nation's gross domestic product? Can local governments interested in economic growth and development influence business locations? These types of questions motivated economists to become interested in the geographic aspects of economic activity.

Location Economics and Role of Transportation

Interest in business and industry location soon led to a coincident interest in transportation issues. Changes in transportation technology prompted the exploration of new issues in urban and regional economics. For example, newly built transportation facilities helped the movement of people and goods from the inner city to the suburbs. This exodus from downtown contributed to the increasingly severe problems of poverty and crime in the core cities. Further, the tax base for financing remedial measures in the core areas eroded as a result of the exodus to suburban fringes.

The exodus of wealthier households left the urban center with a large number of unemployed workers or workers employed at low wages, while at the same time the downtown district attracted many high value-added businesses employing large numbers of high-salaried suburban residents. Another result of the exodus was an increasing demand for public services in the city center, but without the financing available to cover the increasing needs. This led to the strong predilection toward formal analysis and planning, often without measurable results.

Professional concern over the lack of useful and timely information for dealing with the increasingly severe problems of urban and regional communities created a coincident demand for economic data collection and data reporting systems. Effective statistical gathering and reporting techniques were available for many aspects of the national economy. But the federal government does not provide the detailed data for local economies that it offers for the national economy.

The dramatic changes of the last quarter century now call for a close look at the causes and sources of these changes. Both the causes of change and consequences of change differ greatly from one region to another. For example, the post–World War II rural-to-urban migration subsided in the 1970s. There was even a turnaround in some parts of the country during this period. While improved agricultural technology reduced the demand for farm labor, nonagricultural industries in the metropolitan areas readily employed the new migrants from the rural, agricultural areas. The metropolitan areas soon became the hotbeds of new businesses created from the human capital of diverse backgrounds, skills, and capabilities. This mix of new resources brought forth a tremendous variety of readily available products for an increasingly affluent resident population.

Advanced Urban Economies and Exports

The rural-to-urban migration represented a shift in employment and business opportunities from natural resource-based industries and commodity production to many different kinds of consumer and producer services and manufacturing. The high-order, high value-added services and manufacturing became concentrated in the major metropolitan core areas. They represent the principal export-producing sectors of advanced urban economies. Product specialization evolved in industries demonstrating a comparative advantage over similar industries competing in export markets. Intellectual property, rather than land and other forms of real property, eventually became the new measure of power and wealth.

Economic linkages between and within the core area and its periphery are an indispensable part of the well-being of both areas. These linkages take many forms, such as the purchases of material inputs by manufacturers in one area from those in another. They include the purchases of banking and accounting services by businesses in rural areas from businesses located in nearby metropolitan areas, and information transfers by specialized information-generating and information-using enterprises in both rural and metropolitan areas.

Linkages also include the various means of income redistribution, a principal function of governments, particularly the federal government. Income redistribution carries sets of costs, including maintaining administrative structures for revenue collection and the many forms of regulation. Generally, the rural areas and dominantly rural states appear to be the net beneficiaries of these transfers, which occur largely at the expense of the more prosperous segments of our society. The greater ease of collecting direct tax revenues at the federal than at the state and local levels of government, and the increasingly difficult task of sustaining local government functions from property tax revenues, leaves the federal government as the principal revenue-collecting agency. Meanwhile, state and local governments incur an increasingly greater responsibility in the provision of essential government services.

The high-order, high value-added services and manufacturing are now concentrated in the major metropolitan core areas. They are the principal export-producing sectors—that is, the economic base—of these advanced urban economies. They account for an increasingly larger share of the "new dollars" coming into these areas. Much of the manufacturing and many of the services formerly concentrated in the downtown districts of the metropolitan core areas are now moving to the less densely populated suburbs, which offer less expensive sites for the newly relocated facilities. These relocating businesses often face critical labor shortages, however, because of the lack of adequate housing at these new sites. For some areas, "reverse commuting" on the part of the central city labor force thus becomes the only remaining alternative for addressing the emerging metropolitan area labor shortages.

The concentration of the specialized, high value-added services in metropolitan core areas is a tremendous responsibility for local governments, given the importance of the services sector as a new engine of regional economic growth. Yet, a diminishing tax base relative to the growing public costs of maintaining the essential infrastructure of the metropolitan core areas, particularly for the support and reorganization of the local K-through-12 public education systems, imposes severe restrictions on the fiscal capacity of metropolitan area governments. These governments and their constituencies provide an essential economic environment for the new engines of economic growth. They must also share a large part of the available resources with the local and state governments of nonmetropolitan areas and their constituencies.

How We Define a Region

Perhaps the best place to resume our overview is by defining what we mean by "city" or "region." Whether we are talking about an urban region, a rural region, or even a multistate region, some basis for establishing territorial dimensions must be established. Before we can discuss regional processes, regional development, or any of the other issues contained in this book, we must understand what constitutes a region and its important elements and constituencies.

Administrative Regions

A commonly used definition of a region is based on the limits of some administrative authority. A school district is bounded by the limits of authority for a local school board, a city by its city government, a county by some county board, and so on. The administrative region defined by the limits of its administrative authority is a less than satisfactory definition for most economic analyses. By limiting a definition to administrative boundaries, we often ignore the influence of suburbs and peripheral areas, both of which are affected by and exert major economic influences on the administrative city. For example, by limiting analysis to a region inside a state's borders, we ignore the influence of activities in adjoining states. Further, it is not uncommon to find administrative jurisdictions overlapping one another. Such overlapping authorities make clear administrative definitions difficult.

If the definition based on administrative authority is generally unsatisfactory, why do we use it so often? First, many studies are funded by administrative entities. These entities often find it hard to justify an analysis that would potentially help competing regions gain an informational edge. Further, many regional data are collected and reported on an administrative basis, dictated in part by the measures enacted for income redistribution. Analyses dependent on such "secondary" data are often constrained to use administrative boundaries, primarily those of state and local governments.

Finally, except for overlapping authorities, administrative boundaries are among the easiest to draw on a map. The limits of an administrative region are readily explained to the lay public. Administrative boundaries are what the general population thinks of when they think of a region.

The U.S. Census uses several definitions that include administrative criteria. An *urban place* is defined as a region with an incorporated or unincorporated identity with a population of at least 2,500 individuals. There is really nothing more to the definition than this. A municipality is simply defined as a city bounded by the limits of authority for a municipal corporation. Local government is formed by the conditions outlined in the municipal charter (or its equivalent), and this government is charged with providing certain urban services, such as police and fire protection. This is also what we generally think of when we think of a city.

Contiguous Regions

Contiguous definitions fill a larger space, with smaller territorially differentiated entities. Groups of counties often form the basis for regions based on contiguous considerations. We shall soon see that combinations of metropolitan areas are also often used for contiguous-based regional definitions.

The U.S. Census has some definitions based on contiguous properties. Because these definitions are popular with the Census Bureau, a primary source of information for regional analysis, they are also quite popular in regional research and analyses. We will present three of these, namely, the *urbanized area,* the *metropolitan area,* and the *consolidated metropolitan statistical area.*

An urbanized area (UA) contains a large municipality (defined above) and a surrounding region in which there are population densities exceeding 1,000 people per acre. The region must have a total population of at least 50,000 individuals. This definition draws a picture of a municipality surrounded by what is termed the urban fringe. The urban fringe and the metropolitan core are two contiguous regions that make up the larger region, the urbanized area.

A metropolitan area (MA) includes a core population cluster (either an urbanized area or a municipality) that is relatively large. The general cutoff for size is a core cluster containing at least 50,000 individuals. The region is then defined as this core cluster and surrounding, integrated counties. The primary criterion for integration is commuting patterns. In other words, an integrated county is one in which a large number of individuals (at least 50 percent of the population) commute to the core county for their livelihood. There are other criteria as well (population density requirements and percentage of population pursuing nonagricultural occupations, for example). These will not be detailed here. If the core city does not have a population of 50,000 but the total population of the integrated area is 100,000, the region may still be defined as a metropolitan area.

A metropolitan area is defined in terms of contiguous counties. The core county contains the dominant population cluster. Other counties are then included in the definition based on commuting flows, discussed above. The county basis for definition creates problems in some regions, particularly those in the western part of the United States, where counties can be quite large. Because whole counties are included, there may be a substantial rural component to this "urban" definition. Metropolitan areas are important because they form the basis for much detailed economic information provided at the local level by the U.S. Department of Commerce. Studies dependent on already-collected (secondary) data of this type are quick to use the detail provided at this level of definition.

A consolidated metropolitan statistical area (CMSA) is a metropolitan area with a population exceeding one million. It is consolidated if it contains two or more *primary metropolitan statistical areas* (PMSAs). An MSA is simply a metropolitan area that is not a part of a CMSA. No integration characteristics are required if PMSAs are to be included in the broader CMSA. Further, all of the counties in the CMSA are contiguous. In the 1990 census, there were 396 urbanized areas, 284 metropolitan areas, and twenty consolidated metropolitan areas.

We compare the economic structure of three MSAs (including two PMSAs)—Washington, D.C., and environs; Minneapolis-St. Paul; and Detroit—using measures of total employment in "one-digit" industry groups to show differences in the distribution of industries in each area and their rates of change over the past quarter century (Table 1-1). These are the 1994 MSAs. Twenty-five units of local government—five counties in Maryland, eleven counties and six cities in Virginia, and two counties in West Virginia, along with the District of Columbia, formed the Washington, D.C., PMSA. Thirteen counties—eleven in Minnesota and two in Wisconsin—formed the Minneapolis-St. Paul MSA. Six counties—all in Michigan—formed the Detroit MSA. Total population serves as a reference measure for comparing levels and changes in the ratio of employment to population. For two of the three metropolitan areas, total population grew more rapidly than the nation as a whole—a pattern that correlates with above-average employment growth. The District of Columbia experienced

TABLE 1-1 Total employment in specified industry, by place of work, 1969 and 1994

Industry	D.C. MSA		MnStP MSA		Detroit MSA		United States		Total Change, 1969–1994			
	1969 (thou.)	1994 (thou.)	1969 (thou.)	1994 (thou.)	1969 (thou.)	1994 (thou.)	1969 (thou.)	1994 (thou.)	D.C. (pct.)	MnStP (pct.)	Det. (pct.)	U.S. (pct.)
Total	1,622.6	3,021.3	977.6	1,830.6	1,877.3	2,298.5	90,878	144,391	86	87	22	59
Farm employment	19.5	12.8	17.7	16.3	11.2	8.1	3,925	3,001	–34	–8	–28	–24
Agriculture services, forestry, fisheries, and other	6.8	NA	2.4	12.0	4.3	15.1	504	1,694	NA	399	255	236
Mining	2.3	3.1	0.8	2.0	1.6	2.2	733	912	34	151	40	24
Construction	89.1	152.3	50.8	74.6	83.4	91.2	4,460	7,287	71	47	9	63
Manufacturing	68.7	101.3	232.0	279.3	637.1	444.7	20,543	19,025	47	20	–30	–7
Transportation and public utilities	74.0	128.3	62.8	93.1	91.7	97.5	4,788	6,923	73	48	6	45
Wholesale trade	45.1	89.1	64.2	106.6	92.7	124.7	4,097	6,774	97	66	35	65
Retail trade	222.0	428.8	157.0	308.2	289.3	395.4	13,422	24,276	93	96	37	81
Finance, insurance, and real estate	114.5	232.6	76.2	162.8	117.4	171.4	5,894	10,635	103	114	46	80
Services	361.4	1,094.0	179.6	558.2	315.6	706.8	16,430	42,239	203	211	124	157
Federal, civilian	342.6	393.1	20.8	22.6	34.6	31.2	2,919	3,080	15	9	–10	6
Military	124.1	93.1	17.2	13.1	17.9	12.2	3,419	2,371	–25	–24	–32	–31
State and local	152.1	252.9	96.2	181.4	174.8	197.3	9,744	16,174	66	89	13	66
Population	3,150.1	4,466.5	1,991.6	2,688.4	4,476.6	4,307.1	201,298	260,341	42	35	–4	29

Source: U.S. Department of Commerce, Regional Economic Information System, 1996.

exceptional employment and population growth during the 1980s, as we show later, from its private services. Minneapolis-St. Paul, on the other hand, showed above-average growth in every category, except contract construction. These contrasting growth experiences relate to sharp differences in the economic bases of the three MSAs, that is, the export-producing sectors of the local economies that generate goods and services for the purchase of nonresidents, including visitors. For the District of Columbia, the economic base is, of course, the federal government, although its local impact has shifted with the growing importance of its regulatory activities, which has produced a private-sector increase in lobbying resources to influence current and forthcoming legislation. The contrasting growth experiences also correlate with differences in the labor costs, as shown in later tables. The Detroit MSA economy, for example, is characterized by high wages and high unemployment—a legacy of the post–World War II period when U.S. motor vehicle manufacturers, centered in the extended Detroit metropolitan area, dominated their domestic and global markets.

The entire series of introductory discussions focusing on the unique experiences of each of the three MSAs emphasizes the importance of the economic base—its advantages and disadvantages of location, its linkages with other areas and regions, the legacy of its origins, and its history of growth and change. These discussions also highlight the many useful sources of data that are readily accessible for the study of urban regional economies. They document, for example, differences in industry growth rates attributed directly to differences in export markets, both domestic and foreign, served by each of the local economies. For the District of Columbia, the relevant local activity that brings in new dollars is the federal government and the activities that reach into the private sector and, therefore, attract a host of private businesses and organizations. The new leaders of the computer industry, for example, "are finding that prodding government in its direction is a lot more difficult than clicking a mouse."[1] For the Detroit MSA, on the other hand, the local economy is highly dependent on the resident automobile manufacturers—their plants and corporate offices, which also attract some specialized private services that remain hidden in the aggregate numbers.[2] We continue with these comparisons later in this chapter and also in Chapters 6 and 11.

Other Definitions

Other definitions used primarily for analytical purposes are based on nodality or homogeneity characteristics. Each of these will be discussed briefly before moving to a final approach that will prove to be very important for the direction taken in this book.

Homogeneous regions can be defined in terms of units of measurement that have some common feature so that, at least in some respect, the region can be said to be homogeneous. The homogeneous characteristic is generally defined in terms of one common feature. It does not follow that the region must be homogeneous in every respect.

Homogeneity becomes the basis for definition when an analyst wishes to compare regions or define regions in terms of a commonality of culture and values, as in the writings of Frederick Jackson Turner in *The Frontier in American History* and of Howard W. Odum and Harry Estill Moore in *American Regionalism.* The regional concept as a "tool for social research" was reviewed by Rupert Vance during the symposium on American regionalism held at the University of Wisconsin in April 1949 and published in *Regionalism in America,* edited by Merrill Jensen. Vance noted that regionalism is a concept that cuts across many lines, including its use in the early historical studies of American sectionalism as well as the area studies program of the U.S. military during World War II. This concept of a region emerged again in Joel Garreau's *The Nine Nations of North America* (1981) and Darrell Delamaide's *The New Superregions of Europe* (1994).

The nodality approach emphasizes the interdependence of various components within the region. The distance factor is extremely important when using this kind of a regional definition. The important link between nodal regions is *flows*—of products, people, communications, energy, and so on. The heaviest flows are usually between specific points called *centers.* We can measure these flows in any number of ways, but the principle remains that *interaction* guides the regional definition. Emphasis on the *nodes* or *poles* inevitably diverts attention from the delineation of regional boundaries to the center of a region. Boundaries are, therefore, not as easily defined as is true for some of the other approaches.

An important form of nodal region is the *labor market area* (LMA). Simply stated, the LMAs represent the commuting areas of individual county residents as reported in the census of population and housing. We can show clusters of LMAs around a core LMA that serve as an air transportation node in the U.S. air transportation network of air nodes and connecting cities. The entire group of LMAs is a proximity-based economic region.

Core LMAs have similar economic functions and roles in the emerging global information economy. They are the world-class transportation, telecommunications, and distribution centers. They have a rich diversity of industries—the export-producing and import-replacing sectors of manufacturing, transportation, finance, insurance, banking, business, and other producer services and consumer services such as entertainment, recreation, and health care. Most important are the strategic management functions in the core area downtown districts. These functions require one-on-one relationships between information providers and information users. They include the highly differentiated information-related services for achieving and maintaining the competitive edge of local businesses in regional and world markets.

Outside of the core to the LMA are *transitional areas,* which have the advantage of lower site costs and generally lower labor costs. The lower labor costs are due in part to lower costs of living in these areas than would be true for the metropolitan area. The cost of space is less. There is less congestion and loss of time because of it. Another savings is from a lower commute cost, which need not show up in wage demands. Because of these cost advantages, the transitional areas are experiencing relatively rapid growth in manufacturing. In fact, these areas quite often display rates of growth in employment and in income that exceed the metropolitan core areas. The proximity of the transitional areas to metropolitan areas also allows for access to each other's activities. Population in the transitional area can take advantage of the entertainment offerings of the metropolitan area without too expensive a commute. Semifinished goods, branch plants, and related interactions are high between the two areas.

A new designation (different from the official census definitions) might find the metropolitan area represented by the downtown of a central city, the neighborhoods contained in the central city boundary, and the neighborhoods in *ex-suburbia*—suburbs beyond the second-ring suburbs. The economy of this node depends upon its diversity of economic activities and resources. The forces of *agglomeration* thus sustain the region's economy.

Agglomerate forces are of three major types. The first is related to the sheer size of an economic region. Certain types of activities require large, local markets in order to achieve economies from scale in their operations. Once they locate in metropolitan areas, these firms are able to survive because of lower costs stemming from efficiencies in large-scale operations.

The second type of agglomeration is related to the proximity of business enterprises to specialized resources, to research, and to suppliers of intermediate products. Savings occur in searching for such resources and intermediate products, in training specialized labor, in engaging in research and development, and in a host of related requirements for efficient specialized manufacturing and service provision.

A final type of agglomeration is based on flows of communication. Specialized producer services often require face-to-face communication to conduct business. The supplier and demander locate proximate to one another (agglomerate) to facilitate such face-to-face communication.

These agglomeration economies provide core LMAs with certain advantages over outlying regions for the production of high-order goods and services. Details of this concept are presented later in the book.

Different Ways of Thinking about Regions

Textbooks on economics principles—macro and micro—seldom refer to regions and their location, or to their structure. Even a careful reading might conclude that economic principles apply similarly to one region as well as another. This is especially the case when we focus on money markets, interest rates, and the activities of a central bank with a capacity to affect a nation's monetary condition and a central government with a capacity to affect a nation's fiscal condition and, in turn, its monetary condition.

From a regional economics perspective, we start with a market economy and issues of investment and resource productivity among a nation's localities and communities. A regional economics perspective starts with a different common denominator and a different sense about the role of state and local governments. We look beyond the general level of income and employment and the interest rates charged by central banks as we introduce the concepts of space and proximity of one activity to another into our discussions. The cost of living and the productivity of the workforce differ greatly from place to place. Local markets also differ in products handled and prices paid, as does the performance of local and state governments in the delivery of public services.

Perspectives on Economics Principles

Economic principles have much to say about scarcity and how we can allocate limited resources to meet unlimited human needs and demands. *Macroeconomics* concerns mainly aggregates or a national economy as a whole, especially the problems of inflation and unemployment that central governments can tackle. *Microeconomics* deals mostly with the behavior of individual consumers and producers and the incentives that affect their behavior. *Regional economics* relates to both disciplines in concepts and tools of regional analysis.

Macroeconomics

Macroeconomics deals with a nation's economy as a whole. Its focus is largely on problems of unemployment and inflation and the monetary and fiscal polices for dealing with them. Macroeconomics textbooks include additional topics such as the gross national (or domestic) product, the demand for and the supply of money, and the Keynesian model—with and without money. They may include related topics on income distribution and poverty, international (but not domestic interregional) trade, international finance, and aggregate economic growth and development. Regional economics covers the domestic regional counterparts of each of these topics. For example, central to the study of regional economies is a region's economic base, commonly represented by its exports to markets outside the region. Since exports are less than the total regional product and yet account for all of it, the concept of an export, or economic base, multiplier is frequently introduced into discussions of the effects of change in export demand on a region's economy.

Economic base is a demand-oriented concept predicated on an earlier theoretical construct that was developed by John Maynard Keynes. The most fundamental form of the Keynesian model is the assertion that regional income is a function of the level of household consumption and investment, as we show later in this chapter. In the Keynesian model, consumption is a function of the level of income, but not all of the total income is spent on consumption. This proportion is the marginal propensity to consume, or the fraction of any additional dollar of income that is spent, the difference from one being the marginal propensity to save.

Ultimately, levels of income and consumption depend on the level of investment—an exogenous source of income affected not by local consumption, but rather by the economic policies of the central government. These aggregate values presumably represent the results of consumer behavior as individual households have more income to spend.

The reciprocal of the marginal propensity to save is the Keynesian multiplier. Any change in autonomous (nonconsumption related) spending leads to a multiplied change in the region's income level. Saving is an income leakage. This multiplier is used in conjunction with any form of autonomous spending.

If we are speaking of export base logic, the level of a region's income is determined by taking the multiplier times the exogenous (external) level of spending, or exports. A second leakage is thus introduced: imports. Such a model recognizes that individual localities are more open (depend more on trade) than national economies and that internal levels of consumption depend very much on income earned from external sources. Local government spending and taxation, because of small import leakages and the immediate and more direct effects on personal consumption spending, are commonly viewed as the primary concern of the individual local economies and their governments.

Microeconomics

Microeconomics deals with markets and the pricing of goods and services sold in these markets. Its focus is on market incentives, like price changes and price differences, and their effects on consumer and producer behavior. Microeconomics textbooks include topics such as consumer choice and product demand, producer choice and product supply, the product market and its varying forms of competition, and the factor markets of labor and capital. Regional economics covers the same topics and uses the same analytical tools as microeconomics, but introduces the additional effects on economic activity of location and proximity on price changes and price differences from one locality or region to the next.

Before the coming of large discount retailers to rural areas, the local food market and the friendly local hardware store were commonplace for the simple reason that the products and services they offered were worth their costs. The prices were a bit higher locally and the choices were less than in the larger places some miles away, but the savings were not worth the time and trouble of overcoming the extra miles of travel.

Technology and trade changed all this. Rural areas with an agricultural base adopted the new technologies; bought the new cars, trucks, and tracks; and experienced the advantages of low cost and convenient access to more distant product markets for their grain and cattle and the production inputs of farming. The same applied to household purchases. The large retail discount merchants seized the opportunity to establish new, one-stop shopping centers near the rural population centers, catering to farm businesses and offering lower prices, a wider range of choices, and even better service.

Improved incomes of rural people accelerated inevitable changes in the rural markets and communities. Farm consolidation hastened the decline in the number of farm and nonfarm businesses. Most of the youth of these communities left for the metropolitan areas in search of new opportunities and a new way of life. Main Street businesses lost their former advantage of distance to protect them from the price and service competition of metropolitan area retailers. Meanwhile, new forms of market organization emerged in metropolitan areas to protect established businesses—like a dominant international airline at a major U.S. air node—from the competition of new entrants into the metropolitan area product markets.

Perspectives on Regional Economics

A regional economics perspective also has the dual concern of markets and governments, but with a much greater emphasis on local markets and local governments. Differences between the two, however, lead to concerns about the proper role of markets and the proper role of government in local affairs. Both are transient concerns, based on the societal values, institutions, and organizations currently prevailing. We start with the prevailing market economy perspective, focusing on its perceived role when thinking about regions and representing its consequences for the delineation and study of nodal-type economic regions.

Market Economy

Describing the market economy perspective as one of choice and chance puts a heavy burden on the individual decision maker and on the market economy itself. The element of choice occurs in important ways, such as when a young married couple purchases their first home through a realtor, or enrolls their first child in a nursery school or in private grade school rather than public school. The chance that a choice may go astray exists in varying degree with each choice. Nor are local markets entirely free in the sense that institutions and organizations regulating or controlling these markets limit choice.

The labor market model of a local economy contrasts sharply with its historical antecedents in the modeling of local economies—von Thünen's "isolated state," the Weberian location triangle, or the application of "O'Reilly's law" in the delineation of a local trade area. A diverse and expanding local labor market has become an essential requirement for a growing metropolitan area to have the critical labor resources for adapting quickly to changing economic conditions. Analytical tools for the study of metropolitan areas, such as economic base and input-output models, now incorporate ways to measure these important requirements of a viable metropolitan area economy. Christaller's (1966) representation of a hierarchy of local and regional trade areas focuses only on the product markets of local retail stores and private services.[3] Fox's "functional economic areas" and the methodology of their delineation bring us to the current use of the daily trip to work in delineating local labor market areas (Fox and Kumar, 1966).

A market economy perspective emerges also from further examination of the three MSAs and their contrasting economic structures and experiences from the peak of one business cycle to the recovery of another. Table 1-2 shows the total earnings (in current dollars) of employed workers, by place of work rather than household residence, in each MSA and the nation (the latter still serves as a reference area). While the aggregate of private services showed a tenfold or greater increase in labor earnings from 1969 to 1994, selected categories of private services increased even more. Business, health, and legal services, for example, increased by a factor of twenty or more, leaving personal and educational services far behind. Built into the 1994 dollars, of course, is a consumer price index four times its 1969 level.

Thus, a market economy perspective on the internal structure and external relationships of the residents of an economic region starts with the internal structure of the LMA, namely, the daily patterns of commuting and trade. These are internalized, within-area activities, totally resident focused. They include finding and forming favorable learning environments for both family members and corporate employees.

TABLE 1-2 Total earnings of employed workers in specified industry, by place of work, 1969 and 1994

	Total Earnings							
	D.C. PMSA		MnStP MSA		Detroit PMSA		United States	
Industry	1969 (mil.$)	1994 (mil.$)	1969 (mil.$)	1994 (mil.$)	1969 (mil.$)	1994 (mil.$)	1969 (bil.$)	1994 (bil.$)
Earnings, total	12,568	105,229	7,296	54,833	16,908	80,617	619.4	4,084.9
Agriculture, agricultural services, forestry, fisheries, other	121	771	103	310	79	298	21	79
Mining	21	NA	11	44	22	45	6.3	36.3
Construction	750	5,042	562	2,577	1,025	3,332	40.6	218.1
Manufacturing, total	569	4,440	2,133	12,354	7,522	26,626	173.8	747.3
Transportation and public utilities	645	5,806	595	3,659	927	4,425	43.4	272.6
Wholesale trade	443	4,242	651	4,930	991	5,892	36.8	254.8
Retail trade	1,207	7,776	832	4,970	1,685	6,634	67.3	392.2
Finance, insurance, and real estate	592	6,384	456	5,104	734	4,614	33.9	302.9
Services, total[a]	2,653	38,851	1,064	14,060	2,175	20,406	95.7	1,127.0
Hotels and other lodging places	NA	1,477	42	278	NA	231	3.9	38.7
Personal services	154	770	96	505	210	594	7.7	34.8
Business services	544	8,625	144	3,318	329	4,598	13.0	202.7
Health services	412	6,597	317	4,110	617	6,980	26.1	366.4
Legal services	254	4,599	74	1,204	133	1,418	6.0	90.0
Educational services	251	1,600	47	469	70	405	6.7	46.2
Social services[b]	NA	979	NA	652	NA	637	NA	36.6
Membership organizations	350	2,709	103	460	216	591	8.5	35.7
Engineering and management services	NA	9,050	NA	1,513	NA	2,902	NA	144.3
Government and govt. enterprises, total	5,566	31,550	890	6,816	1,710	8,336	100.6	654.8
Federal, civilian	3,774	20,206	170	1,024	306	1,282	25.9	131.4
Military	804	2,795	28	123	45	129	14.4	47.9
State and local	988	8,549	692	5,670	1,359	6,926	60.3	475.5

Source: U.S. Department of Commerce, Regional Economic Information System, 1996.

[a]Services total includes all two-digit services industries and non-disclosures.

[b]Social services is new under the 1972 standard industrial classification. It consists of establishments previously classified under hotels, health services, educational services, membership organizations, and miscellaneous services.

The economic base of the area is its exports, including visitor spending, in all its detail. On a regional scale, however, the exports are, in part, internalized within the region. The local labor market and local product markets thus function internally. Like capital markets, the product markets accessible to local businesses may also function on a global scale. This perspective thus focuses on issues that extend beyond the labor market area and the region, such as introducing new business technologies and practices to increase worker productivity and in the process create new, well-paying jobs for local residents.

This text has a market economy perspective in a proximity-based regional delineation that focuses on the 29 U.S. air nodes in the contiguous 48 states. Chapter 3 provides an extended discussion of this system as a framework for the study of urban regional economies.

Public Economy

Words and phrases often used in describing the public economy include "budget-maximizing bureaucrats," "excessive and costly regulations," "self-serving legislation of politicians," and so on. These descriptions of the public economy originate from the public-choice school of economic thought, as represented in the writings of Buchanen, Niskanen, and Tulloch. They contrast sharply, of course, with the early marginalists, like Gossen, who equated the work of a government's civil service with the maximization of the public interest. At least one of these expressions also describes the actions of some businesses, especially those engaged in the exercise of their own local monopoly power.

A common characteristic of, perhaps even synonymous with, the concept of a public economy is the use of administrative versus market approaches in the pricing of products and services. While the civil service has a high degree of job security, its legislative members operate on two- to four-year election cycles. The public economy approach thus lacks the pricing discipline of competitive markets. It tends to work on a short-term planning horizon. Its outcomes, therefore, are not necessarily restricted by readily understood economic considerations, perhaps any more than its voting districts represent functional economic communities.

We continue in Table 1-3 with comparisons of the three MSAs by first recalling the importance of the public sector in each area. Only in the District of Columbia, however, is the public sector an important part of the economic base, that is, the part that generates the goods and services bringing new dollars into the local economy. The Minneapolis-St. Paul MSA is marked by a diversity of export-producing industries. The automobile industry in the Detroit MSA, on the other hand, accounts for very few additional services.

TABLE 1-3 Proportion of total earnings of employed workers in specified industry, by place of work, 1969 and 1994

Industry	Proportion of Total Earnings							
	D.C. PMSA		MnStP MSA		Detroit PMSA		United States	
	1969 (pct.)	1994 (pct.)	1969 (pct.)	1994 (pct.)	1969 (pct.)	1994 (pct.)	1969 (pct.)	1994 (pct.)
Earnings, total	100.0	100.0	100.0	100.0	100.0	100.0	100.0	100.0
Agriculture, agricultural services, forestry, fisheries, other	1.0	0.7	1.4	0.6	0.5	0.4	3.4	1.9
Mining	0.2	NA	0.2	0.1	0.1	0.1	1.0	0.9
Construction	6.0	4.8	7.7	4.7	6.1	4.1	6.5	5.3
Manufacturing, total	4.5	4.2	29.2	22.5	44.5	33.0	28.1	18.3
Transportation and public utilities	5.1	5.5	8.1	6.7	5.5	5.5	7.0	6.7
Wholesale trade	3.5	4.0	8.9	9.0	5.9	7.3	5.9	6.2
Retail trade	9.6	7.4	11.4	9.1	10.0	8.2	10.9	9.6
Finance, insurance, and real estate	4.7	6.1	6.2	9.3	4.3	5.7	5.5	7.4
Services, total[a]	21.1	36.9	14.6	25.6	12.9	25.3	15.5	27.6
Hotels and other lodging places	NA	1.4	0.6	0.5	NA	0.3	0.6	0.9
Personal services	1.2	0.7	1.3	0.9	1.2	0.7	1.2	0.9
Business services	4.3	8.2	2.0	6.1	1.9	5.7	2.1	5.0
Health services	3.3	6.3	4.3	7.5	3.6	8.7	4.2	9.0
Legal services	2.0	4.4	1.0	2.2	0.8	1.8	1.0	2.2
Educational services	2.0	1.5	0.6	0.9	0.4	0.5	1.1	1.1
Social services[b]	NA	0.9	NA	1.2	NA	0.8	NA	0.9
Membership organizations	2.8	2.6	1.4	0.8	1.3	0.7	1.4	0.9
Engineering and management services	NA	8.6	NA	2.8	NA	3.6	NA	3.5
Government and govt. enterprises, total	44.3	30.0	12.2	12.4	10.1	10.3	16.2	16.0
Federal, civilian	30.0	19.2	2.3	1.9	1.8	1.6	4.2	3.2
Military	6.4	2.7	0.4	0.2	0.3	0.2	2.3	1.2
State and local	7.9	8.1	9.5	10.3	8.0	8.6	9.7	11.6

Source: U.S. Department of Commerce, Regional Economic Information System, 1996.

[a]Services total includes all two-digit services industries and non-disclosures.

[b]Social services is new under the 1972 standard industrial classification. It consists of establishments previously classified under hotels, health services, educational services, membership organizations, and miscellaneous services.

TABLE 1-4 Earnings per worker in specified industry, by place of work, 1969 and 1994

Industry	Earnings per Worker (current dollars)							
	D.C. PMSA		MnStP MSA		Detroit PMSA		United States	
	1969 (thou.$)	1994 (thou.$)	1969 (thou.$)	1994 (thou.$)	1969 (thou.$)	1994 (thou.$)	1969 (thou.$)	1994 (thou.$)
Earnings, total	7.7	34.8	7.5	30.0	9.0	35.1	6.8	28.3
Construction	8.4	33.1	11.1	34.5	12.3	36.5	9.1	29.9
Manufacturing	8.3	43.8	9.2	44.2	11.8	59.9	8.5	39.3
Transportation and public utilities	8.7	45.2	9.5	39.3	10.1	45.4	9.1	39.4
Wholesale trade	9.8	47.6	10.1	46.2	10.7	47.3	9.0	37.6
Retail trade	5.4	18.1	5.3	16.1	5.8	16.8	5.0	16.2
Finance, insurance, and real estate	5.2	27.4	6.0	31.4	6.3	26.9	5.8	28.5
Services	7.3	35.5	5.9	25.2	6.9	28.9	5.8	26.7
Federal, civilian	11.0	51.4	8.2	45.2	8.8	41.1	8.9	42.7
Military	6.5	30.0	1.6	9.4	2.5	10.5	4.2	20.2
State and local	6.5	33.8	7.2	31.2	7.8	35.1	6.2	29.4

Source: U.S. Department of Commerce, Regional Economic Information System, 1996.

Labor earnings, when divided by the total number of workers, reveal still another set of differences among MSAs—whether or not the area, the economy, and its industries are viewed as being *high wage* or *low wage*. Compared to the United States, both the District of Columbia and Detroit are high-wage local economies, as shown in Table 1-4.

The federal government is clearly the economic base of the District of Columbia and environs. Indeed, it is the sole economic base, even more so than manufacturing of motor vehicles is the economic base of the Detroit MSA. In the District of Columbia, the federal government attracts a variety of private services, ranging from business services to legal services, membership associations, and engineering and management services.

Table 1-5 illustrates the general tendency of all MSAs to converge toward above-average earnings per worker in most, if not all, industry categories. Only the federal military is consistently below average in the Minneapolis-St. Paul and Detroit MSAs—a result of the dominantly part-time employment in the National Guard.

The separation of the market economy from the public economy on a territorial scale has important implications for the residents of the metropolitan region. It offers certain opportunities for political trade-off without readily ascertained economic costs and penalties, given the division of economic constituencies within most voting districts, particularly in the metropolitan core area.

TABLE 1-5 Relative earnings per worker in specified industry, by place of work, 1969 and 1994

	Area Earnings per Worker Relative to U.S.					
	D.C. PMSA		MnStP MSA		Detroit PMSA	
Industry	1969 (pct.)	1994 (pct.)	1969 (pct.)	1994 (pct.)	1969 (pct.)	1994 (pct.)
Earnings, average	13.6	23.1	9.5	5.9	32.1	24.0
Construction	–7.5	10.6	21.7	15.4	35.1	22.0
Manufacturing	–2.2	11.5	8.7	12.6	39.6	52.4
Transportation and public utilities	–3.8	14.9	4.6	–0.2	11.7	15.2
Wholesale trade	9.1	26.5	12.8	22.9	18.9	25.7
Retail trade	8.4	12.3	5.7	–0.2	16.2	3.9
Finance, insurance, and real estate	–10.1	–3.6	3.9	10.1	8.7	–5.5
Services	26.0	33.1	1.7	–5.6	18.3	8.2
Federal, civilian	24.2	20.5	–7.7	6.0	–0.4	–3.6
Military	53.7	48.6	–61.8	–53.5	–40.7	–47.9
State and local	4.9	15.0	16.1	6.3	25.6	19.4

Source: U.S. Department of Commerce, Regional Economic Information System, 1996.

For example, river basin planning in the United States began with the Tennessee Valley Authority (TVA). The TVA represented an effort by the federal government in the 1930s to build dams and generate low-cost electric power. A secondary goal of the TVA was to create local jobs in the economically lagging states drained by the Tennessee River (Selznick, 1966). In western states, river basin planning meant the Columbia Basin Project in the Pacific Northwest (McKinley, 1952), the Central Valley Project in California, the Missouri River Basin Project covering much of the old Louisiana Purchase of 1803, and others that introduced an even broader set of purposes with the sale of water. River basin planning continued through the post–World War II period until the 1980s, when it was terminated by the U.S. Congress.

This regionalistic emphasis on central planning meant that problems of urban sprawl, rural depopulation, and fiscal impoverishment of core area municipal governments generally were not addressed. Nonetheless, individual state legislatures, especially those with large rural memberships, cooperated willingly with the river basin planning authorities for the federal largess that they provided. Meanwhile, the forces of technology were reshaping both rural and urban America with continuing rural depopulation and abandonment of metropolitan core areas by those seeking the more homogeneous environment of the surrounding new suburbs.

The market-driven economies of downtown districts, with temporary exceptions such as Detroit, St. Louis, and Pittsburgh, retained their viability and even enhanced the importance of their regions through a shift to new, high value-added tenants. These new tenants were better able to pay rents than the stores and shops that sold out or that moved to suburban shopping malls. Thus the public economy perspective of the metropolitan core area began focusing on issues of infrastructure development, especially for the downtown district and its business and government employees.

Infrastructure development included factors such as constructing high-speed highways and downtown parking ramps, facilitating access to outside market destinations and supply sources, financing major business convention and sports facilities, and building sustainable neighborhoods and communities. These efforts were not geared, however, to reducing disparities in household income and wealth between the core area immediately beyond the downtown district and the surrounding suburbs that have become the new places of residence for the departing core-area, high-income households.

Understanding Regional Economies

Regional economies are integral parts of a national economy. Each region is blessed by its individual uniqueness that contributes to the economies of other regions through its exports and imports of goods and services. Location, linkage, and legacy are the essential elements of each region's uniqueness. Differences persist because of differences in nature and sustainability of the economic base of each region and its capacity to generate savings or otherwise support continuing investments in the region's productive capabilities.

Location, Linkage, and Legacy

Comparing the industry profiles and measures of economic outcomes in the three MSAs points out the importance of location in accounting for their contrasting histories. This applies not only to the current location of basic (that is, export-producing) industries that are the local markets for the supporting industries but also to the location of basic industries relative to their export markets. The comparisons thus point to the importance of inter-industry linkages in accounting for the unique economic profile of each area. Critically important also are the legacies of the past as they affect the likely economic futures of each of the areas. A downsizing of the federal government in terms of direct employment, for example, means a reduction in the local demand for housing and consumer service in the District of Columbia MSA. If the regulatory role of the federal government were reduced, the demand for producer services also would be reduced, with further repercussion on the local economy. A shift in the dependence of the Detroit MSA manufacturers from a relatively high-wage local labor force to a low-wage labor market elsewhere is a comparable alternative for segments of the Detroit area economic base. Whether or not these shifts occur in the private sector depends on the sorts of inter-industry linkages that may exist or develop at a new site.

Important determinants of business location and development—apart from where the business entrepreneur lives—are the costs of labor, energy, land and space, environmental protection, and industry regulation, all of which vary from site to site. Costs of maintaining local public facilities and services also vary from place to place and from one time period to the next. In addition, transfer costs—not simply the costs of transportation but also the costs of the many different producer services required to move a product from its production site to the place of its final use—are increasingly important. These include the costs of producing, distributing, evaluating, and using highly differentiated market and product information essential in completing each business transaction, whether negotiated face-to-face in a downtown office or electronically in a distant commodity exchange. Both von Thünen's (1826) land use model and Weber's (1909) plant location model address these sorts of location issues. Neither model, however, addresses directly the investment decision and its importance in industry location.

Every location decision, recognized as central to the study of urban regional economic systems, involves the decision to invest. Investment could be in a manufacturing plant at an existing site or a new site, in a new home in a new residential area, or in a government office at a new site. Rather than building a new facility, a manufacturer may opt for an existing one at a new site or modify existing production methods to improve productivity and increase output at the facility's present location. Much depends on the expected profitability of the manufacturing operations at the alternative sites, not only in the near future but also in the years ahead as individual product cycles approach diminishing rates of market growth and increasing pressures from competitors.

Business location decisions are demand driven as well as supply driven, with population and its workforce being a major location determinant on both sides. Use of highly aggregated numbers on industry production and related labor skills obscures the specific product markets and supply sources affecting the location of the production facility and its supporting infrastructure and services.

Some students of regional industrial systems at least imply that regions, rather than individual businesses and communities, compete. For example, regions with outstanding air transportation and education, among other assets, display competitive advantages that filter down to the businesses located there (Porter, 1990). High-valued manufactured products, together with the high-valued services that people seek or provide, become the new clients of every world-class transportation system. Education, particularly technical education, has its own market tests of relevancy, such as the monitoring of graduates' careers and the correspondence of these careers and their locations with their training and the sources of support. This requires the capacity for coping with the changing markets—local and global—and acquiring the skills for successfully participating in the new market economies.

Activity linkages bring together many different participants in an urban regional economy. The most commonly occurring forms of economic linkages center on income and product disbursements along with receipts of businesses, households, and governments. The automobile industry's input purchases from and output disbursements to various product markets illustrate these. Individual manufacturers in the industry purchase primary inputs of labor and capital; inputs of automobile parts and components from other manufacturers; and business, financial, transportation, and other producer services. The automobiles are then shipped to dealers throughout the country and abroad. They finally reach their end users in households, businesses, and governments.

One measure of the likely inter-area trade is the importance of each industry in the area relative to its importance in the nation. If automobiles accounted for 15 percent of an area's total industry employment but only 5 percent for the nation, then the difference between the two percentages is a measure of the likely exports of automobiles to the rest of the nation. If this measure were to drop sharply to 10 percent, the area would now share less of the nation's automobile production. If this area were totally dependent on the export of autos for its economic base, then it would have every reason to expect a proportional reduction in total employment and economic activity. Changes in this proportion over one or more time periods provide a measure of both the area's changing economic prospects and the importance of having more than one export-producing industry in its economic base.

The legacy systems, like those of transportation, that sustain and enlarge a region's base economy develop over many product cycles that initially play an important part in the repeated success of a region's export-producing enterprises. They also play a part in the ultimate decline of a once-successful industry by burdening new businesses with unnecessarily high taxes and misdirected public outlays. The goal of a regional transportation system should be the same as that of a successful business. Paraphrased for the state or regional transportation agency, "It is not to create value for users of the transportation system, but to mobilize users to create their own value from the system's various offerings" (Normann and Ramirez, 1993, p. 71).

TABLE 1-6 Total personal income per capita, by place of residence, 1969 and 1994

	Total Personal Income per Capita (current dollars)							
	D.C. PMSA		MnStP MSA		Detroit PMSA		United States	
Income Source	1969 (thou.$)	1994 (thou.$)	1969 (thou.$)	1994 (thou.$)	1969 (thou.$)	1994 (thou.$)	1969 (thou.$)	1994 (thou.$)
Total personal income	4.7	28.8	4.4	25.2	4.4	24.7	3.8	21.7
Earnings, total, by place of work	4.0	23.6	3.7	20.4	3.8	18.7	3.1	15.7
Less: Personal cont. for social insur.[a]	0.2	1.6	0.2	1.6	0.2	1.4	0.1	1.1
Plus: Adjustment for residence[b]	–0.1	–1.0	0.0	–0.2	0.0	–0.3	0.0	0.0
Equals: Net earn. by place of residence	3.7	21.0	3.5	18.7	3.6	17.0	2.9	14.6
Plus: Dividends, interest, and rent[c]	0.6	4.2	0.6	3.6	0.5	3.8	0.5	3.4
Plus: Transfer payments	0.4	3.6	0.3	3.0	0.3	3.9	0.4	3.7
Earnings, by source:								
Wages and salaries	3.6	19.7	3.2	16.8	3.2	15.4	2.5	12.4
Other labor income	0.1	1.9	0.2	2.0	0.3	2.1	0.1	1.5
Proprietors' income[d]	0.3	2.0	0.3	1.6	0.3	1.2	0.4	1.8

Source: U.S. Department of Commerce, Regional Economic Information System, 1996.
[a]Included in earnings by industry but excluded from personal income.
[b]Net inflow of the earnings of inter-area commuters.
[c]Includes capital consumption adjustment for rental income of persons.
[d]Includes inventory valuation and capital consumption adjustment.

We return, again, to the three MSAs to illustrate the persistence of industry legacy systems and their consequences—in this case, for personal income levels of area households. The Detroit MSA, for example, maintained above-average per capita personal income levels over the 1969–94 period despite a sharp downturn in manufacturing and population growth (Table 1-6). Also illustrated is the expansion of the commuting areas demonstrated by the "adjustment for residence." In 1969, the 1994 MSA delineation encompassed the commuting area, except for the District of Columbia. This was no longer true by 1994. The MSA delineation generally lags behind the expansion of local commuting from place of residence to place of work.

The almost unbelievable complexity and richness of the legacy relationship means that any restructuring of business not accompanied by acute sensitivity to the importance of these relationships is likely to fail. This is increasingly well documented in the recent literature (von Hippel, 1988). It is evident also from the perspective of an individual seeking employment in another regional cluster of the same industry. Interpersonal ties among professionals in the same industry in a new locality often help job seekers find satisfactory employment.

TABLE 1-7 Relative total personal income per capita, by place of residence, 1969 and 1994

	Area Personal Income per Capita Relative to U.S.					
	D.C. PMSA		MnStP MSA		Detroit PMSA	
Income Source	1969 (pct.)	1994 (pct.)	1969 (pct.)	1994 (pct.)	1969 (pct.)	1994 (pct.)
Total personal income	22.4	32.6	14.6	16.3	16.5	13.8
Earnings, total, by place of work	29.7	50.2	19.1	30.0	22.7	19.3
Less: Personal cont. for social insur[a]	44.9	49.6	20.4	45.0	17.5	34.2
Plus: Adjustment for residence[b]	NA	NA	NA	NA	NA	NA
Equals: Net earn. by place of residence	26.9	43.7	17.9	27.8	22.3	16.1
Plus: Dividends, interest, and rent[c]	8.7	22.0	9.7	4.3	2.4	12.1
Plus: Transfer payments	4.9	–2.0	–5.5	–18.3	–11.7	6.3
Earnings, by source:						
Wages and salaries	40.2	58.8	24.2	35.6	27.6	24.2
Other labor income	–24.8	26.5	25.6	33.8	86.3	45.3
Proprietors' income[d]	–19.1	10.4	–17.0	–11.7	–31.5	–35.4

Source: U.S. Department of Commerce, Regional Economic Information System, 1996.
[a]Included in earnings by industry but excluded from personal income.
[b]Net inflow of the earnings of inter-area commuters.
[c]Includes capital consumption adjustment for rental income of persons.
[d]Includes inventory valuation and capital consumption adjustment.

Technology, often viewed as the driving force in a region with internationally competitive industries, is nurtured and sustained by a region's core competencies. The core competencies, in turn, are supported by a region's infrastructure, both public and private. These include education, health care, and transportation facilities, particularly air transportation.

Identifying a region's core competencies is much like identifying the core competencies of a business. Prahalad and Hamel (1990, pp. 81–82) propose three tests for a company intent on building its core competencies. Core competencies should provide "potential access to a wide variety of markets." They should make "a significant contribution to the perceived customer benefits." They should be "difficult for competitors to imitate." State and local governments successful in building affordable transportation, and technical education systems that address critical market and resource access requirements of their export-producing businesses, have a "step up" on their competitors in building the regional infrastructure of a globally competitive regional industry cluster.

The final comparisons in this series (Table 1-7) further document the persistence of past industry patterns. The Detroit MSA, for example, shows above-average per capita labor earnings, with the exception of proprietors' income, which remained well below the national average. At the same time, transfer payments in the form of Social Security and social welfare payments to local recipients increased sharply, contrary to the trend in the two other MSAs.

The metropolitan core area and its air transportation node provide their industry clusters, as well as nonmetropolitan clusters linked to the core area, with a global reach. These new engines of economic growth (Jacobs, 1984) are powerless, however, without the strong, productive business and governmental linkages that extend well beyond the metropolitan core area. In fact, manufacturing in rural areas was the fastest growing segment of the globally competitive, export-producing industry clusters of U.S. metropolitan core areas in the eighties in large part because of these linkages (Maki and Reynolds, 1994).

Geography and Trade

Krugman (1991) comments on the lack of his own awareness of the close connection between international trade and economic geography despite a professional lifetime of reading and writing about it. He notes (p. 2) "the tendency of international economists to turn a blind eye to the fact that countries both occupy and exist in space—a tendency so deeply entrenched that we rarely even realize we are doing it." He suggests further that "one of the best ways to understand how the international economy works is to start by looking at what happens *inside* nations" (p. 3).

The location of most new businesses is usually determined by the location of its founder. Metropolitan core regions spawn a large proportion of new businesses. Close proximity to information sources and business and government decision centers facilitates business foundings. New businesses account for much of the employment growth in recent years, as documented by the numerous studies cited by Reynolds and Maki (1990a, b; 1994a, b) in their recent reports on business foundings and job growth in 382 U.S. labor market areas (see Chapter 3).

Export-producing industries are the vital links between a state's economy and national and global markets. Production obviously is for markets in a market-oriented economy, but production of goods and services in any region depends on imports of those goods and services not produced in the region. The export-producing industries thus derive the required income payments for purchasing imports as well as compensate labor and capital for the primary inputs used in production. Interstate and interregional trade allows each state or region to exploit its comparative advantage and thus increase its gross domestic product.

Import replacement is a phrase borrowed straight from Jane Jacobs' *Cities and the Wealth of Nations* (1984). Take, for example, the majority of the counties in the Northern Transportation Corridor specializing in the production of agricultural, forest, and mineral products (Reynolds and Maki, 1990a). Opportunities for import replacement in the counties that depend on natural resources are few. For the core areas, imported goods and services are essential inputs in production and consumption. The more diverse the area, the more likely that intermediate rather than final markets account for the larger share of total imports.

The export-producing activities make these purchases possible. As exports expand, imports also increase. At some point, however, opportunities for import replacement attract new locally produced products for local consumption. These opportunities occur because of the innovations and improvisations—mostly from small businesses—that result in import-replacing products entering local markets.

Investment—Public and Private

Investment decisions—private versus public, short-term versus long-term—represent the bottom line in urban regional growth and change. Private investment decisions focus on expected net return on a given capital investment, taking into account standard discounting of risks and uncertainties relating to the cost of investment financing and the leveraging of owners' equity in the investment. Public investment decisions focus on the perceived societal or interest group benefits, less the expected direct costs, also discounted for perceived political as well as economic risks associated with the investment. Most investment is based on a short-term planning horizon.

Both private and public capital investment projects contribute immediately to regional economic growth during the construction of the facility and subsequent economic activity generated by the spending of construction payrolls and the purchases of construction employers. These effects continue recycling in the local economy over a several-year period in the case of major construction projects.

Completion of the construction project marks the potential beginning of the operation of a facility that not only produces a particular product for purchase locally or for export to more distant markets but also provides a new or added income source for local residents. The spending of this income may take many forms: Some of it may be expended for immediate personal satisfaction, less for saving and investment. Whether or not the new or added income source represents a net addition to total income depends on whether the income originates from outside the area—from exports of the goods and services produced at the new facility—or from the purchases of visitors. Both bring new dollars into the area.

An important attribute of a preferred location for a business enterprise is the local infrastructure—the physical facilities and economic resources shared, in varying degree, by all local businesses. This infrastructure includes the regulated industries—transportation, communications, and public utilities—as well as banking, finance and insurance companies, management consulting agencies, and research and development laboratories (Moss and Brion, 1991; Noyelle and Stanback Jr., 1984). It represents the macroeconomic entity essential to the individual export-producing businesses in the local economy.

By definition, the export-producing businesses are part of the local base economy. Typically, the largest employers in this category are branch plants or headquarter offices of multinational companies trading in global markets (Daly, 1991). Corporate decisions based on national and global—rather than local—considerations particularly affect the branch plants of large corporations in these localities. The quality and availability of local training and education in public schools and postsecondary educational institutions also are decision variables relating to the likely productivity of the local workforce.

Support industries, serving the region's residential sector as well as the local transportation and telecommunications infrastructure, help strengthen a region's economic base. Local governmental efforts, along with the local macroeconomic environment, directly affect both support industries and local infrastructure. Support industries produce goods and services for local intermediate and final markets. Local industries purchasing semifinished products are the intermediate markets, while households, businesses, and governments purchasing finished products are the final markets. The location attributes of support industries are simple and straightforward in their implications for new business formation: Their markets are local. Imports from outside the LMA fulfill any excess product demand. Therefore, economies of scale in production and production knowledge are the critical limiting factors facing entrepreneurial efforts in the establishment of strongly competitive new business ventures tapping into existing local markets.

Increasing infrastructure expenditures is a popularly advanced proposal for creating jobs and reducing the "infrastructure gap." Much macroeconomic analysis supports this proposal (Aschauer, 1991). On the other hand, region-specific and mode-specific studies of existing transportation systems emphasize productivity-improving alternatives that point to cost savings that, in turn, are targeted to high-priority projects. The likely outcome of this controversy is a serious "credibility gap" for proponents of new federal funding for transportation infrastructure.

Federal funding for transportation infrastructure, driven by conventional "pork barrel" politics, seeks the macroeconomic analyses for its justification. Unfortunately, the allocation of the federal funding to states and local governments on the basis of votes and political constituencies turns the priority-setting process on its head. An alternative approach to infrastructure spending is targeting federal funding for the highest-priority projects without the imposition of congressional district constraints. Some states and regions would be winners in the alternative approach, and some states would be losers. Both parties are likely to pressure their congressional delegations into a more costly alternative—another likely possibility not included in macroeconomic studies.

The widespread reshaping of regional infrastructure systems is both a cause and a consequence of increasing global competition. State and regional transportation is thus no longer an activity isolated from events beyond state and regional borders. Yet, much of transportation systems planning is held captive by the durability of its own legacies in maintaining costly, but unneeded, roads and bridges and sustaining a world-class transportation network to almost every corner of administrative jurisdictions.

Infrastructure investment is only one of several emerging tasks of regional governance. Capacity-building and intraregional cooperation involve state, local, and regional organizations.

Some Key Issues

Given the tools and concepts for measuring a region's performance, the next step is one of understanding how a region's performance changes over time and the sources and causes of these changes. What are the additional improvements, if any, in a region's economic performance that we can expect from deliberate efforts of society to intervene in the underlying market processes?

Endogenous Economic Development

The Commission of European Communities proposes a four-point program of endogenous measures under Section 5b on rural development: (1) Develop partners for cooperation to stimulate small and medium-size entrepreneurs' volatility, (2) cooperate with university systems to access results from technological research, (3) educate and train within the enterprises to develop skills in business and production processes, and (4) advise and assist on prototype development, feasibility studies, investment project studies, and economic and finance studies.[4] These measures focus on improving the productivity of labor, capital, and entrepreneurial inputs in production. They point to a renewed appreciation of the importance of the endogenous factor in economic development, but only in the context of metropolitan core area linkages with selected transitional and peripheral areas of a nodal economic region.

Domestic economic policies of the United States and many of the European Community's twelve countries emphasize productivity improvements in export-producing industries and the competitive position of all local industries in their respective markets. One proposal for improving industry productivity in the

European Union (E.U.) is to redefine the principal aim of regional policy toward the restoration of national and E.U. competitiveness in export markets. This would be accomplished by targeting aid on a small number of eligible districts rather than broad regions, in short, " . . . allow localities with the greatest potential for growth within existing assisted areas to be pinpointed" (Armstrong, 1996, p. 207). Such a proposal, of course, runs counter to the notion that governments should not target, but upgrade, all industries, given the inherent difficulties in choosing winners coupled with inherently high propensities to favor noncompetitive industries.

What matters for improving a region's standard of living is the productivity of its industries, not *what* industries it competes in, but *how* it competes in them (Porter, 1996, p. 67). One perspective is that regional policy should promote specialization, upgrading, and trade among regions by encouraging the formation of industry clusters. For example, in Massachusetts, the business and investment climate has been "dramatically improved" with this approach (Porter, 1996, p. 88). Implementation of such policies puts a heavy burden on access to place-specific information and the appropriate methodologies for making sense of this information in resource targeting for cluster formation purposes.

Education is commonly viewed as a key contributor to a trading region's core competencies and its endogenous economic development. It provides—according to the current rhetoric on supporting increased educational funding—the essential means for workers in an information-driven local economy to acquire the new skills that are an integral part of an increasingly technology-intensive global economy. The educational delivery system, particularly at the postsecondary level, now faces its most critical challenges, however, in providing new learning environments for its students to successfully acquire the core competencies sought by employers and society generally. Competency standards for each set of skills presumably would facilitate both the learning process and the placement process. Once established, standards for skills competency could provide a common basis for accurate and honest information exchange among employers, prospective employees, and educational institutions preparing students for these skills. New occupational classifications would then emerge from this collaborative effort of setting and maintaining these new skill standards. These are some of the often-cited goals, rather than the reality, of our current educational efforts.

Vulnerable Local Economies

Criteria for assessing the vulnerability of an area's economic base—risk, costs, productivity, and flexibility—vary by the location of an area in the city-region settlement system, in which the core area is the principal transportation and communications center. Neither one of the two steps cited earlier in the delineation of the economic base—export producing and income generating—sufficiently differentiate the industry and activity composition of a local economy to provide useful and practical measures of its vulnerability to economic shocks. Unanticipated changes in export markets and exogenous income flows are endemic with vulnerable rural areas and natural resources–based economies. A third step under the rubric of risk and costs, productivity and flexibility, takes into account the differentiating characteristics of vulnerable and viable regions.

Strong linkages between labor market core area businesses reduce a rural area's vulnerability to unanticipated economic change by reducing the risks and costs faced by its individual businesses. The decentralized markets of economic theory function well because of the variety of contractual arrangements between buyers and sellers. Recurring negotiation sustains and improves these contracts. Manufacturing businesses in core areas, for example, initiate numerous contractual arrangements with businesses in rural areas to produce their product in peak periods or in periods of tight labor supply. They may initiate other contractual arrangements for integrating rural input supply sources into their production scheduling through professional and technical training programs in supply management, inventory control, and production scheduling (Ley and Hutton, 1987; Scott, 1988).

Vulnerable rural areas may learn from viable rural areas through an assessment of the differences between the two focuses on productivity and flexibility. Businesses forming, locating, or expanding in transitional rural areas adjacent to the metropolitan core area of a large region have an important advantage. They have proximity to world-class information systems and producer services for adapting rural area businesses to changing production technologies and global markets (Beyers, 1991). In these areas, business volatility—changes in total employment due to births and deaths, expansions and contractions of business enterprises—correlates with economic growth (Reynolds and Maki, 1990a,b). Vulnerable rural areas of metropolitan-focused regions are, in large numbers, transitory. Those close to metropolitan core areas experience economic and social pressures of rapid population growth. A new location equilibrium for manufacturing enterprise, driven by lower site and production costs, transforms many rural communities into the expanding urban frontier of the metropolitan core area (Scott, 1988).

Most of the commonly used criteria for sorting vulnerable rural areas from metropolitan areas that can provide important information for their survival are lost in a strictly administrative delineation of regions. Rural improvement measures typically exclude some sort of proximity criterion for delineating functional rural economic communities. Voting districts, while including parts of both rural and urban areas, have no requirements in their delineation for including these critical rural-urban linkages.

Sustainable Economic Growth

Both the traditional and the new theories of economic growth fail to fully account for observed differences among regions in economic performance. The traditional model of economic growth emphasizes capital accumulation—the combining of labor and capital to produce the output of an economy (Plosser, 1992). This output depends on the productivity of its primary inputs, that is, its production technology. However, the traditional model ignores sources of added productivity.

The new models of economic growth provide for internally generated growth through processes affecting societal rates of savings and investment. Government tax policy, for example, influences long-term rates of capital accumulation, which, in turn, generates beneficial external effects that potentially justify further government intervention. The same rationale justifies government subsidy of education, resulting in a more productive workforce. A more productive workforce, in turn, leads to improved economic growth and individual well-being and justifies changes in tax laws to stimulate investment in productivity-improving equipment (De Long and Summers, 1992). Neither traditional nor new models of economic growth account for differences in resource productivity from one location to another resulting from proximity to the core LMA of an economic region.

A new generation of models relating to regional economic growth, on the other hand, explicitly includes measures of regional economic organization in accounting for differences in the various measures of regional economic performance (Treyz, 1993). They focus on local economic linkages to product and labor markets, and also local factors, such as the quality of entrepreneurship and technical education. They build on regional economic base theory, input-output analysis, production theory, consumption theory, investment theory, location theory, urban agglomeration theory, central place theory, interregional trade theory, and related paradigms and concepts pertaining to the structure and performance of firms, industries, and markets within a regional economic system.

Even with this abundance of theoretical approaches to modeling regional economies, their complexity exceeds the scope of any one or combinations of several of these approaches in providing an effective framework for sustainable economic development. The American Economic Development Council, for example, defines economic development as the "process of creating wealth by the mobilization of human, financial, capital, physical, and natural resources to generate marketable goods and services." Bingham and Mier (1993) extend this definition by differentiating the roles of the private and public sectors in local economic development. They acknowledge that regional economic development has many meanings. They believe they capture these meanings in a series of "generative metaphors," such as ". . . problem solving, running a business, building a growth machine, preserving nature and place, releasing human potential, exerting leadership, and a quest for social justice." Baumol (1967; 1985), on the other hand, views the inherently nonprogressive nature of government and its lagging productivity as a source of its fiscal problems.

A "bottom-up" regional policy perspective calls for special understanding of the linkage between regional organization, regional governance, and regional performance.[5] The value of product disbursements, for example, varies by industry and by location. They are high value added in the metropolitan core area, low in the rural periphery. They correlate closely with levels of industry investment per worker. Where expected profits are high, with little risk of failure, investment and production follow, along with jobs that generate above-average earnings. The location of production activity in an economic region, whether in the core area, the periphery, or somewhere in between, has much to do with the earnings of its labor and capital resources and its long-term viability. Yes, it does matter where you are.[6]

We address some implications of the various forms of regional organization for businesses and government agencies concerned with investment in growth-influencing resources and facilities and their location. These include (1) the nurturing and support of globally competitive business enterprises, (2) the emergence of self-reinforcing industry clusters, (3) the shift to sustainable economic development focusing on a location-dependent economic base, and (4) the formation of functional economic communities of regional governance that make the difference between a vibrant and viable region and one that is on its way down.

The first of the four themes, globally competitive business enterprises, brings "new dollars" into the local area. It brings these new dollars into the area because its products compete successfully with the products of other business, both domestic and foreign. The commuting area of the local labor market establishes the functional economic community of export-producing business enterprises.

The second theme, self-reinforcing industry clusters, provides the essential elements of what Saxenian labels as a "regional network-based industrial system" (1994, p. 4). The industry clusters nurture close working relationships with one another and depend on the countless numbers of informal exchanges among their workers. By industry clustering, we mean the bringing together of a group of companies that rely on an active set of relationships among themselves for individual efficiency and competitiveness. These become, as noted by von Hippel (1988), important sources of product and process innovation for a business enterprise. The individual business clusters are an integral part of a larger complex of business enterprises that form a regional industrial system.

Providing the infrastructure for the emerging industry clusters and the regional industrial system puts a heavy burden on the third theme—sustainable economic development. The intent of such claims is, in the final analysis, to modify existing institutional arrangements in order to provide a new and different structure of conventions or entitlements. The central city of the metropolitan core area now seeks a rearrangement of these claims that in the future will provide for a fair and equitable sharing of the costs and the benefits of the essential urban infrastructure supported largely by the central city. This situation calls for institutional innovations, as suggested by Bromley (1989, p. 111), ". . . that will give expression to particular interests, which then show up as claims against the prevailing institutional structure." From these institutional rearrangements, an entrepreneurial government can emerge that considers its principal task the management of its space economy as a "good place" in which to do business and to live.

Entrepreneurial local governments join with one another to form a functional economic community of regional governance—the fourth theme. The commuting area of the local labor market defines the community. Ideally, the individual communities would maintain local environments favoring successful business enterprise and well-kept residential neighborhoods. They would collectively provide the community resources for reducing the heavy burden of poverty and aging on any one community. They also would share in the support of the infrastructure of transportation and education that is essential to the viability and competitiveness of local business enterprise.

Emerging from this cursory examination of urban-centered regions and urban regional economies are a series of key issues faced by residents (businesses, government, and households), economic regions, and metropolitan core areas. Central to the study of urban regional economies is the location decision, which ultimately relates to the investment decisions of this trinity of economic players. Also important are activity linkages. Critical for all these issues, of course, is the capacity and desire to effectively use these understandings in building the economic base and quality of life in each of these urban regional economies and to courageously explore each of their alternative futures.

Plan of the Book

We bring together a number of leading topics in urban and regional economics, such as location and land use, in Chapter 1 (Introduction). We also cite the role of information and its production, interpretation, and use as an increasingly important location factor for certain types of businesses and localities. The remaining individual chapters we present under three topical headings—Concepts (Part 1), Tools (Part 2), and Applications (Part 3).

Concepts

In Part 1 we discuss linkages between firms that make up various industries, in different areas and regions. We assess the importance of local resource supplies, especially labor, to the location decision. Of particular interest in the emerging global economy is the presence of local labor markets with the skills required by industries and commanding wages that can be afforded. Without the essential public and private infrastructure to provide market access for the export-producing enterprises of each labor market area, workers and employers would lack the income to sustain their households and businesses from one year to the next.

The topic of location and land use in Chapter 2 is of historic importance to students in agricultural economics and related fields of study. However, relevant concepts have shifted in importance with the growth of the service sector, particularly the information-related activities. Transfer costs still remain as important location determinants, but with technology and information transfer moving to the top of the list. Land use now becomes a central concern of local governments in metropolitan core areas. This involves largely urban land uses and the preservation of green spaces in the metropolitan area, rather than only agricultural land uses in the urban periphery. Even in the peripheral rural areas of an economic region, the preservation of original habitats often tops concerns relating to the production of agricultural, mineral, or forest products for human use.

Local retail markets define the central places of natural resource-based local and regional economies in Chapter 3. In the early years of our nation's history, farm-to-market roads provided access to local markets for farm products as well as household goods and services. Local labor markets now define the central places of the new technology-based economies. The commuting areas of these central places form the new economic regions. Their core areas are the largest and highest-order central places in each region. These also serve as the principal transportation and communication (that is, information transfer) centers of the emerging global economy. Because of their size and importance, the owners of these critical high-order services are in a position to assert monopoly power in their product pricing in many of the largest metropolitan areas.

Exports (that is, transfers of goods and services to buyers outside the place of residence of workers in a local labor market, described in Chapter 4) form the economic base of each LMA. Area-to-area specialization and the successful exercise of an area's competitive advantage in local and regional markets brings opportunities for much inter-area trade. Thus, the total sales of goods and services to buyers outside a region is much less than the aggregate of exports within a region. Central to an area's competitive advantage, however, is the productivity of its resource use—natural, material, and human. But its producers must have quick and affordable access to their market destinations and supply sources to maintain this advantage. For distant rural areas, this means low-cost transportation of locally produced and processed commodities. For downtown districts and central cities, market access is measured in milliseconds rather than days and weeks. Dependable air transportation and opportunities for face-to-face contact, however, remain the essential requirements for successful access to both product markets and supply sources in information-intensive, knowledge-based local economies.

The infrastructure cited in Chapter 5 is the common assets of a locality, the facilities and institutions that serve all its residents. According to the *American Heritage Dictionary,* it includes "the basic facilities, services, and installations needed for the functioning of a community or society, such as transportation and communications systems, water and power lines, and public institutions including schools, post offices, and prisons." Increasingly important are the services for technology transfer—the schools, colleges, and research institutions. Also high on the priority list is availability of, and access to, affordable housing for an area's resident population and workforce. Also important is the active participation of small and medium-size enterprises in local economic development and access to available systems for the effective transfer of commodities, people, information, and knowledge.

Tools

In Part 2 we turn to tools of analysis and measurement. The first chapter in this section (Chapter 6) explores methods for forecasting the likely course of local and regional growth and change. Specific tools, such as shift-share analysis, input-output modeling, and econometric modeling are presented. The second chapter (Chapter 7) addresses the building and use of social and economic accounts. Concepts without the tools for their application become a burden rather than an asset to a discipline and those who profess it. Each of the concepts cited earlier leads to the forecasting of local and regional economic activity and a need for the sorts of tools presented in the next two chapters.

To forecast, as suggested in Chapter 6, is to estimate or calculate something in advance, like the revenue forecast for a state agency or a sales forecast for a business enterprise. Most forecasts depend on some form of trend analysis and extrapolation. Some become extremely complicated in the number of variables and assumptions involved in making the forecast. Forecast complexity is usually associated with forecast accuracy, that is, precision and lack of bias. Precision is a measure of the closeness of the forecasts to the actual values. Bias is undervaluation or overvaluation of trend as revealed by a consistently low or high forecast of change. One test of forecast accuracy is the comparison of its result with that of a random draw. Another is comparison with a naïve model, for example, saying that next year's revenue is the same as this year's revenue, plus or minus some calculated trend value. In Chapter 6 we present the most commonly used forecast methods in urban regional economics under the two categories of custom-made and ready-made models. We illustrate the preparation and use of these models with brief examples of their application.

Accurate forecasts depend on accurate and readily accessible data, as argued in Chapter 7. The most complete data are national in coverage of the economy and population in totality. Much detail is available for both the economy and its workforce. Population statistics have many categories of detail, with much available for individual counties and minor civil districts. The economic data are much less complete for the smaller areas. Even less complete are data covering individual counties and localities. Even if the small-area statistics were as complete as the national statistics, the separation of place of residence from place of work and workplace ownership calls for an even larger database than at the national level to fully account for the likely local effects of a changing national economy. Chapter 6 focuses on the local statistical sources for building and using the forecast methods presented earlier under the rubric of social and economic accounts.

Applications

Part 3 deals with some applications of the concepts and tools of urban regional economics covered in the preceding chapters. It asks the question, What economic and social problems have emerged out of the processes explored in Part 1? The question of central cities with attending poverty, crime, and high site costs are among the issues addressed. The idea of the central city as a driver of economic development is presented, along with various methods and data for measuring its successes and failures. Finally, the future of the central city and its relation to the rest of the economy is explored.

Chapter 8, the first chapter in Part 3, focuses on downtown districts of metropolitan core areas—the central cities of economic regions—such as Boston, New York City, Cincinnati, and Minneapolis-St. Paul. The downtown districts provide high-order business services, such as finance, insurance, real estate, and management consulting. Various studies have noted that a downtown location usually implies membership in a dense and very important network of business linkages that involve many face-to-face meetings. These sorts of businesses can pay the increasingly high rents that drive out former occupants, such as retail stores and personal services. Downtown districts may house one or more professional sports teams, mostly in publicly financed facilities that remain occupied at the discretion of their occupants. The local sports team owners enjoy the privilege of membership in an exclusive national club of team owners, operating outside existing rules pertaining to the regulation of business monopolies.

Strong urban neighborhoods and communities, discussed in Chapter 9, are extremely critical to the viability of downtown districts and central cities. Urban neighborhoods are defined by proximity. Geography, rather than communal ties, brings neighbors together. An urban community also can be defined by its geography, namely, its commuting area. When the urban neighborhoods lack essential infrastructure and services, like good schools, safe streets, and affordable housing, they experience population decline. Their problems eventually spill over into surrounding neighborhoods and the downtown district.

Understandably, neighborhood survival and revitalization become important issues for urban regional economics to address. Jobs for many neighborhood residents are disappearing, starting with the downtown districts. Here the new businesses acquire a new workforce that largely commutes daily from suburban homes. Many of these neighborhoods still retain their long-established institutions of health care and higher education. These become the new resource centers for neighborhood renewal and revitalization. They also may provide favorable locations for technology-intensive businesses, such as a manufacturer of medical devices.

The rural communities covered in Chapter 10 are, like urban communities, the commuting areas' central places, but smaller ones. Some rural communities provide the new work sites for the expansion of many manufacturing facilities engaged in fierce price competition in their respective commodity markets. These are the transitional rural areas outside the central cities and other metropolitan areas. Many of these areas face limited opportunities for business expansion resulting from a limited workforce. Lack of affordable housing limits expansion of the current workforce.

The more distant peripheral rural areas remain dependent on their traditional economic bases of agriculture, mining, and forestry. The high-amenity areas attract seasonal residents and occasional visitors. For the most part, however, these are the declining areas of an economic region. Historically, they have provided the new workers for the expanding industries of the region's growing metropolitan areas. Millwork manufacturing is one industry that is found in both the metropolitan core area and the peripheral rural area. An important issue for the peripheral rural areas is the impact of free trade measures, like the North American Free Trade Agreement, on local export-producing industries. Sugar beet production and processing in western Minnesota is an example of this sort of industry, presented as another case study in Chapter 10.

Chapter 11 discusses the use of labor market areas to build economic regions. The labor market areas closest to a given metropolitan core area form a proximity-based economic region. The urban neighborhoods and communities, along with the rural areas and communities of a given economic region, share in varying degree the resources of the same central city and its downtown district. This sharing occurs because of the many day-to-day (and longer) linkages between businesses and households in both the core and surrounding areas. Inter-area trade is a visible manifestation of these linkages. Through trade—the exporting and importing of goods and services between areas—each area is challenged to exercise its comparative advantage, whatever it is. Air transportation and the global transportation networks link each of the metropolitan core areas into the global economy.

Economic regions with continuing and growing linkages between their metropolitan core area and rural area activities, and the same for other metropolitan areas in the region, have strong possibilities for continuing growth and development. The core area, especially, serves as a "growth pole" for the entire region, with its important spillover effects on surrounding areas within the region and trade linkages with the more distant ones. How metropolitan core areas generate economic regions requires something more than exports, namely, the capacity to replace wide ranges of its imports. We focus on the role of imports, as well as exports, in the delineation of an economic region and its structure and activities.

Notes

1. John Simons and John Hayward, "*Gates Opening:* For the High Tech Industry, Market in Washington is Toughest to Crack." *Wall Street Journal,* March 4, 1998, p. 1. This article notes that high-tech industry officials are slowly becoming convinced that they have no choice but to respond to the pleas of their political friends to get involved in Washington politics. It offers the thought that the "very qualities of individual initiative, rational intellect and bristling self-confidence that have allowed Silicon Valley pioneers to flourish in business have contributed to industry's fumbles in the slow, cumbersome, consensus-driven arena in which political business gets done." The article notes that the "clash of cultures, in fact, goes a long way toward explaining why the booming high-tech industry remains something of a bust in the world of politics and government."

2. The Detroit experience is reminiscent of the German experience, with its high wages and generous worker benefits following World War II, now dampened by high unemployment, along with the collapse of East Germany. Yet, the political party expressing free market views is losing, while the opposition party that still promotes high unemployment solutions calling for "higher wages to increase consumer spending and spreading around the jobs that exist by shortening the workweek" is gaining in public popularity, as reported by Greg Steinmetz. This article notes, however, that the debate is growing over whether or not to relax labor standards and risk letting what are viewed

as bad jobs crowd out good jobs, a "debate critical to Europe's largest country as it tries to cope with unemployment that has hit 12.6 percent." See Greg Steinmetz, "*Bottom Rung:* A German Who Offers Low-Pay Service Work Dismays Countrymen—They Want Only 'Good Jobs'; But High Unemployment Starts to Change Minds," *Wall Street Journal,* March 3, 1998, p. 1.

3. One of the more widely used books in urban economics, for example, lacks even one chapter (out of twenty-two chapters) on the urban labor market. Its topical coverage, however, is very broad, ranging from the "first cities" and the "demise of small stores" to the "single tax" and the "optimum amount of crime." Discussion of economic base and input-output models, which applied economists often use in predicting urban economic change, is limited to nine pages, and then only to highly aggregated representations of these models that show roughly the same technical coefficient values for all urban areas rather than the sharply different values derived from detailed industry data now readily available to students of urban economies (O'Sullivan, 1993 pp. 138–146).

4. The commission gives priority to "development of endogenous potential" in the following: (1) Development of the primary sector in order to help agriculture adapt to the reform of the Common Agriculture Policy, and to ensure that it plays a positive role in the economic and social development of rural areas; (2) development of other activities, including small and medium-size enterprises in particular; (3) expansion/promotion of tourist and leisure activities and the creation of nature parks; (4) respect for the environment; and (5) European Social Fund assistance for training programs in support of agricultural and nonagricultural activities (from Commission of the European Communities, Annual Report of the Implementation of the Structural Funds, Brussels, 1991).

5. Tom Redburn, writing in the *New York Times* about the slow recovery of the New York region from its three-year recession (June 12, 1994, p. A12), links regional economic organization to regional performance when he notes that "Regional assets, in many cases, have become liabilities. Job gains will be retarded by New York's heavy dependence on the mammoth corporations now slashing their work forces for greater productivity and competitiveness." He also notes that "Among big cities, only Detroit has a lower working age population than New York City. Various explanations are offered, including generous welfare benefits, poor education in troubled urban neighborhoods and infestations of illegal drugs and crime. But a lack of promising jobs that require more strength and stamina than study is also important."

6. A recent lead article in *The Economist* (July 30, 1994, pp. 13–14) starts: "The cliché of the information age is that instantaneous global telecommunications, television and computer networks will soon overthrow the ancient tyrannies of time and space." It is quick to note, however, that through history, companies in a fast-growing field tend to concentrate regionally and that in reality "history counts: where you are depends very much on where you started from." The most advanced use of the Internet system, for example, has been "to strengthen the local business and social ties among people and companies in the Silicon Valley"—a reality that "offends not just the techno-enthusiasts but also neo-classical economics: for both, the world should tend towards a smooth dispersion of people, skills and economic competence, not towards their concentration."

PART 1
Concepts

2

Location and Land Use

The location decisions of individual businesses and households are critical to the growth and development of urban regional economies. For example, Detroit's future was set when the automobile industry decided on that region as opposed to Chicago or New York. The locations that followed, the labor force that was developed for the industry, the finance that was required and forthcoming, and the infrastructure support offered by local government all combined to establish Detroit as the auto-making center of the United States for many decades.

Understanding the location decision of a business or household is the first step in our journey toward understanding urban regional economies. Early attention by economic researchers focused on the location of trading establishments. For example, grain elevators and their sideline businesses catering to farmers and retail stores, and personal services catering to the needs and desires of the farm family. Retail and service businesses prospered in central places that maximized access to customers.

The size and type of central place depended on the surrounding economy and its people and products, moving to and from the centers of local commerce. The driving force in the growth of these central places was the shift from an economy based on self-sustaining farms to one of producing more than household needs. The surplus was then traded for other products not available in the immediate household. We still carry the legacies of the past, even to this day, as we demonstrate by starting with the topic of location and land use as an entry to the field of urban regional economics.

Other early efforts identified location patterns, mostly for manufacturers, that were dependent on factors such as the location of nontransferable resources, local factors affecting the costs of production, transfer costs for resources that were transferable, and distribution costs of the final product. Weight-gaining and weight-losing products became the focus of attention along with characteristics that would make possible the location of economic activity at points in between nontransferable resource locations and final markets. More recently, the emergence of technology-intensive manufacturing and producer services as the principal exports of metropolitan areas has brought into prominence local labor markets and information transfer costs as major location determinants for cutting-edge industries. By labor market area, we mean the daily commuting area for the larger urban places characterized by a resident labor force that is somewhat smaller than the total number of local jobs. Information transfer costs refer to the costs associated with the transfer of information from place to place, from the producer or the owner of the information to its buyer.

Location Determinants and Structures

Low site cost is an important factor in accounting for the recent rapid growth of manufacturing in many rural areas. Transfer costs are important in retaining technology-intensive businesses in metropolitan core areas, while local entrepreneurship and new business formation are important in the growth of both the metropolitan core areas and the adjoining rural areas within close proximity to the core areas. Finally, political boundaries constrain certain private actions while facilitating others. The net result is diverse spatial patterns, some of which are more efficient than others in allowing consumers and producers to build the transportation and communication systems for reaching their respective destinations.

This section begins with a discussion on the price and value of land. Much of the discussion on rents relates to an era in which agriculture was the dominant industry and transportation was relatively difficult. The section ends with some transitional statements leading to more modern theories of location. Central places are discussed later in the book but are briefly introduced in this section. The emerging role of manufacturing is also presented, as is a short discussion on the now important role of other site costs, such as the regional cost of labor or additional costs imposed by environmental protection policies.

Our attention then turns to the costs of transfer. Emphasis is placed on the transfer of information and its effects on location. Hypotheses and data are presented to illustrate the points made.

Site Costs

Site costs affecting the location of economic activity include those relating to a particular piece of land. Also included are specialized labor skills, industry regulation, and environmental protection. All affect the costs of producing a particular good or service, including anything from a standardized commodity, like the No. 2 corn traded worldwide, to a highly differentiated product, like timely and accurate information for making business decisions. Timely and accurate information is especially important to export-producing firms seeking access to carefully targeted domestic and global markets. Thus, site costs may enhance the competitive advantage of some rural areas with low labor and environmental costs while reducing it for others that lack timely access to markets. Conversely, site costs may erode the competitive advantage of some urban areas while increasing it for others characterized by large concentrations of skilled labor and timely access to markets and to market information.

How do we determine the value of land? The value of land is equal to the discounted value of the returns from the employment of that land over time. A well-established principle tells us that a dollar tomorrow is not worth the same as a dollar today. The proverb, "A bird in the hand is worth two in the bush," speaks to this principle.

If I have a dollar today and the interest rate is equal to five percent, what is that dollar worth one year from now? The answer, of course, is \$1.05. But we can turn that question around and ask, What is \$1.05 a year from now worth today? The answer is \$1.00. We found that answer by dividing \$1.05 by 1 + 5 percent.

What is a dollar today worth in two years? If the interest on that dollar is compounded on a yearly basis, the dollar today is worth \$1.0525 after the two years have passed. What is \$1.0525 two years from now worth today? The answer is \$1.00. We found this by dividing \$1.0525 by $(1 + .05)^2$. If we were analyzing three years from now, the discount factor would be $(1 + .05)^3$, and so on, as long as the income stream is expected to exist.

The formula for the value of land, therefore, is

$$V = \sum_{t=1}^{n} \frac{R_n}{(1+i)^n}$$

$$V = \frac{R_1}{(1+i)} + \frac{R_2}{(1+i)^2} + \frac{R_3}{(1+i)^3} + \ldots + \frac{R_n}{(1+i)^n}$$

where V = present value of the land, t = time period measured in discrete units (day, month, year, etc.), i = rate of discount (interest on an alternative investment plus any risk factor), R = rate of return from the land in any one particular year, and n = number of time periods into the future, potentially running indefinitely toward infinity.

Relative Productivity

Site value is also determined by the economic rent its occupancy generates. In years past, the value of a site usually referred to its agricultural productivity compared to other sites equally distant from their product markets. The higher the relative productivity, the greater the value and the less added costs of transportation. The rent paid on the most productive land was based on its advantage over the least productive land. The rent accrued wholly to the landowner. The two more noted proponents of this view were David Ricardo (1817), who emphasized the differential fertility of land, and J. H. von Thünen (1826), who developed a framework for measuring the influence of location differentials on land rents.

Ricardo's theory saw rent as a residual reward to land's productivity. In today's economic view, we assume pure competition with three units of land—the first highly productive, leading to low per unit costs; the second moderately productive, with intermediate per unit costs; and the third the least productive, exhibiting the greatest costs. Figure 2-1 shows the cost and demand conditions facing the firms located on each of these units of land.

If the price of the product (in this case, corn) is at level P, and the average costs of the three sites (1, 2, and 3) are as shown, it is clear that the first land to be cultivated will be at site 1, where economic profits can be made. Now the question is, Who gets these profits? If we assume that the average cost curve includes all production costs except those for land, the excess return is available to bid for land. We have assumed pure competition. Therefore, the producers will bid against one another for site 1 until all of the excess return accrues to the landowner. The producing firm will make only normal profits once the rent is paid.

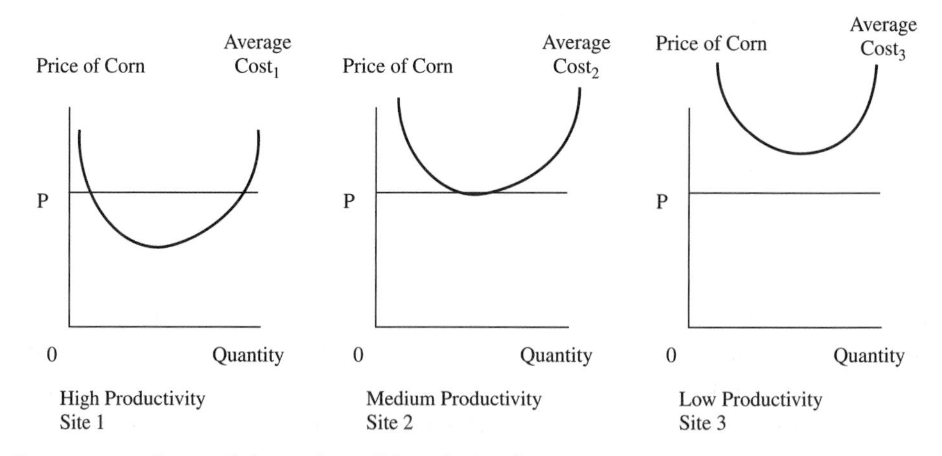

FIGURE 2-1. Cost and demand conditions facing firms.

Site 2 can also be farmed. However, because of its lower productivity, no rent will be bid. The firm makes normal profits and the landowner receives no rent. Site 3 will not be farmed under current conditions since no economic return can accrue to either the farmer or the landholder.

If prices rise, site 2 will go into full cultivation and a rent will be earned by the landowner. If prices rise to a high enough level to make site 3 profitable, it, too, will be cultivated and rents will be earned. Of course, as the price continues to rise, the rent on the most productive site will rise also as profitable farmers bid against one another for its use.

Location Access

Von Thünen's theory took a slightly different twist. Assuming a flat and homogeneous plane with transportation equally possible in every direction, von Thünen saw land rent as a function of access to the region's core, where population and the bulk of economic activity takes place. The highest rents go to land that provides this access. So the core area houses the commercial and processing space, both of which require dense populations and face-to-face bargaining and communication. These processes, in turn, use land intensively. The next ring contains relatively high-value truck crops that are both fragile and perishable and therefore expensive to transport to the core, where they can be sold. The second-highest rents occur within this ring of the defined region. The next ring contains grain farming. In this ring, land is used more extensively than in the previous rings, and rents are lower. Beyond this last ring are peripheral activities, such as logging and mining. Land rents are the lowest in this last ring, and land is used most extensively.

The resulting landscape was predicted to be a series of concentric circles, beginning with the core and ending with the logging and mining operations. As one moves from the core toward the periphery, land rents fall and the extensiveness of land use increases.

The value of a parcel of land to the owner is the discounted value of rents over future time periods. The value of land to the producer is the discounted value of profits (net returns) from the use of that land over future time periods. Assuming the market is competitive (a common but often bothersome assumption) land will go to its highest and best use, that is, the use netting the greatest returns. When agriculture was the basic exporting industry of most areas, the location of economic activity turned around the role of land as property and the returns it provided to its owners. Land was considered a fixed resource. The economic challenge facing its owners was to find its highest and best use, usually some form of crop production or livestock production, or a combination of the two.

By the turn of the nineteenth century, the economic activity worthy of study included industry. These were the manufacturing concerns concentrated in emerging urban places. They had no means of support other than the excess production these places generated for shipment to other areas. Soon, the location of an industry using more than one input in production became the focus of a new generation of location scholars, exemplified in the works of Alfred Weber (1909) and his location triangle (discussed later).

The term *industry* is no longer confined to manufacturing alone. It now includes the full range of remuneratively productive economic activity. Nonetheless, the agricultural legacy left by the early writers on land and location economics persists in the location literature on rural trade and service centers. Theories of a central place and an urban hierarchy (discussed in the next chapter) refer to the dispersed population around each central place and the retail stores and service establishments this population supports. The frequency of purchase and value of goods and services are found to result in an ordering of central places. The central place studies of Christaller (1966) and Lösch (1964) show the existence of such urban hierarchies.

Studies of industry location within an urban place depart sharply from the rural and dispersed population emphasis of many contemporaneously completed studies of central places and urban hierarchies. Particular kinds of retailing, for example, concentrate in certain districts, while others agglomerate among dissimilar establishments that are nonetheless complementary in consumer purchasing behavior. Location within specialized industrial parks, shopping districts, or malls thus becomes a function of the drawing power of these microcenters of industry and commerce and the compatibility of a given activity with those around it. The activity survives as long as it pays the rent and provides an acceptable discounted rate of return to its owners.

Increasingly important to both existing and new businesses are labor costs and availability at a given site. The attraction of low labor costs provides an incentive for industry expansion in some rural areas, especially when the low labor costs accompany low land and environmental costs. These are the areas experiencing growth in manufacturing—contrary to the experience of long-established metropolitan areas currently losing much of their former industrial base. However, the availability of large pools of skilled labor remains an important location attraction for the metropolitan areas marked by large concentrations of technology-intensive industry. This attraction exists even in the face of high labor costs. For these areas, the existence of an active and diverse local labor market reduces the costs of business layoffs and expansions for both employees and employers—an important location attraction for start-up businesses entering a new market with a new product.

Other site costs include those imposed by government regulations that affect businesses differentially from place to place. The costs of pollution abatement and environmental protection affect the expected rate of return from a given site and, therefore, business location. This is especially true in urban areas that once were the centers of heavy industry; these areas now have only the vacant land left behind that is condemned for urban uses until its toxic wastes are removed. The removal of these wastes imposes exceptionally high time and money costs for most manufacturing businesses, reducing these businesses' expected net returns over time and precluding their future expansion possibilities.

Transfer Costs

Transfer costs are associated with the movement of primary and secondary inputs to production sites and of the product outputs to their respective markets. They pertain to a wide range of transfer activities—from trucking and warehousing to air transportation and satellite communications. They also include the added rent incurred in face-to-face communication in the production, interpretation, and distribution of unique, high-margin, decision information. Thus, for some activities, transfer costs correlate inversely with the level of population concentration and traffic congestion. These activities find rural sites more attractive than urban sites. For others, transfer costs become prohibitively high in rural areas. Such activities do best in the downtown districts of major metropolitan agglomerations.

Pricing of Transfer Services

Ground transportation and air transportation are the common forms of transfer of people and products from one site to another within and between urban regional economies. With both forms of transportation, the pricing of the service provided is an increasingly contentious issue. High levels of ground transportation in heavily populated areas result in high levels of highway congestion and air pollution—external costs that are not assessed directly to their source, namely, the individual motorist. In the case of air transportation, both the airlines and downtown business travelers prefer a nearby airport location, but for different reasons. The airlines prefer a limited supply of slots to contain competition and maintain pricing dominance, while downtown businesses favor the shorter distance to the airport over having larger and more distant airports, simply to enhance worker productivity. The monopoly pricing may present a win-win outcome for the two parties, but not for other business travelers and the general public. These are outcomes that result from a rationalization of the air transportation system in behalf of the dominant airlines at major international airports.

CONGESTION PRICING — Road congestion results in large hidden costs to both the commuter from place of residence to place of work, and the nearby resident. These costs include the losses of time and wasted fuel and the increase in pollution and related environmental and health damages. Just the time and fuel losses exceed $75 billion in the United States, according to findings of the Texas Transportation Institute. These losses are even greater in Europe and Asia.

One solution is to put usage meters on cars, with different peak and off-peak charges. Gary Becker, the 1992 Nobel laureate in economics, believes that every vehicle should be equipped with a radio transponder that emits an automatic vehicle identification and the time of day for recording as it passes by an electronic toll collector (ETC). This would be a better solution than building more highways (Becker, 1998). Everyone owning a vehicle would be required to have a transponder. Anyone failing to do so would be caught by the ETC and reported to the central office, which would regularly send the accumulated charges to each driver.

Because the usage charges would be higher for peak hours than off-hours, Becker believes that drivers would respond by shifting to the off-peak hours and even reduce the total amount of transportation by carpooling and other ways of reducing the direct charges. A major difficulty exists, however, in the lack of public acceptance of congestion pricing and the reality of strong opposition to it on the part of downtown businesses and the daily commuter to workplaces in these businesses.

MONOPOLY PRICING—The pricing of a limited supply of a product or service is readily demonstrated in the ever-popular price-quantity charts of microeconomics textbooks. In the case of a single, or dominant, service provider, for example, the effective market price is much higher than the competitive market price. For some services, the economies of size are so large, or the control of the marketplace is so entrenched within the jurisdiction of a single service provider, that effective price competition is impossible. This is the sort of problem faced by users of interregional air transportation at major U.S. air nodes in our largest metropolitan areas.

Solutions for dealing with monopoly pricing in air transportation again derive from the microeconomics studies, specifically findings on how individual airlines, passengers, and freight shippers respond to various market incentives. One solution is to offer airport slots to competing airlines at periodic auctions. The highest bidder would retain rights to a slot for a prescribed time period, with a certain maximum number available to any one airline. The income derived from the slot sales would be used to maintain airport premises and develop new markets for airport services.

An alternative solution is to move the existing airport to a new and larger site. This approach also involves an overview of the airport under different sets of assumptions about airport and airline operations and the likely businesses and personal responses to various market incentives. A realistic planning horizon for a new airport would extend over many years, involving establishing a likely location for the airport and initiating a process of land purchases relating to the new airport facilities and their space requirements. Plans must include recycling the old airport, which is probably more valuable to local governments for its possible residential, commercial, institutional, and specialized industrial use than as an airport.

Either solution requires a well-informed understanding of the likely consequences of changes in airline and air transportation user behavior as a result of the proposed changes. Again, a major difficulty exists in the lack of widespread public acceptance of both solutions and the reality of strong opposition to either solution on the part of the dominant airlines and many downtown businesses.

Information Transfer Activities

Johansson (1987) identifies three types of activities that generate contacts: (1) production and handling of knowledge, (2) service activities and goods, and (3) dispersal of standardized information. The first category is connected with the renewal activities of the firm. The second category involves the service and goods-producing activities necessary for the proper functioning of the firm. The last category contains market activities. We can then add a fourth descriptor: activities directed toward the control and regulation within firms.

Hence, from a firm's perspective, four types of activities can be observed, which result from the different needs associated with business meetings. The frequency of business meetings is determined by the general expected level of uncertainty during the exchange of information. This further implies that the level of competence, education, and experience needed increases with the level of uncertainty.

Table 2-1 presents a brief description of each of the four business activities involving the transfer of information. These activities, in turn, depend on unique local conditions for their optimal transfers. Transfers of information between renewal activities and routine business service and production activities, for example, occur within the firm. A high level of uncertainty characterizes them, and the information requires a high level of competence in its most effective uses. On the other hand, information transfers between routine business service activities and market activities occur outside the firm. The market activities in this transfer involve lower levels of uncertainty and competence.

TABLE 2-1 Differentiation of business activity in the firm by type of contacts and level of uncertainty and competence

Higher level of uncertainty and competence for contacts within the firm	
Renewal activities	Routine business service activities
• Innovation • Knowledge • R & D, etc.	• Financial services • Marketing • Advertising, etc.
Lower level of uncertainty and competence for contacts outside the firm	
Production activities	Market activities
• Control • Regulation, etc.	• Selling • Purchases, etc.

Source: Modified from Hugoson and Karlsson, 1996, p. 7.

Information transfers characterized by a high level of uncertainty and competence involve a high frequency of business meetings. Conversely, those with a low level of uncertainty and competence require a low frequency of business meetings. The high-order firms engaged in these activities locate in the downtown districts of large metropolitan areas, but their face-to-face contacts may extend over a much larger geographic area. The distant contacts generate long-distance business travel and the demand for cost-effective air access to the distant contacts and markets.

Studies associated with information transfer face a common difficulty of limited data. For example, the database developed by Jönköping, which is the most complete for the study of long-distance travel, is built on a new survey each year based on monthly interviews with approximately 2,000 respondents. The surveys provide data on long-distance trips between 1,940 regions (Hugoson and Karlsson, 1996). The survey regions are aggregated into functional regions or labor market areas (LMAs). For the 1991–92 period, the database had over 84,000 respondents, with 5,552 reporting business trips within Sweden between 70 LMAs, generating 7,892 two-way main trips of more than 100 kilometers. Expanding the survey numbers to the entire country of 9.5 million population yields an estimated 110 million long-distance trips yearly, of which 24 million were business trips.

Renewal activities occur within the firm as well as in cooperation with other firms or the public sector. The activities involved in this category concern the creation of new business ideas, the imitation of existing products, and the creation of new basic knowledge. These kinds of activities demand high competence to be able to find, create, analyze, or implement new technology and knowledge as new business ideas. When an external link is established, a high frequency of meetings is often required because of the complex interpretations of nonstandardized information and the knowledge that is inherent in these types of activities.

Routine business service activities occur outside the firm and contribute to the support of the firm's needs from the service sector. Examples of business contacts in this category include consultants in areas such as financial, insurance, juridical, marketing, advertising, accounting, transportation, headhunting, hotel, and restaurant services. Prevailing uncertainties and the demand for competence in the business service network imply that the frequency of business meetings is high.

The last categories are production activities and activities related to the production processes that occur within the firm. These activities keep the production process going, dictate the level of production, and give business services to the firm in case they are not supplied from the market. These activities also cover the distribution of products through the firm's network. Business meetings involved in these kinds of activities consist of regulation, control, and distribution decisions within and between the firm's geographically dispersed production, administrative, selling, and buying units.

The Jönköping study provides an analytical framework for determining the sensitivity of individual sectors of a region's economy to air transportation (Hugoson and Karlsson, 1996). It is highly instructive in the design and construction of a comprehensive database and model for understanding and predicting the demand for long-distance business travel. It also addresses the importance of a comprehensive survey of all long-distance travel by all transportation modes. These data would be available for studies of long-distance transportation. The findings support a series of hypotheses about business travel, of which two are particularly important to this study: (1) business travel is primarily connected with meetings between white-collar workers at higher levels within organizations, and (2) business trips intensify within growing regions. That is, other things being equal, growing regions will exhibit a higher order of travel than nongrowing regions. Long-distance business trips were found to be generated particularly by geographically dispersed firms and in situations connected primarily with the distribution of knowledge.

Industry Location and Activity Clusters

We start the present examination of economic activity location by focusing on businesses that form clusters of interdependent industries, even in the case of retailing. They, in turn, depend on large, diverse, and active local labor markets and efficient transportation to and from these activity clusters. The new technology-intensive businesses depend also on evenhanded approaches by all levels of government to the regulation of economic activity that otherwise could jeopardize their contributions to the economic well-being of their localities by unfairly reducing their competitive advantage in global markets.

Forming Activity Clusters

Similar businesses, catering to the local resident population, form clusters of competing enterprises, each sharing part of a total customer base that is larger than the aggregate of customers when the same businesses are dispersed and isolated from one another. Dissimilar businesses, if depending upon each other for proximity-sensitive production inputs or output markets, form clusters of market interdependence. Printing and publishing businesses, among the last remnants of a once prominent manufacturing sector in our major cities, still locate close to downtown business customers. Technology-intensive manufacturing businesses also locate in the downtown district to maximize access to the local labor market. These establishments form local industry clusters. They depend on a local labor market that extends well beyond the boundaries of the central city. Finally, the almost ubiquitous urban/regional transportation infrastructure provides the dispersed workforce daily access to the industry clusters.

Business establishments are of many sizes and shapes, catering to intermediate markets of other businesses and final markets of consumers and investors. *The Standard Industrial Classification (SIC) Manual,* published by the U.S. Department of Commerce, lists several thousand different types of businesses that are part of one or more industry groups. These listings are based on a classification scheme of one to four digits. For example, a four-digit livestock slaughtering business (SIC 2011) is a part of the three-digit (SIC 201) meat products industry. The meat products industry, in turn, belongs in food products manufacturing (SIC 20)—one of twenty two-digit manufacturing industries. All manufacturing (SIC 20 through SIC 39) forms a one-digit industry group classification.

One establishment may be a part of several four-digit industry groups, depending on the diversity and range of its product line. The same establishment may be a part of a multiestablishment firm. A large meat packing plant, for example, may slaughter hogs and also process them, producing bacon, ham, and half-carcasses that are further processed in a retail outlet into smaller retail cuts. The hog slaughtering plant, because of weight loss of live animals in transit and processing, typically locates in an area of concentrated hog production. The same company may own several facilities. These specialized facilities typically locate in areas of population concentration near to food distributors or retail stores.

Transportation Costs and Requirements

Accounting for and predicting the location of economic activity is a central concern of urban regional economics. As mentioned earlier, past efforts focused primarily on farming and the processing of farm products. The type of production depended on the agricultural productivity of the given site and its distance from markets. The processing site was close to the production site as determined by the weight loss in processing, the proximity of the site to the nearest competitor, and the relative transportation costs for material inputs and product outputs. In the case of manufacturing, a product using inputs from several sources might find several sites available, depending on relative transportation costs for the material inputs and the product outputs. The emphasis in both cases, whether a given site or a given product, was on standardized commodities readily traded in domestic and international markets.

Commodity-producing businesses, such as a livestock slaughtering plant, typically locate in areas of concentrated production of the primary raw material, but some distance from competing processing facilities. Thus, each processing facility seeks its own supply area, protected by the cost of transportation to a competing processing plant. Active price competition exists largely along the perimeters of supply areas, with prices being higher at greater distances from a given facility. In effect, some more distant producers benefit from access to more than one processing facility, depending upon the intensity of competition among these facilities for a limited input supply.

A processing plant with more than one raw material input, such as a steel manufacturing facility, faces a more difficult location decision than the single input plant. It may incur a different transportation schedule not only for each of its raw material inputs but also for its product output. The least-cost location is one that minimizes the total transportation bill, assuming the same processing costs at any location. The Weberian triangle (Weber, 1909) illustrates the decision problem for the two-input, one-output facility by introducing the various transportation costs as relative weights with the weight-minimizing location being inside the triangle.

The processing of the raw materials into a finished product may also incur a physical weight loss, as in the case of a steel manufacturing facility that converts iron ore and coal into steel. This would bias the facility location toward the input source. The location of the steel manufacturing facility also takes into account the use of a third raw material input, namely, limestone. Its early location, in Pittsburgh or Cleveland, was closer to the coal fields and limestone quarries of Pennsylvania and Ohio and the steel-using manufacturers in the Midwest than the iron mines near Lake Superior.

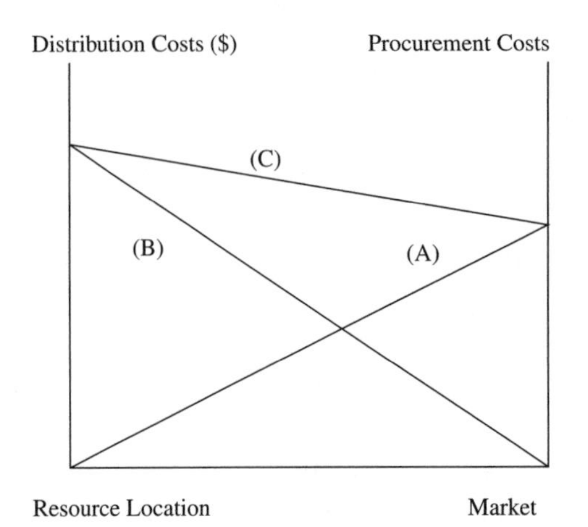

FIGURE 2-2. Simple one-input, one-output model.

One notion of firm location is associated with the location influences associated with weight-gaining versus weight-losing production processes. First mentioned by Edgar M. Hoover (1948), the basic premise is that weight-gaining processes locate near the market, while weight-losing processes locate near resource supplies.

Take the example of iron ore. The first stage in processing iron ore takes place near the iron ore supplies in northeastern Minnesota. Why? Since the process involves a great deal of waste, why ship the waste when it can be discarded at the mine, leaving the much lighter and less bulky iron ore to ship?

On the other side, take the example of soda drinks. The inputs to soda are essentially syrup and water. Water is relatively ubiquitous (although some people in the western United States may dispute this). So why not ship the syrup to the market and add the weight of the water there?

Figure 2-2 shows how this simple one-input, one-output model works.

Assume the firm has one resource input and one output. The resource is in a location that is separated from the market where the single output is sold. Assume further that production costs are equal at the two locations. Finally, assume that transportation costs are linear functions of distance (that is, the costs increase at a constant rate over distance) and are based on the distance and the weight of the commodity being shipped. Under these conditions, the firm's choice will stem from the costs associated with transporting either the raw material or the final product.

Assume that line A in Figure 2-2 represents the cost of procuring the resource. If the firm is located at the market, the costs of procurement will be at their maximum. The same is true for distribution costs (line B) when the firm is

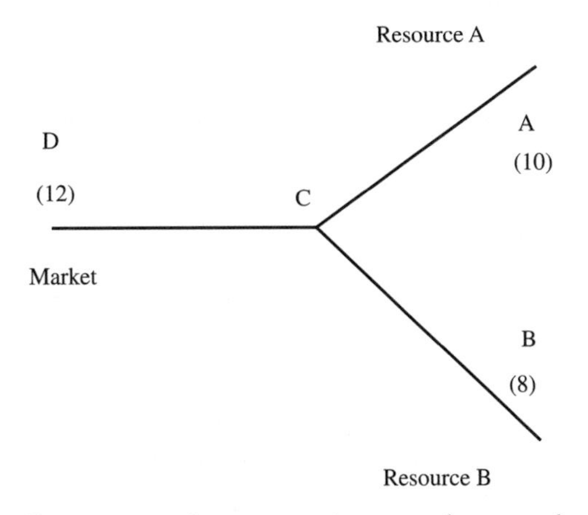

FIGURE 2-3. Assume two inputs and one market.

located at the resource location. Line C is the total transportation costs, or A + B. When the firm is located at the market, the total transport cost equals procurement cost since distribution is zero (or at least minimal in this relatively simple example). The same is true for distribution costs, which are equal to total cost when the firm locates at the resource location.

Figure 2-2 shows that procurement costs increase at a lesser rate than distribution costs. This is demonstrated by the fact that line A has less slope to it than line B. Total cost is at its lowest point at the market. Therefore, this is where the firm will locate under the specified assumptions. Note also that an intermediate location is not likely since total cost is above its minimum at these intermediate points.

This example is consistent with that of a weight-gaining process. Adding water to cola syrup at the resource point adds weight and therefore transportation outlays for the finished product. Thus, the distribution costs are high relative to the procurement costs. Such a firm would be considered market oriented.

Under what conditions is an intermediate location suitable? The classic argument along these lines was put forward by Weber (1909) (Figure 2-3).

Assume two resource inputs and one market. Assume further that the firm can locate anywhere it chooses; that is, there are no roads or rail lines that dictate the location. Assume finally that production costs are equal at every location, so that transfer costs are the sole determinant of the firm's location.

Under these conditions, an intermediate location becomes likely. To explain this, let us introduce the concept of *ideal weight*. Ideal weight is the weight of the commodity times the cost per unit of distance and per unit of weight to ship that commodity. For example, if the commodity weight is ten pounds, the transfer costs are $1 per pound per mile; the ideal weight is equal to $10 per mile.

As we move toward a particular location (A, B, or C in Figure 2-3), we save the ideal weight (indicated in the parentheses). However, as we move away from the other possible locations, we have to pay the ideal weights of the other two locations in additional transportation costs.

For example, if we move toward location B (having the lowest ideal weight), we save $8 per unit of distance traveled but we incur $22 ($12 at the market plus $10 at resource point A) per unit of distance in additional procurement and distribution costs. This is not a particularly good deal.

What if we try to save on the highest ideal weight? If we move toward location D, we save $12 per unit of distance traveled, but we incur an additional $18 ($10 at A plus $8 at B) in additional procurement costs. This is still a loser.

In fact, there is no extreme location we can choose that would minimize our transportation costs. So we choose an intermediate location. Where is that intermediate location? If transportation is equally possible in every direction, the intermediate location will occur at a point of balance between points A, B, and C. Imagine that the ideal weights at each extreme are located on a series of strings knotted at point C. Each would pull C toward its particular location. While it is true that point D has the highest ideal weight and will therefore exert the strongest pull on that string, it is also true that the ideal weight of twelve pounds does not overwhelm the other two. The balance between these weights might favor the market somewhat, but the balance will be somewhere in the middle.

In many instances commodity transportation costs are not the principal determinant of a firm's location. Improvements in transportation technology, production processes aimed at making lighter and therefore more transportable products, and increased market dependencies all serve to make the commodity transportation theories less relevant to today's firm and its location decision. While the locations of natural resources are still important to some firms, other resources are increasingly becoming the focus of firm location decisions.

Chapter 3 deals much more extensively with labor and labor markets. Here, suffice it to say that evidence is mounting that indicates the labor resource is becoming a major location determinant for many industries. Highly skilled labor, in particular, is becoming more important as an input to the location decisions of "high-performance" firms and industries.

For example, steelmaking facilities have historically been oriented toward the inputs of iron ore and, more importantly, of coal. Early steelmakers were often oriented toward water, either for transportation or as a source of energy. As steel manufacturing technology advanced, orientation toward cheap sources of electricity became more important. But with modern ease of energy transportation, with decreased emphasis on iron ore and coal transport, and with improved technology in steel manufacturing emphasizing scrap metal inputs, the technology-intensive steelmaking and steel-using facility is no longer as dependent on the energy, water, and coal as it once was. The new emphasis is on both the availability of scrap iron, which is primarily located at large market locations, and the labor resource, also located near larger metropolitan areas.

For a technology-intensive steel-using (as opposed to steelmaking) facility, the location-determining input is neither the steel inputs nor the energy sources, but labor. Its initial location preference is the metropolitan area with a large, diverse, and active local labor market.

Eventually, however, even the technology-intensive manufacturer can no longer incur the increasingly high labor and other site costs of a growing metropolitan area location. It faces an even more difficult decision: relocating to a lower cost site or restructuring its manufacturing by specialization in the most technology-intensive processes, which produce the highest-margin products. It may still retain the less technology-intensive processes by relocating them at lower cost sites, outsourcing the production of intermediate inputs, or depending upon multiple supply sources for these inputs.

Most metropolitan areas have many examples of businesses, such as Honeywell in the Minneapolis-St. Paul metropolitan area, that once maintained their manufacturing facilities in one area but have dispersed new facilities to other metropolitan and/or rural areas. In the case of routine manufacturing processes, the presence of a low-wage labor force becomes a major location determinant.

Intra-urban Location Processes and Linkages

The competition for land and space in an economic region differs from place to place because of proximity to a metropolitan core area and the development of its location-dependent activities. These activities account for the changing place-to-place geography of production in a region. Their type and intensity are among the high-order determinants of the economic viability of any regional agri-industrial system. Political boundaries, however, mark the territorial scope of government influence in business location. They can serve as trade barriers between countries but not between states and regions within the United States. Restraint of internal trade is prohibited by the U.S. Constitution. Nor is the mobility of labor from one political jurisdiction to another impeded by political boundaries. Interstate differences in social programs may, in fact, encourage household migration from one state to another. Similarly, some states and even cities and counties provide tax and other business incentives to encourage business migration from one political entity to another.

These, however, are not the decisive differences among state and local governing institutions that influence economic growth. More important is the investment, or the lack of it, in public infrastructure that facilitates business formation and expansion and provides an economic environment for successful business enterprise and a productive workforce.

Bid-Rent Functions for Firms and Households

Assume a monocentric city serving as a transportation node. Assume also that the firm cannot substitute among productive resources; that is, the firm cannot substitute capital for labor at any location. Assume that production costs (except for transportation) and revenues are equal at every location. Finally, assume that the firm must pay transportation costs at a fixed rate per mile in order to get its product to the city center.

Under conditions of equal costs and revenues at all locations, it is intuitively obvious that the greatest profit will be at the city center, where there are no transportation costs to subtract from revenues. Since revenues and production costs are equal at every location, total profits must fall as the firm moves away from the city center because transportation costs are increasing.

Remember that rent is a residual. The rent that a firm could bid for a particular location depends upon its profits after other production costs are deducted from revenues. Therefore, as a firm moves from the city center, its ability to bid rents for a particular location falls in proportion to the increase in total costs of transportation. Figure 2-4 shows a typical bid-rent function for such a firm.

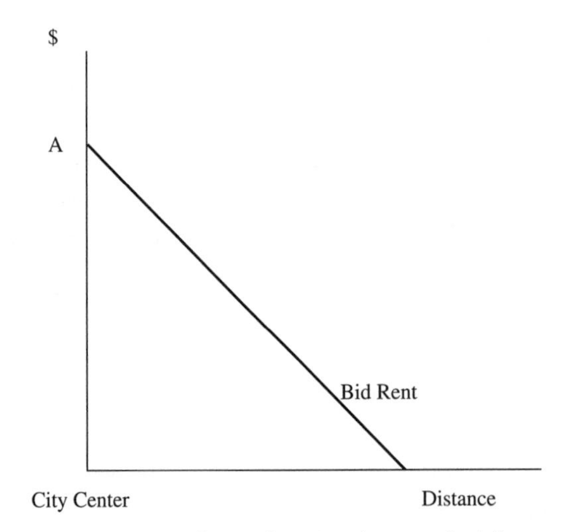

FIGURE 2-4. Bid-rent function for a typical firm.

The slope of the bid-rent curve is determined by the transportation rate per unit of distance. The maximum bid rent is at the city center (point A), where transportation costs are equal to zero. At that point, whatever the excess of revenues over production costs, this excess is available for rent. Remember microeconomic theory: When we say the entire excess is available for rent, we mean that the firm could pay this rent and still make a normal profit. Under competitive conditions, profits will eventually be reduced to normal levels. In fact, firms in a competitive market will bid for land until the rent absorbs any potential profits above normal levels.

To get to this last point, we need to realize that there are several bid-rent functions, each associated with another line of activity. Some firms will experience higher returns at the city center than other firms, but they will also have a steeper bid-rent function because of high costs of transporting their goods or services. Others will be able to make much lower profits at the city center but will have a flatter bid-rent function because their products are less costly to transport over distance. Figure 2-5 shows three activities under these conditions.

Office functions require face-to-face communications. Higher-order office functions, such as corporate headquarters, bank branches with decision-making authority, research and development, and high-order business services (legal, accounting, etc.) require such face-to-face communication. Locating away from one another greatly increases the transportation costs of obtaining such communication.

Most manufacturing processes do not require the same levels of communication. Rather, products must be transported, usually by truck, to the city center. The order of the manufactured products is such that they can't compete with the high-order office functions for locations at the city center. However, they can

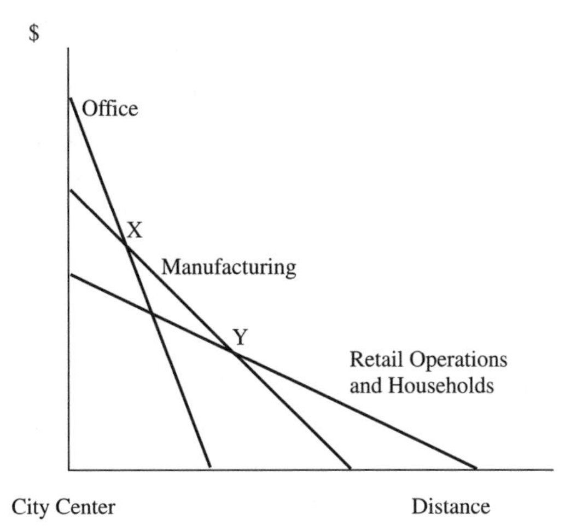

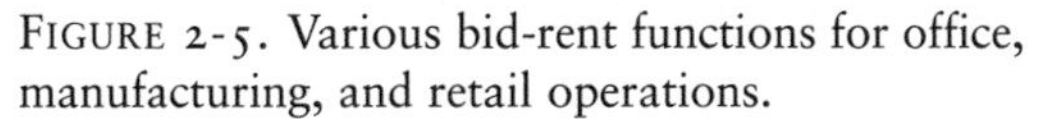
FIGURE 2-5. Various bid-rent functions for office, manufacturing, and retail operations.

compete for sites between points X and Y (Figure 2-5), where they outbid rivals for these locations. In a similar manner, retail operations, with relatively low transportation costs but also lower revenues over costs, find their competitive advantage for sites existing beyond point Y.

If we were to rotate this entire diagram in a circle, leaving X and Y as pointers to breaks in activity, we would have concentric circles—zones of specialization—much as were described by von Thünen. The center circle would contain office functions, the second circle would contain manufacturing, and the outer circle would contain retail operations and households. Beyond the last circle would be agriculture, forestry, mining, or other basic resource-dependent activities.

Remember that we have made rather restrictive assumptions to build this model. Of course, in the real world, retail operations exist both downtown and in the outer suburbs. But the principles for these locations are still contained in the model. In fact, the retail operations located downtown are generally of a different type than those located some distance from the center; the downtown retail operations offer higher-order, office-employee types of goods and services. The revenues over costs are higher in the downtown location for these types of retail firms, and they take on more of the characteristics of the office function listed above.

The point of this model is that, under competitive conditions, any parcel of land will go to the highest bidder. While a number of sometimes controversial assumptions go into this highest- and best-use statement, it certainly has at least some element of truth. The evidence of that is that the highest rents tend to be at centers of activity and decrease as one moves from these centers.

One final point: The bid-rent curves take on a slightly different look if we drop the "no resource substitution" assumption made at the beginning of this model. Without going into a great deal of detail on this issue, if substitution is possible, the firm will substitute labor and capital for land when land rents are higher, and land for capital and labor when rents fall. Therefore, the output per unit of land will be greater in the city center, where the rents are highest, and will fall as capital and labor are substituted for land when one moves from that center.

Bid Rents and City Structures

We will talk about intra-urban markets for land in this section. Why? The internal structure of a city in a free market depends upon the location decisions of private firms and households. These decisions are made, in part, in response to the rents necessary to obtain the services of land. Location decisions, in turn, influence the level of land rent for any particular location in the city region. Finally, not all locations are equal. Therefore, some locations will exhibit higher land rents than others. Firms that can thrive in these high-rent areas are those that gain the greatest competitive advantage from a particular location. All this results in patterns of land use, with areas specializing in some particular activity (manufacturing, retail, business services, or households, for example).

Our initial assumption is one of *pure competition.* Pure competition as a concept is best used as a point of departure. Two important and useful characteristics are associated with this assumption: (1) Typical firms will not make economic profits (profits beyond the minimum required to remain in business) in the long run, and (2) factors like zoning, covenants, transportation systems, and property taxes are ignored. Our initial theories are those of a perfectly functioning market. The influences of these other factors affecting location can best be understood in terms of their influences (positive and negative) on perfectly functioning market outcomes.

We note in Chapter 3 that the sizes and distributions of cities depend upon the functions they perform. Functions are defined in terms of the orders of goods or services based on minimum geographic sizes of markets necessary to sustain a particular activity. The same is true for the internal geography of a city. The central business district tends to house office and specialized retail functions. It is no accident that this has happened. Many of these spatially concentrated activities are the result of land rent influences on location decisions.

One interesting point is that, once an urban structure is in place, changes in structure tend to be slow in coming. Structures are in place that need to be either converted or demolished in order for new activities to enter. This means that cities that emerged as long as a hundred years ago have land uses, or at least remnants of land uses, that are still in place.

The cities that emerged as dominant in the nineteenth century, such as New York, Boston, Philadelphia, Chicago, New Orleans, San Francisco, and Los Angeles, were cities that depended very much on their ports for economic development. Other cities, such as St. Louis, Denver, Dallas, Wichita, and Kansas City, developed because they were on major paths of household migration, were origins or destinations for agriculture products (cattle, for example), or had major rail spurs coming through the region. In other words, our old friend transportation became a major determinant of city location and specialization.

What happened within the cities themselves? Major industrial development surrounded the terminals of major transportation nodes. When the railroad came into being, it generally ran its early lines to the previous transportation centers—near the ports of such cities as Chicago and New York—emphasizing even more the location influence of transportation nodes.

The early U.S. city (prior to the 1920s, when the internal combustion engine came into play) tended to be monocentric; that is, the focus of economic activity tended to be in the center and declined as one moved from the center toward the periphery. Industry and retail activity located near the transportation centers to ship and import goods and services. The population located around this central band of activity to minimize the time and expense of the trip to work. Agriculture located in the bands surrounding the households to minimize the costs (especially of spoilage) of getting food to the household table. And timber and hunting activities tended to be farthest from the city center.

This meant that much of urban industrial and residential infrastructure was built in the central city. As mentioned earlier, much of that infrastructure remains today. Our early discussion of the structure of a city will assume this monocentric city structure. We will modify that particular assumption at the end of the chapter in order to note differences that result.

Continuing with the assumption of a single city center, we will make a couple of even more restrictive assumptions that are inconsistent with Figure 2-5 (we will relax these later). We assume that all trips are commuting to jobs. We assume that all jobs are in the city center. In other words, the household must trade off their preferred location against the time and expense involved in commuting to the center.

It is important to note that a large component of commuting costs is the value of time spent commuting. That value is expressed as opportunities foregone during the commuting period. Many studies indicate that the value of commuting time is equal to up to half of a person's salary. This means that commuting is most expensive for high-paid individuals.

We also assume that this is the only trade-off considered by the typical household. Preferences for schools, tax levels and government services provided, and shopping considerations are not factors in household location, at least for this initial analysis. Distance from work is the only consideration.

At a given level of utility, the household will be trading land rents for commuting expenses. As one moves farther from the city center, land rents fall because bid rents fall. However, commuting expenses increase as distance from the city center increases. Therefore, the household will bid lower rents for a unit of land that is farther from the city center. The bid-rent function for households will take on the same characteristics as the bid rent functions for business firms noted earlier. What is more, as is the case for business location, the bid rent function for households will be a straight line if there is no capital for land substitution and will be convex to the origin if substitution is allowed (Figure 2-6).

Households have a preference for land. In other words, the income elasticity of demand for land is greater than one. Under these conditions, the household will move away from the city center until it reaches a point at which the extra (marginal) value received from larger land parcels equals the extra (marginal) costs associated with moving one unit of distance farther out.

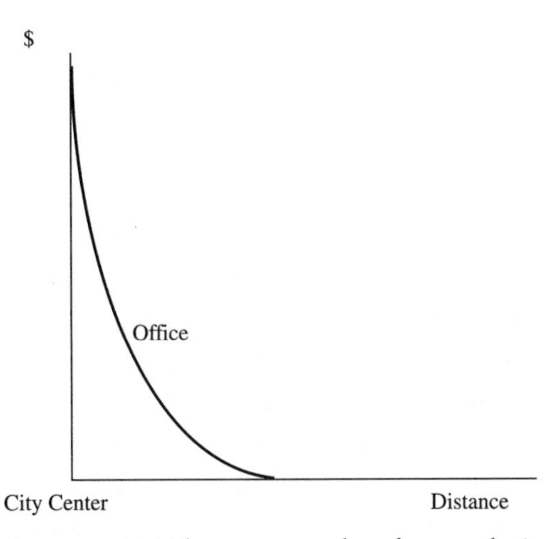

FIGURE 2-6. Bid-rent curve when factor substitution is possible.

What would be the predicted effect of a new or expanded highway on the conclusions from this analysis? Either a new highway or the construction of a new lane on an existing highway has the effect of reducing time spent in commuting. This reduction in time reduces the costs of commuting. Other things being equal, the reduction in cost should encourage households to move farther from the city center. We are oversimplifying here, but the construction of multilane highways to the city centers of older, established cities would be expected to lead to an increase in suburban development.

If there is competition both in the product market and for the use of resources (land in this case), then the firm will only make normal profits in the long run. Any profits made over the costs of production will be used to bid for land. Land will go to the highest bidder. Earlier in Chapter 2, we showed that, under such conditions, each parcel of land in a city goes to its highest and best use. The result is the most efficient allocation of land according to the economics definition of efficiency.

Households and firms bid against one another for the use of land. The resulting land pattern, assuming transportation is equally possible in every direction, takes on the shape shown in Figure 2-5, with households and retail operations sharing the space furthest from the core.

The same principles hold if there is competition, but transportation is not equally possible in every direction. However, the concentric circles may look more like polygons overlaid one on top of the other. Wherever there is a road that is traveled by a number of people, land rents are at their peak near the road and fall as one moves away.

A special case is that of the beltway highways that surround many U.S. cities. This system, coupled with the rise of the truck as a major transportation mode, allowed manufacturing to move from close to the city center out to the beltway. For reasons listed above, these beltways also allowed households to move farther out from the center.

Manufacturing, when it was in the city center, used a vertical process in which hooks and belts took the product from the top floor, through the production process, down to the first floor, where it was loaded on the train, ship, or what have you for transportation to markets. Rents were high, so multistoried plants arose as the firm substituted capital for land. With the internal combustion engine and the highway system, manufacturers were free to move their locations to the suburbs, where more of their workers resided. What is more, because of the lower rents, land was substituted for capital. Horizontal production lines were developed, the forklift was invented, and production became much more efficient.

With improvements in communications through the personal computer and the World Wide Web, some office functions are moving out of the city center. Improved means of communication allow the home office to remain downtown, where big deals are cut and cutting-edge research and development occur, while allowing more routine operations such as bookkeeping, household banking, and even specialized retail operations to move where site costs are lower. All of these processes are leading to rates of economic growth that are much higher in the outer rings of the city than in the city center. Some manufacturing operations move even farther out, to rural (peripheral) areas. These operations are taking advantage of both cheaper land and a cheaper labor force. Most manufacturers locating in the peripheral areas are characterized by routine operations requiring a labor force with moderate to low skills. If the labor market becomes tight in domestic peripheral areas, the firms often move to other countries, where labor surpluses exist.

From a land use and land rent point of view, the monocentric city becomes multicentric. Land rents rise around the beltway highway as business concerns bid for the available space. Figure 2-7 shows a comparable situation to the one just described. Offices will reside both in the city center and close to the beltway. Manufacturing may actually leave the city center, depending on whether the second peak rents overlay the second ring of the circles. But manufacturing will surely be located in the suburban area. Households and retail operations will occupy the remaining space to the east of the manufacturing sector. And some manufacturing, retail, households, and office functions will move to the more rural, peripheral areas.

Neighborhood trade and service centers, when associated with the secondary peaks of manufacturing and office functions, provide new options for reducing the time and cost of the daily journey to work. These centers continue to face

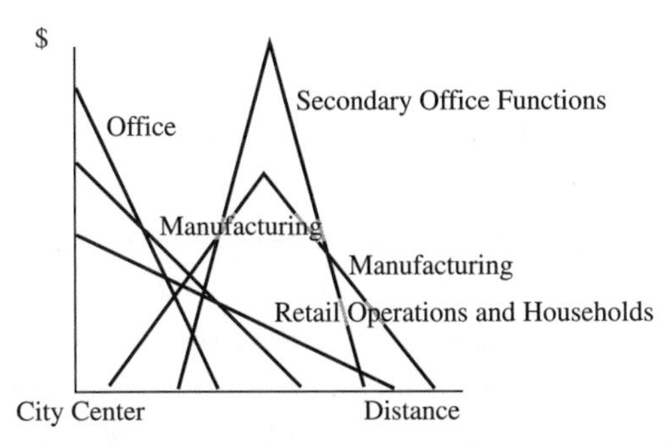

FIGURE 2-7. Secondary peaks in bid rents create a second center around the beltway highway.

stiff competition, however, from large shopping malls, like Minnesota's Mall of America, and superstores, like Walmart. Again, the lessons learned in building and maintaining active and viable centers for the delivery of essential services in rural areas also apply to the urban neighborhoods of central cities.

Urban land prices gradually decline from the city center, except for the suburban shopping districts and high-amenity-value properties. Commercial property values in the Minneapolis downtown district, for example, range from more than $200 per square foot to $5 per square foot for the most marginal properties. Residential properties in the suburbs range from more than $20 per square foot to less than $2 per square foot, or an equivalent range in acres from more than $1 million to less than $100,000. Location, in terms of access to markets, services, and amenities, is the critical determinant of urban commercial and residential real estate values. Chapter 3, Local Markets and Central Places, continues this discussion in the context of the variety of local markets found in an urban regional economy.

The construction of important local highway linkages between rural communities and their metropolitan core areas follows the urbanization of metropolitan regions. This process is accompanied by the gradual expansion of the urban periphery, the conversion of rural into urban land uses, and the disassociation of place of work and place of residence for a large proportion of the metropolitan area population.

The devising of roadway development options at this stage of a metropolitan area's outward expansion means finding and implementing the one option out of several that best serves the interests of both local residents and the larger economic community. Concerns about urban sprawl would be addressed at an earlier stage of the metropolitan area's expansion. At this stage, the people are in place, have jobs, generate above-average levels of disposable income, and support a growing number of local businesses, thus reducing their dependency and their neighbor's dependency on more distant shopping centers and places of work. The reconciling of development options with other local concerns thus begins with an awareness of what is already in place that may be affected by the proposed roadway development.

The reconciling of roadway development options often starts in a real-world setting with the immediate concerns of adversely affected property owners. Assessing the economic effects of the development options (although this addresses the concerns of local residents) is more global and distant, both present and prospective, than the concerns of adjoining residential property owners.

One case study on the reconciling of development options comes from a recent study of roadway development for an approximately 20-mile segment of a principal highway corridor in a newly urbanized part of the Minnesota-St. Paul labor market area. This segment of highway borders the west-side boundary of the metropolitan urban services area. The Minnesota Department of Transportation initiated two sample surveys—one of households in a dozen minor civil divisions and municipalities, the other of business establishments in the area's principal trade centers.

This summary was based on two surveys—one of 600 resident households (representing a resident household population of nearly 20,000), and the other of 200 local businesses (representing a resident business population of nearly 2,000). The household survey covered a geographic area of 19,321 households with a total adult population of 42,171—an average of 2.18 adults per household. The survey also provided an estimate of the total work trips of respondents and one additional adult trip per household for households with more than one adult. This procedure thus underestimated the total trips of households with more than two trip-generating adults.

These surveys provided the factual context for assessing various roadway development proposals affecting the economic and social well-being of local residents. The findings address the local economic conditions that ultimately determine the overall benefit and cost of roadway development in which individual local residents share in varying degree. They focus on the intermediate level of economic impact assessments—not on the individual property owners in the focus area (the Corridor) or those in the entire multicounty Metropolitan Region, but rather, on the local economy represented by the business and household activities and their participants.

The distribution of the Corridor workforce in Table 2-2 illustrates the different sets of relationships between place of residence and place of work in an expanding metropolitan area. More than 84 percent, or 29,600, of the total Corridor workforce of 35,100 commuted to jobs outside their place of residence. Less than half of the commuting—15,200—was to destinations inside the

TABLE 2-2 Comparing specified resident survey workforce and daily trips to work with business survey jobs in specified areas, Trunk Highway 12 Corridor, January 1995

			Household Survey						Business Survey	
			TH 12 Corridor Trips							
Corridor Subarea	Total Work Force (thou.)	All Trips (thou.)	Total (thou.)	East (thou.)	Central (thou.)	West (thou.)	Down-town (thou.)	Other Met. (thou.)	Total Corridor Jobs (thou.)	Exc. Def.[a] Work Force (thou.)
Total	35.2	29.6	15.2	9.7	3.2	2.3	3.8	9.3	28.4	6.8
East	17.2	14.3	6.6	6.2	0.3	0.1	2.5	2.9	11.2	6.0
Central	9.5	8.1	4.5	2.2	2.2	0.2	0.8	3.2	12.3	–2.8
West	8.6	7.2	4.0	1.3	0.7	2.0	0.4	3.1	4.9	3.7

Source: Minnesota Center for Survey Research, TH 12 Corridor Surveys, January 1995.
[a]Exc. Def. = excess or deficit.

Corridor. The journey to work in the two downtown districts of Minneapolis and St. Paul and metropolitan area suburbs accounted for the larger share of total trips. Moreover, the total workforce exceeded total jobs by 9,700 in two of the three subareas, namely, East and West. The deficit local workforce in Central subarea was supplemented by in-commuters from outside the Corridor, as suggested by the small resident, non-commuting local workforce of 1,500 (i.e., 9,600 total – 8,100 commuting) and 3,200 total commuters to the Central subarea. Thus, the actual workforce deficit was closer to 7,700 (i.e., 12,300 – 4,600) than the –$2,800 listed in the table as the deficit workforce. The local workforce includes multiple jobholders who reduce the larger deficit by approximately 10 percent. Finally, large subarea differences occur in the relative importance of commuting and the destinations of commuters.

Table 2-3 summarizes the baseline tabulation of the estimated 135,600 weekly trips of an estimated 19,321 Corridor households to places of shopping (groceries and clothing), entertainment, recreation (golf, tennis, and other participation sports), children's activities, and personal health care. A large number of trips contribute to the duration of travel on local and regional roadways. Duration of trip varied from ten minutes one way for convenience grocery shopping to 33 minutes for a sports event and 34 minutes for a cultural event in downtown Minneapolis or St. Paul. Trip frequency is thus only part of the problem in affecting access to work, shopping, personal care, recreation, and entertainment.

We can derive the trips to place of work from the Resident Survey estimates of trips per employed person. We first convert shopping and personal care trips from a weekly basis to a daily basis, assuming a six-day week for this purpose. We then reduce nonwork trips to account for combined trips, in this case, by

TABLE 2-3 Shopping, entertainment, recreation, children's, and personal care trips per week of Trunk Highway 12 Corridor households, January 1995

Subarea	Total (thou.)	Shopping (thou.)	Part.[a] Sports (thou.)	Dining Out (thou.)	Sports Event (thou.)	Children (thou.)	Movie (thou.)	Health Care (thou.)	Cultural (thou.)
Total	135.6	59.0	23.4	19.7	9.8	8.9	5.9	4.7	4.2
East	67.9	30.0	11.2	10.3	4.6	4.4	2.9	2.3	2.2
Central	36.4	15.2	7.1	5.3	2.7	2.2	1.5	1.3	1.1
West	31.3	13.8	5.1	4.1	2.5	2.3	1.5	1.1	0.9

Source: Minnesota Center for Survey Research, TH 12 Corridor Surveys, January 1995.
[a]Part. = participation.

approximately one-third. We thus derive a top daily trip estimate of 46,900 for the Corridor study area—29,600 work trips for five days a week and 15,000 all other trips for six days a week.

Density and Differentiation of Economic Activity

The density and differentiation of economic activity also affects the costs of doing business and the location advantages of particular areas for various types of businesses. When economic units are located close to one another, per-unit costs of doing business decrease. These economies stem from scale in attracting customers to the place of business, both internal to the individual business and external to other businesses. These costs differ from the costs of information transfer among information providers cited earlier. Agglomeration cost savings come from the clustering of economic activity, that is, an attraction for other firms affected by noncollusive proximity economies once a cluster begins to form.

Agglomeration Economies

There are several sources of agglomeration economies:

- *Internal economies:* These are the economies from larger-scale production discussed in the typical principles of economics textbooks. As a firm becomes larger, resource specialization, standardized inputs, robotics, and other factors lead to lower per-unit costs, at least up to a point. The typical downward-sloping, long-run average cost curve is the result.
- *Localization economies:* Localization economies are external to the firm when it locates near its suppliers or in proximity to other firms producing the same product. These economies are internal to the industry, but external to the firm.

The classic example of a localization economy relates to the garment industry in New York City. Firms in the garment industry tend to be small, produce products that undergo frequent style changes, and produce relatively high-margin products. A skilled labor force exists in the region, which makes it easier for firms to increase or decrease their workforce as market conditions warrant. What is more, by locating near one another, the firms are in a position to observe the new styles and methods used by competitors. Frequent contacts over lunch or during leisure-time activities increase both communication and speed the flow of information, which were identified as important location determinants in the previous section of this chapter.

Finally, while no one firm in this industry can produce an important input (e.g., buttons) at a scale allowing for internal economies, the presence of such a large number of firms makes possible the establishment of a new supplier specializing in button manufacturing. Such a manufacturer can achieve internal economies to scale not possible for the other shops. At least a portion of these savings can be passed on to the button manufacturer's customers. The internal economies to the button manufacturer become external economies to the rest of the industry.

Such external (localization) economies do not lead to a movement along the dress manufacturer's long-run average cost curve. Rather, they result in a downward shift of the entire curve, making per-unit costs at every level of production lower.

Of course, many other examples can be given. The computer industries in the Silicon Valley, Boston, and the Twin Cities in Minnesota enjoy many of the characteristics of localization economies.

- *Urbanization economies:* These economies are internal to metropolitan areas, but external to both the firms and industries that are located in the metropolitan area. Urbanization economies stem from the dynamic nature of a metropolitan area. They include (1) the diversity and skills of the metropolitan labor force, (2) the presence of large numbers of people out of which tomorrow's entrepreneurs evolve, (3) the frequency of communication between "ideal people" in the region, and (4) the agglomerating influence of research and development activities that spawn new and innovative processes and products. Furthermore, the provision of public goods and services are subject to economies to scale. Such savings are passed forward to the industries and firms in the region in the form of more efficient provision of public goods.
- *Economies from industry linkages:* These economies involve savings in transportation costs, provision of just-in-time inventory, and provision of

specialized products; all result when input providers and demanders locate proximate to one another. All are sources of internal economies to industry clusters and external to the firm.

- *Shopping agglomerations:* All of the external cost savings discussed above accrue to the firm. Shopping agglomerations are of a slightly different type. When you shop at a store, the true price is greater than that posted on the price tag. It is the shopper who searches for just the right product, travels to the store to secure that product, and brings the product home. Search and travel costs are part of the true cost of the product from the shopper's point of view.

These costs are reduced when retail operations and personal-service providers locate proximate to one another. Search costs are reduced when providers of shopping goods are located together. Shopping goods are those that generally require comparison shopping before purchase, such as automobiles, shoes, and clothing. In fact, when providers of related goods, such as shoe and clothing stores, locate near one another, shopping is made easier. Of course, this principle is the basis for the mall concept, which allows shoppers to do all or most of their shopping at one location. One trip to the mall and one trip home reduce the travel costs associated with shopping as well. Shopping agglomerations tend to work against the tendency of competing stores to locate away from one another in order to monopolize geographic markets.

Most resident households of both rural and metropolitan areas depend on personal transportation for access to place of work, shopping, and personal services, as well as schooling. Very few, if any, areas are self-contained, functional economic communities. Even a central city with a minimal share of its total economic activity originating or terminating outside its boundaries is still an open economy, with much trading of goods and services to and from consumer and producer markets outside the area. The distinguishing characteristic of the population agglomerations formed by the resident households of metropolitan areas is the performance of the resident labor force and the attributes that contribute to this performance. Thus, we focus on the resident labor force and its qualitative attributes rather than the resident population as an "agglomeration."

We cannot even imagine the resident labor force or the resident population being a "jumbled mass" as the term agglomeration might suggest. Within the central city, the resident population forms a wide variety of multiblock neighborhoods, some more functional than others. Beyond the central city boundaries are the several "rings" of suburbs, differentiated by age and quality of housing stock, the quality of schools and schooling, and the safety of person and property. For some metropolitan areas, the outer boundaries of the surrounding suburbs form a "municipal urban services area" beyond which urban development may not occur. Reality quickly confronts efforts to contain "urban sprawl," as metropolitan authorities are faced with court orders prohibiting such restrictive practices. Failure in limiting urban development by restricting the availability of essential municipal services means, of course, a continuing expansion of the metropolitan area's outer boundaries.

Another distinguishing characteristic of metropolitan population agglomeration is its continuing expansion beyond existing municipal boundaries. This expansion, in turn, further accentuates the disassociation of place of residence and place of work as well as place of shopping, recreation, entertainment, and personal care.

The resident households of rural areas face contrasting futures depending upon their proximity to metropolitan areas and environmental attributes. Residents close to the metropolitan area have the option of commuting to jobs either in the city or in nearby rural areas experiencing the opening of new businesses and/or the expansion of existing ones. Local landowners benefit from increases in land prices resulting from the growing demand for residential, commercial, and industrial space.

On the other hand, rural areas more than an hour or two from the metropolitan core face continuing declines in population and employment. The exceptions are, of course, the rural areas within commuting distance of the job growth in secondary metropolitan areas, such as Milwaukee outside the Chicago metropolitan area, or Fargo-Moorhead outside the Minneapolis-St. Paul metropolitan area. As jobs increase in the secondary metropolitan areas, so does the labor force, thus adding to the size and vitality of their local labor markets and their attractiveness to new and existing businesses. In the declining areas, however, out-migration to primary and secondary core commuting areas is the common mode of household adjustment.

The unknown looming on the horizon is the effect that telecommunications technology will have on residential locations. The rapidly advancing communications software and the chips to support software are both in rapid advance, making home employment a viable option, especially for high-skill consumer and producer service providers. There is some evidence that the freedom this technology affords is leading professionals to seek even more rural locations than those characterized by suburbia. Amenities and other quality-of-life attributes of the small town are again becoming attractive to these professionals. Futurists guess as to where this movement will end. We won't join in those speculations here.

Public Facilities and Infrastructure

Government facilities and local infrastructure comprise still another set of economic units engaged in clustering their activities with those in the private sector.

The regional clustering in the 1960s was prompted, in part, by the desire of state governments to consolidate their services in larger urban centers. State offices in these centers could serve the household and business population of a multi-county area rather than of one municipality or county, the rationale being the minimizing of travel time and costs for the total population.

Local infrastructure and local economic development go hand in hand. Without local economic development, demand for local infrastructure would not exist. The determinants of local economic development are, firstly, endogenous to the area. These determinants include local entrepreneurship and the productivity of the local labor force engaged in producing goods and services for shipment or sale to buyers located outside the area. Local infrastructure is a necessary but not sufficient condition for local economic development.

Even wilderness areas located considerable distances from large population concentrations depend on local infrastructure for access to and from specific recreation sites. Nonresident users are beneficiaries of highly valued wilderness experiences that become essentially transfers or a subsidy from the resident population that supports the local infrastructure to the nonresident population that benefits from its availability. In this example, local infrastructure allows economic development to occur, provided nonresidents reimburse the resident population for the derived benefits from access to the recreation sites due to availability of local infrastructure.

Government regulations, however, impose added costs on start-up businesses in metropolitan areas, especially small and medium-size enterprises (SMEs). The costs of living as well as the costs of doing business are higher in these areas than in small cities and rural areas. Yet a progressive tax system may fail to account for these differences, thus further eroding the initial competitive advantages of metropolitan areas, particularly the central cities and their SMEs. In addition, local government control of the critical air transportation infrastructure may be compromised by a locally dominant airline. Such an airline can assert its government-sanctioned monopoly power by imposing a price premium on the airfares of originating passengers. The high fares may be occasionally reduced for defensive purposes when a new airport tenant threatens the dominant airline's monopoly position by offering lower airfares for its passengers.

A more subtle but nonetheless real deterrent to the growth and viability of emerging information-based local economies is the belief held by many public sector leaders that only large companies, along with government, create the "good" jobs that pay well and provide satisfactory levels of worker benefits (Friedman, 1996). The SMEs, on the other hand, can offer only low wages and dead-end nonunion jobs with minimal benefits, according to this large-company bias. Such attitudes work to the detriment of both the SMEs and the metropolitan areas that provide the most fertile ground for their birth and regeneration, and an optimal location for their continuing growth and expansion.

Land Use Control

Land use control is a long-established prerogative of state and local governments. They assert this control by establishing an urban service boundary to achieve higher development densities in existing urban areas, granting building permits only to approved building proposals, and adopting zoning ordinances to segregate incompatible or undesirable land uses.

Metropolitan Urban Service Area

A well-defined metropolitan urban service area (MUSA) (as noted again in Chapter 5) is one means of controlling the deployment of metropolitan services within the jurisdictions of a metropolitan area governing authority. The actual boundary depends on a host of factors, starting with the existing metropolitan roadway system and residential development patterns. If the goal is to contain the infrastructure costs of urban development, new infrastructure construction must be contained as well. The immediate issue is one of drawing the MUSA boundary and establishing appropriate means for enforcing compliance with its provisions.

Theoretical underpinnings of the MUSA boundary exist in the evaluation of the added infrastructure costs of metropolitan area urban expansion outside the boundary and the higher costs of urban land acquisition when development is contained inside the boundary. Winners and losers differ in each case. Winners of urban expansion beyond the boundaries are the new home owners now able to build or buy within the MUSA boundary. Some local governments may experience a larger tax base without commensurate increases in costs. Losers are state and local governments burdened with higher infrastructure costs but without the corresponding increases in tax base to cover these costs. For many local governments in the built-up areas, urban expansion beyond the boundary may result in an immediate loss in local tax base. The added commuter highway congestion and air pollution are added hidden costs of the low-density urban expansion.

The Twin Cities (Minneapolis-St. Paul, Minnesota) Metropolitan Council established a MUSA boundary and encourages development within the boundary with the use of its approval authority over the capital budgets for sewers and transportation. A 1977 report of this agency's Fully Developed Area Task Force emphasized an intensive reuse of the existing public and private investment structure and a regionwide sharing of the costs of such redevelopment. A proposed metropolitan reinvestment fund was viewed as a major tool for the implementation of the redevelopment plans.

This Metropolitan Council growth management strategies policy of the 1990s established three growth options—current trend, concentrated growth, and growth center—as the basis for the regional forecast assumptions. The current trend option reflected a projected growth pattern shaped by market demands but also assumed that the council's development policies would be maintained (orderly economic development through the management of regional systems).

The Metropolitan Council's focus is long term. Local government would devise the shorter-term staging plans. The second of its three options—concentrated growth—holds the current MUSA boundaries until 2020. Most of the region's growth would occur within the current MUSA at higher densities. The growth center option encourages most of the growth to locate within the current MUSA, focusing on "high-density, pedestrian- and transit-friendly, mixed-use nodes and corridors" (Metropolitan Council, 1997). The Council's statements have only limited reference to and discussion of specific policy incentives—market or nonmarket—and land use controls that have established track records in effectively restricting development inside or outside the MUSA boundaries, consistent with "due process" safeguards.

Building Permits

Local communities affect their rate and costs of local population growth through the issuance of building permits. More stringent building requirements mean fewer new residential and commercial buildings and eventually fewer people and fewer jobs. This occurs only insofar as the permitting process entails time delays and related costs of development that eventually appear in higher prices for the new buildings. Higher prices for new buildings may encourage longer use and repair of existing buildings, and higher prices and repair costs for older buildings. The issuance of building permits thus provides local governments with an enforceable means of maintaining minimal standards in local building design and construction, but these incur added costs to the owners of the new buildings.

A theoretical justification for the issuance of building permits is the added net benefits it brings to the local community through compliance with quality standards that enhance the values of existing buildings. Again, the results are mixed in terms of winners and losers. Winners are the owners of existing buildings and the local municipalities on two counts of higher prices for current property owners: short term in higher values for existing real properties and local tax base, and long term in sustained high property values for the municipalities. Losers are new property owners, who pay the higher prices, and the renters (some may have preferred being owners), who now pay the higher rental values.

Zoning

Zoning ordinances fall under the categories of nuisance, fiscal, and design. Nuisance zoning separates incompatible land uses; for instance, industrial zoning separates industrial activities, which generate air pollution, from residential areas. Fiscal zoning excludes activities that would not pay their full share of local government costs. Design zoning deals with an urban area's overall architecture by directing residential and commercial building to certain areas that can be served by existing or projected infrastructure and by preserving open spaces.

One alternative to nuisance zoning is the use of effluent charges. A well-designed schedule of effluent charges would vary spatially as well as by the amount of pollution. Higher levels of pollution would incur higher effluent charges. These charges would vary also with proximity to residential or other pollution-sensitive areas. Another alternative is performance zoning, which provides performance standards for each zone. This is a compromise between traditional zoning and effluent charges. It allows for mixing land uses, given the assured compliance of a potentially incompatible use with performance standards that protect nuisance-sensitive areas. Residential nuisances, like off-street parking in high-density residential areas, exist that can be controlled with performance zoning.

Fiscal zoning provides local governments with a means for excluding households or land uses that would impose a fiscal burden on the locality. These include high-density housing, fringe area land uses, and new commercial and industrial development.

Design zoning imposes overall architectural standards for a locality that may prove inequitable for some property owners. A city's landscape design, for example, may call for open space or low residential densities that would be grossly unfair for some current property owners and a potential bonanza for others. These sorts of inequities would be addressed within the framework of transferable development rights that allow for the sale and purchase of these rights. Thus, the owners of development rights for the open space area may sell these rights to the owners of land zoned for high-intensity use. The transfer of development rights thus makes possible a more equitable settlement of the initial preservation zoning issue than was the case under traditional zoning laws.

Winners and losers under new zoning ordinances parallel those in the issuance of building permits or in the drawing of MUSA boundaries. The goal of zoning, like much public legislation, is to promote public health, safety, and welfare. This may or may not transpire, depending upon the participant's vigilance and understanding of zoning processes and their consequences for those affected by them.

Tracking Land Use Changes

The conceptualizing of the urbanization process is an essential part of the task of understanding the evolving shape and direction of the core area's economic growth and development. It also is an essential step in understanding the converting of prime agricultural land into urban uses under different urban growth scenarios. Modeling the urbanizing process would involve, among other steps, the following: (1) identifying and measuring the economic activity in the urban fringe, before and after the converting of prime agricultural land into urban uses in terms of its economic contributions—jobs, output, value added, and economic base—to the local economy and (2) identifying and measuring residential development options associated with the before-and-after scenarios of economic activity. From an academic perspective, modeling the urbanizing process would involve building a theory and framework for analyzing and predicting the urbanization process and its consequences for both the rural and the metropolitan areas of the expanded economic region. But this is not enough. It must also involve building a theory and framework for implementing a democratically accepted, adopted, and supported regional strategy for guiding metropolitan area growth.

Focusing on Key Issues

The topic of competition and development in metropolitan and rural areas covers a wide and varied range of issues, even when limiting this overview to the U.S. economy. To simplify, we focus on three key economic issues facing transitional rural agricultural areas. These include, besides the loss of agriculturally productive land to the urban periphery, the attraction of alternative employment opportunities for farm households and the declining importance of farming in creating jobs, income, and investment opportunities. These are local realities that, in turn, have both positive and negative consequences for the rural agricultural economy and its stakeholders.

The loss of agriculturally productive land to the urbanizing periphery of metropolitan areas is a continuing concern of metropolitan area planners, land subdividers and realtors, state and local legislative bodies and administrative agencies, and environmentalists. Each is a stakeholder in the transformation of rural agricultural areas into metropolitan area suburbs and open-country settlement. Many current owners of the agriculturally classified land, on the other hand, view themselves as livestock breeders for nearby residents, producers of horticultural specialties for metropolitan area residents, or hobby farmers, as well as low-profile land speculators and developers with a long-term investment return.

For the urban planners, farm-based legislators and agencies, and environmentalists, restraining urban expansion remains a lost cause.

The attraction of alternative employment opportunities is the continuing saga of rural areas everywhere. For the United States, however, the vast reservoir of potential rural-to-urban migrants in rural areas no longer exists. In fact, many rural areas now experience labor shortages because of expanding new and existing businesses, including input suppliers for technology-intensive manufacturers in the metropolitan areas.

The declining importance of farming is manifested in series after series of monthly employment figures showing fewer farm and more nonfarm jobs in most localities—a result of applying both traditional and brand-new technologies to the agricultural enterprise, recently accelerated by renewed efforts in land conservation and preservation. Agriculture now shares more of the economic base with manufacturing in a growing number of rural areas. In fact, nonmetropolitan labor markets in many regions now have a larger share of the basic workforce in manufacturing than in agriculture.

Building a Region's Social Capital

Finding common ground in managing urban metropolitan growth undoubtedly is difficult. Its marketplace participants, however, share at least one common value in their propinquity or proximity to one another. Proximity, as many studies show, is the essential condition for the existence of a central city downtown district, a viable and functioning local labor market, or place of gathering for business or pleasure (Hutton and Ley, 1987; Saxenian, 1994). Uncontrolled urban sprawl erodes proximity-based values. Residents of the outlying areas compensate for the loss of proximity by extraordinary increases in the frequency and duration of trips for shopping, schooling, recreating, caring for the health and safety of self and family, and especially the daily journeying to work. All this generates corresponding pressures for more and bigger access roads, with more congestion and pollution that add to the indirect and often hidden costs of urbanization.

Participants in the urban marketplace also share in the consequences of having, or not having, a strong sense of community. Compare northern Italy with southern Italy as Robert Putnam (1993b) does in his book *Making Democracy Work: Civic Traditions in Modern Italy.* Propinquity clearly has had something to do with the civic traditions in modern Italy, but it is more than mere proximity of one leading family to another in Padua, Venice, or Milan. Success in dealing with problems of development and competition in the urbanizing periphery of a metropolitan region depends on the region's social capital. Social capital includes traditions in civic engagement, like the organizing of successful consumer and producer cooperatives, forward-looking labor unions, other civic ventures, and norms of reciprocity, both specific and generalized. Putnam (1993b) refers to the work of Elinor Ostrom in *Governing the Commons* (1990) when noting that communities in which the norm of generalized reciprocity is followed can more efficiently restrain opportunism and resolve problems of collective action.

Closely associated with the sense of community is equity. The lack of equity in paying for the costs of the expanding urban periphery, in large measure, sustains it. New residents are attracted to the urban periphery because of its relatively low land costs, property taxes, and crime rate. The central cities are left with the financing of high-cost services from an eroding property tax base.

Even if metropolitan area residents were willing to pay the full cost of developing the urban periphery, the perceived benefits would outweigh the perceived costs, given the understandable delay in billing the new resident for the annualized full cost of residential relocation. In short, achieving equity in the financing of metropolitan area growth is a problem of public education in associating the benefits of an expanding urban periphery with its costs, as well as establishing the rule of equity in its financing.

Economic Effects of Urban Regional Development

Important in assessing the economic effects of roadway development options is the buying power of local residents, including its locality distribution as well as its magnitude. The findings of the Minnesota Department of Transportation surveys show, for example, that the more distant households (those in the urbanizing portion of transitional rural areas) had a higher proportion of households with less than $35,000 total income and fewer with more than $75,000 total income. The urban-to-rural personal-income gradient reveals important consequences of the personal and environmental attributes of proximity to the metropolitan core area. An increasing awareness of the importance of job access is coupled with increasing difficulties in improving access to most distant job seekers. With a high household income, driving alone in one's own car is the common transport mode in the daily journey to work. This was the case for nearly 88 percent of the employed workforce in the Corridor localities. Only 8 percent depended on a car (or van) pool. This was true for both the respondent and the other adult in the household.

Duration, distance, and frequency of individual round-trips determined the total volume of trips generated by the resident households in the Corridor study area. These in turn varied by household, depending on its employment status, total income, and position in the family cycle. Ultimately, the nature and location of employment of individual household members accounted for the source of traffic-generating activity.

While the elements of propinquity, equity, and community may undergird the successful managing of metropolitan growth, the conversion of rural agricultural land to urban uses will likely continue. This could occur, however, in ways less environmentally and economically damaging than before through various producer and consumer incentives operating in the marketplace. These would be accompanied by community efforts to increase population densities. Examples of possible efforts include enriching the downtown cultural and recreational amenities, supporting community-based child-care and protective services, funding multipurpose neighborhood schools and social centers, and building new, readily accessible rapid-transit facilities serving both downtown and neighborhoods. The job-creating synergism of core area activities that generate high-paying jobs in the core area's downtown district, coupled with easy access to the amenities of both the exclusive residential areas of the urbanized periphery and the downtown theaters and sports arenas, drive the conversion of agricultural land into urban uses. In a real-world context, however, the attempted regulatory control of metropolitan area land uses rapidly becomes an anachronism of its own outlived and outmoded forms of public sector planning.

3

Local Markets and Central Places

A central place is easily reached from many directions. In the agrarian economy familiar to early regional writers, a central place was no more than a team haul away, roughly six miles from its farthest customer. It was the local market for farmers with grain and cattle to sell and then ship to distant markets. Building farm-to-market roads soon became a primary responsibility of local governments. As the motor vehicle and hard-surfaced roads replaced the earlier means of transportation, a few central places graduated to another level of importance, serving not only their immediate customers but those of smaller central places, some of which were a quantum leap beyond the old team haul. The service areas of the larger places were a hundred times their former size. With the coming of many forms of rapid travel and communication, even larger central places grew, not only in size and areas served, but also in the variety and value of the services offered to their customers.

Textbooks on urban and regional economics do not generally combine the two concepts discussed in this chapter. We develop the concepts here, in part, because we use their expressions and language in chapters to follow. They fit within the earlier discussion of location and land use. Moreover, they help us move gradually from a location and land use perspective to a location and labor market perspective. We can do no less than acknowledge the profound changes occurring in urban regional economies worldwide. These changes start with the transformation of the sources of income and wealth from natural resources and physical capital to human capital, and from real estate and various forms of landed property to intellectual property. We focus on some underlying determinants of these changes in Chapter 4, Exports and Economic Base.

We start this chapter with the central place paradigm of the earlier periods of farm-to-market roads and country grain elevators, many with sideline businesses serving farmer customers. This paradigm is associated mostly with a purely market explanation for the emergence and location of trade centers and their interrelationships. We then look at the consequences of the expanding trade and industry on the value of real property and, more recently, of intellectual property in the central places. We talk about "bid rents," local real estate markets, and bridging the "lab-to-market gap." We conclude by returning to the earlier discussion of local labor markets and labor market areas—now among the central concerns of expanding business enterprises, especially those engaged in interregional competition, but increasingly constrained by limited access to the local labor market.

Central Place Theory

The first principles we present are commonly known collectively as central place theory (CPT). CPT is a purely market explanation for the emergence and location of cities. The second set of concepts involve land rent, location decisions, and resulting urban geographic structures. The implication of bid rents on location decisions within an urban area and the impact on the way our cities look and emerge are very important to some of our later analyses. We illustrate these concepts graphically and by real-world examples. These examples summarize more detailed findings on inter-industry linkages and behavior of consumers and producers in specific market situations. As economist Robert J. Samuelson comments, "God is in the details."[1] We begin with CPT.

Origins of Cities and Central Places

Central place theory has less to say about the origin of cities and central places than, for example, their emergence on an extended plane in a largely agricultural economy. Cities now exist for many purposes, but in years past they started as places of worship, defense, or trade. They appeared on hilltops, near rivers and streams, or on ocean estuaries with access to both a continent's interior and other continents across the ocean.

Why Do Cities Exist?

Peter Hall (1966) lists seven world cities—London, Paris, Ramstad Holland (Amsterdam, Rotterdam, and Utrecht conurbations), Rhine-Ruhr (Bonn, Cologne, Düsseldorf), Moscow, New York, and Tokyo. These seven cities grew from a total population of 3.9 million in 1800 to 70 million in 1960—an annual growth rate of 1.8 percent. The population of their seven countries increased from 159.7 million to 644.5 million during this period—an annual growth rate of 0.9 percent. The growth rate for the seven cities, which was twice that of the seven countries, pushed the city share of total population from 2.4 percent to 10.9 percent. Hall (1966) attributes this phenomenal growth of the seven cities to the emergence and expansion of white-collar occupations of all kinds. "It was a shift of interest away from the physical process of production, and towards questions of financing, decisions to produce, and marketing: in other words, from the factory to the office" (p. 26).

We limit our discussion of world cities to six—Athens, Rome, Paris, London, New York, and Tokyo. The early history of these six cities cites the importance of a "high place" in the defense of the early settlements. Once established, the high places and surrounding sites, and later, the walled, fortified cities, became centers of worship, trade, and governance. With the gradual dispersion of knowledge and skills among increasing numbers of inhabitants of these central places, new forms of governance emerged that led to a corresponding dispersion of control over daily activities and individual choices. Centralization of governance gave way to decentralization, and regulation of commerce and industry to deregulation. More local control brought with it, of course, new challenges of authority and governance and new rules for protecting public health, safety, and welfare. We start with the early classics—Athens and Rome.

Athens was developed next to the Acropolis around 1300 B.C., when Theseus was uniting the townships of Attica under Athens.[2] The fortification of the rocky hilltop of Acropolis, which had been inhabited since 3000 B.C., took place about that time. Athens soon spread to the northwest into an area that became known as the Agora (or marketplace) of Theseus. Seven centuries later, during the time of Solon the Lawgiver, much public building was completed in this area. The Acropolis had become the spiritual center of Athens while the Agora had become the center of commercial life.

The Athenian Golden Age began with the defeat of the Persians in 479 B.C. Athenian democracy flourished, with Pericles as its leading statesman during much of the Golden Age. It ended in 431 B.C. with the outbreak of the Peloponnesian War (431–404 B.C.). Athens fell to the Romans in 146 B.C. The Romans treated the city with respect, viewing it as the cradle of civilization. It was burned in A.D. 267 and remained much less than its former self, not achieving significant growth until the establishment in A.D. 330 of Constantinople as the eastern capital of the Roman Empire.

Rome started as a village of shepherds on an easily defended hill close to the Tiber River. Its early history falls into three periods that span nearly eleven centuries—from 753 B.C. to A.D. 330. Romulus, who became Rome's first king, walled the city in 753 B.C. Rome became a republic when it ousted the monarchy in 510 B.C. It remained a republic until 27 B.C., when Augustus became its first emperor, which also marked the beginning of its short-lived Golden Age.

Imperial Rome grew rapidly on the floodplains of the Tiber River and around its seven hills to a population of 1.2 million by the second century. Problems of housing, water supply, and sewage disposal were overcome with its marvels of civil engineering—the many public buildings, the forum that served as a central market and meeting place, and the elevated aqueducts that brought clean water to the city—eventually 200 million gallons daily. In A.D. 330, Emperor Constantine abandoned the city and moved the capital to Byzantine Constantinople.

Rome declined to a mere shadow of its former self until the beginning of its Renaissance in the time of Pope Nicholas V (1447–1455). Modern Rome began in 1870, when Rome became the capital of a unified Italy.

Paris goes back to around 300 B.C., when a Gallic tribe set up a few fishing huts on the Ile de la Cité. The settlement spread to the left bank of the Seine about the time the Romans occupied Gaul in A.D. 53. It lacked the stability and freedom to grow until the twelfth and thirteenth centuries, when its growth soon left the city with narrow winding streets and no large open spaces. Considerable development occurred in some parts of the city between the sixteenth and eighteenth centuries, but its transformation on a grand scale was not achieved until the 1850s under the extraordinary control of all building by Georges-Eugene Haussmann, who reported directly to Napoleon III.

The Romans built London, first known as Londinium (from the Celtic Ilyndin) in A.D. 43 on a crossing of the Thames River. It began as a small Celtic settlement on the north bank of the river, roughly at the site later taken by the London Tower. No remains exist of this early village. By the middle of the third century, Londinium reached a total population of roughly 50,000 within an area of about 320 acres, surrounded by a wall eight feet high, with bastions, towers, and a fort. Recognizing its importance, the Romans gave it the name Augustus in 288. It was already the largest city in Britain.

London was abandoned by the Romans in 410, sacked by the Danes in 851, and largely destroyed by fire in 1666. The rebuilt London and its environs had a total population of 674,000 by 1700. It became the first world city to reach a population of two million. As the city suffered from increasing congestion in the early 1800s, new bridges—Waterloo and London (later in the 1800s, Hammersmith and Tower)—were built across the Thames. The London and Greenwich Railway was established in 1839, followed by the world's first underground railroad in 1854. London was now the largest and most influential city in the world, serving as the center for world commerce and finance. The Great Exhibition, held in Hyde Park in 1851, marked the high point of the British Empire, with London as its central place.

Tokyo was originally a small fishing village on the Sumida River where, in 1456, a Japanese warlord built a castle. In 1657, fire destroyed the castle and most of the city (named Edo), which then had a population of only 107,000. The city was rebuilt on a more open plan to safeguard against fire and with 80 percent of the land occupied by a new castle, samurai estates, temples, and shrines. It again suffered considerable damage in 1707 with the eruption of Mount Fuji. The mid-1700s was the high cultural period of Edo, known as Edokko. In 1867, when the isolationists forced the abdication of the Shogun Yoshinobu in Kyoto, the Meiji Emperor became the new civil ruler. The capital was moved from Kyoto to Edo, now renamed Tokyo, and the castle became the Imperial Palace. The samurai then moved to the provinces; their estates were subdivided into small farms.

New York owes a part of its past to Peter Minuit, a Dutch trader, who bought Manhattan Island from the Algonquin Indians for "a few trinkets worth no more than $24." The Dutch established a settlement on the southern tip of the island, which they called New Amsterdam. The British took over the city, with its population of only 1500, in 1664. British troops took over the city during the War of Independence. Following two large fires and the moratorium on land speculation, its population dropped by one-half, from a peak of 20,000. It was the first capital of the United States from 1785 to 1789—the year George Washington became the country's first president. New York achieved a huge boost in industry and trade with the completion of the Erie Canal in 1825, which now linked the Atlantic Ocean and the Hudson River to the Great Lakes. New York became the "gateway to the nation" as the major point of entry for immigrants—more than seven million between 1855 and 1890 and twice that number in the next 70 years.

Where Do Cities Develop?

There are many explanations for why cities came to be, why they located where they did, and why some urban places grew to be very large while others remained small. Explanations for the sizes and locations of cities include such factors as (1) the impact of large numbers of people on the protection and defense of a particular region, (2) the natural need for humans to interact with one another, (3) the notion of trading centers along major trading routes, and (4) the dependence of location on natural harbors or other transportation factors.

Proponents of the market as an explanation for urban development point to three crucial factors: (1) technology in agriculture providing the food base and the labor force for urban centers to survive, (2) the presence of economies from scale, which would make trade and specialization possible, and (3) a level of competition sufficient to allow firms to enter the market when economic profits are being made. Central place theory assumes the first of these and then concentrates on the other two. Walter Christaller (1966) and August Lösch (1964) are generally credited with the development of CPT. They assumed monopolistic competition, that is, many firms selling slightly differentiated products with free entry and exit into and from the market. In this case, the product differentiation is based on separated locations leading to territorial monopoly power. They also assumed economies from scale. Finally, for simplicity's sake, they assumed a flat and homogeneous plane with transportation equally available in every direction.

We will give the conclusions of the model first. According to CPT, the growth of a city depends upon its specialization in urban service functions. The levels of demand for urban goods and services over the geographic region a particular city dominates will determine both how fast a city will grow and how large it will become. A city's primary function is to act as a service center for the hinterland around it. The city provides this hinterland with central goods and services. These central goods and services can be ranked in higher or lower order, depending upon the size of market necessary for a firm providing this good or service to survive.

What brings these conclusions about? We turn to the assumptions in the monopolistic competition model. *Monopolistic competition* is a market situation in which any profits, beyond the opportunity costs of the entrepreneur, are eliminated by new firms entering the market. The entry of new firms reduces the market share of existing firms making similar products. The *threshold* of a firm is the minimum size of market necessary for the firm to survive, that is, for the firm to make normal profits. In CPT, the threshold is stated in geographic terms. What is the minimum geographic market area required to allow the firm to continue to operate?

The notion of "delivered price" is another principle that space introduces into the traditional theory of monopolistic competition. Most prices at stores are FOB (free on board) prices. The purchaser pays all of the search costs, the transportation costs to and from the store, and even the transportation costs within the store. This means that the true delivered price to the customer is the FOB price plus the search and transportation costs associated with securing the good or service. The farther the store is from the consumer, the greater are these costs and, therefore, the higher is the delivered price.

The normal demand curve is downward sloping. As delivered costs increase, the quantity demanded of the good or service decreases. People near the establishment will purchase greater quantities from the reference store than will people far away (Figure 3-1).

The total area served by the good or service provider is called the *range* of the good. If the range is shown to be greater than the threshold, the firm in question is said to be making economic profits, that is, profits greater than those required for the firm to remain in business. If the range is smaller than the threshold, the firm is losing money. The firm may remain in business in the short

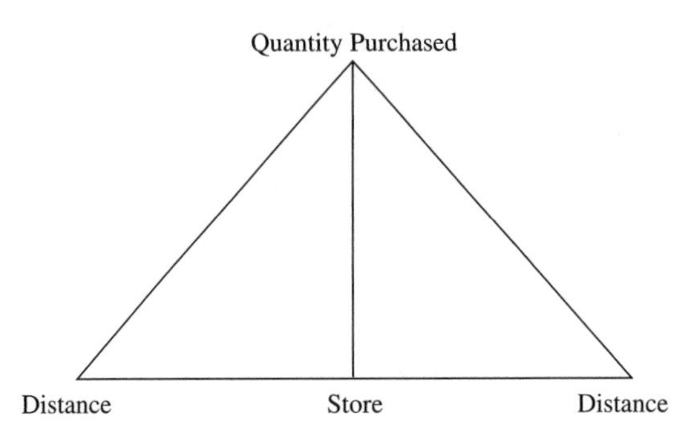

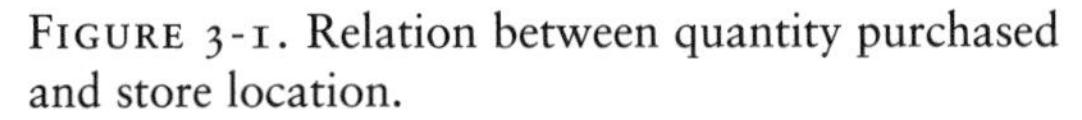
FIGURE 3-1. Relation between quantity purchased and store location.

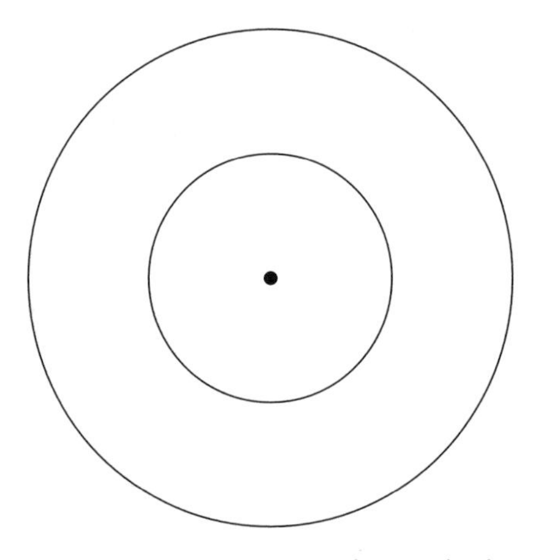

FIGURE 3-2. The range of a good relative to the provider.

run under this latter condition if it can cover its variable costs. However, the firm will leave the market in the long run unless market conditions change to the firm's advantage.

Figure 3-1 shows the range of the good or service from east to west. Of course, if we assume a flat and homogeneous plane with transportation equally possible in every direction, the range also runs from north to south, from northeast to southwest, and so on. If we rotate Figure 3-1 to take into account all directions and then look down on the range of the good from a cloud, we would see that the range takes the shape of a circle (Figure 3-2).

In fact, Figure 3-2 shows the firm making economic profits because the range (the outer circle) is greater than the threshold (the inner circle). It will soon be shown that these economic profits will lead to new firms entering the market, reducing the market share of the original firm.

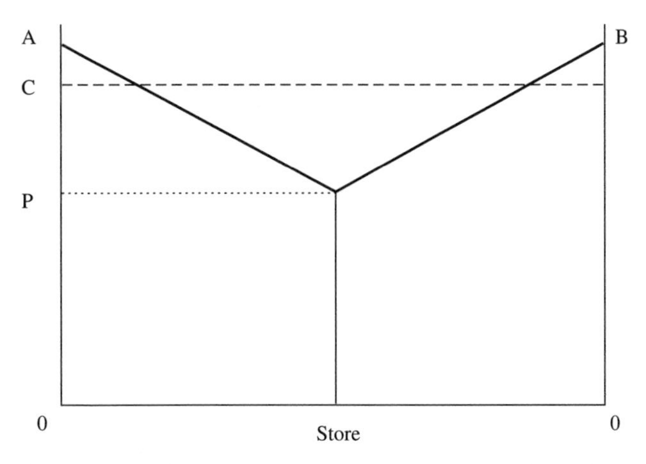

FIGURE 3-3. Store location and delivered price for a firm making economic profits.

The Competitive Process

The geography of monopolistic competition can be demonstrated through the use of so-called *demand trees*. Figure 3-3 shows such a tree. The price at the store is equal to 0P. The delivered price then increases as distance from the store increases, shown by the slanted lines A and B. When the quantity demanded falls to zero, the boundary of the market is determined. The outer axes of the diagram indicate this boundary.

Now assume that the market area associated with 0C defines the threshold of the firm. In other words, the range is greater than the threshold and economic profits are being made. Such a condition encourages new entrepreneurs, who face opportunity costs similar to those encountered by the original firm, to enter the market. Market entry reduces the market share of the initial firm. Monopolistic competition concludes that new firms will continue to enter the market until all economic profits are eliminated and the market is in long-run equilibrium.

If we rotate the delivered price diagram (Figure 3-3) much as we did with the demand cone (Figure 3-1), we again have circular market areas (Figure 3-2). Assuming the purchaser will always buy from the location where the delivered price is lowest, linear market boundaries are drawn between the firms (Figure 3-4).

Of course, the firms would not only encroach on one another from east to west. They would also come from below and above the schematic in Figure 3-4. The original CPT model argued that encroachment would continue to occur until the final market areas are as close to a circle as possible while using up all the available space. The polygon that serves this purpose is the hexagon. According to CPT, the plane would eventually be filled by hexagons of equal size, looking not unlike a honeycomb.

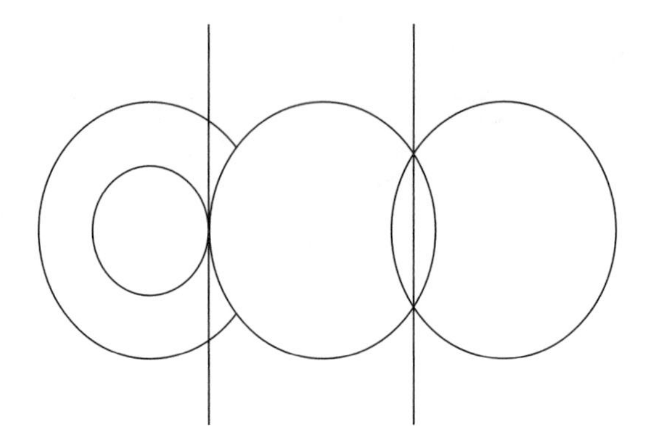

FIGURE 3-4. Schematic of market areas for firms encroaching on one another's markets.

Central Place Theory Once Again

Lösch (1964) developed a theory of urban growth and development using the demand cones and trees in conjunction with the notion that monopolistically competitive firms want to move away from one another so as to maximize their monopoly power over geographic space. Lösch argued that economies from scale lead to firms specializing in different products. Many of these products have similar scales leading to threshold hexagons of similar size. However, firms experiencing larger economies from scale will have larger threshold sizes in order to realize the scale economies. The larger the geographic threshold, the higher order the good or service. Thus, we have a pattern of higher and lower ordered goods and services depending upon the size of scale economies and resulting threshold sizes to be realized.

Let's say that the highest order good or service requires a population of one million to achieve its optimum scale economies potential. Let's say further that the next lower good or service requires 100,000 individuals to achieve its threshold. Finally, let's say that the lowest order good or service requires 10,000 individuals to achieve its threshold. Let's call the highest order good major league baseball, the next order good car dealerships, and the lowest order good restaurants. Further, let's assume our flat homogeneous plane has a total population of 1.5 million individuals and that the baseball team locates in the center of the plane. It is also true that the next two orders will have a firm located in the center to minimize the transportation costs to its customers who, in turn, locate in the center both to work and to enjoy the services of the baseball team (Figure 3-5).

Point A in Figure 3-5 contains all goods and services and the baseball team, car dealership, and restaurant. With 1.5 million individuals located on the plane,

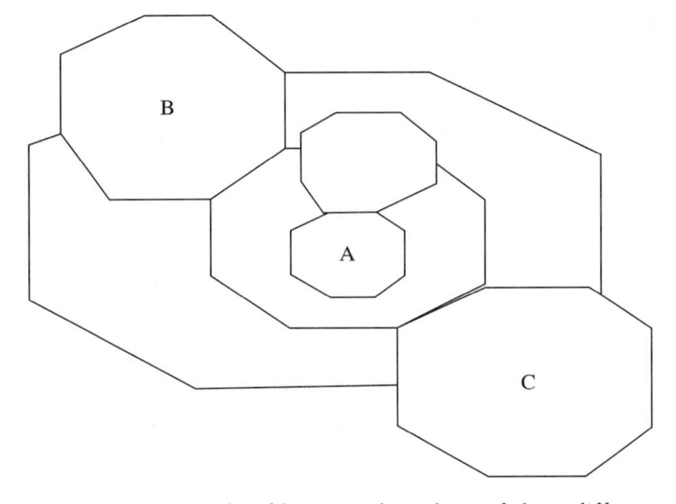

FIGURE 3-5. Sample of hexagonal markets of three different sizes surrounding point A, the geographic center of the flat and homogeneous plane.

only one baseball team is viable. However, there can be 15 car dealerships and 150 restaurants located across the plane. Point A will have the largest population as people locate near to their places of work and entertainment because the largest number of jobs will locate where the greatest level of economic activity takes place. This is a central place of the highest order and constitutes the largest city in the region. Points B and C are both located within the market boundaries of the highest order place and become cities of the second order. Finally, there are many of the smallest towns, each of which is located within the market boundaries of the first and one of the second order places. Each of these lowest order places will have a restaurant.

Of course, there may be several goods and services with the same scale economies and resulting threshold boundaries. For example, grocery and clothing stores might have the same size hexagon as the restaurant's. The largest central place would, under these conditions, house the baseball team, car dealerships, grocery stores, clothing stores, and restaurants. The smallest central places would have grocery stores, clothing stores, and restaurants. The populations of these towns are assumed to shop in the nearest second order place to purchase their cars and travel to the highest order place to attend a major league baseball game.

A hierarchy of central places emerges. The orders are reflected both in terms of economic activities provided and in terms of city population levels. The places are located an equal distance from the next competing central place offering the same level of goods and services. The functions of the various cities depend on the nature of the goods and services they provide, which, in turn, depend on the required threshold geographic area to keep the firms in long-run equilibrium (zero economic profits or losses).

The most obvious criticism of this theory is that we do not exist on a flat and homogeneous plane. We have mountains, streams, and oceans with which to contend. The uneven typography of our landscape would mean that the even patterns of cities of similar sizes would be distorted by other factors. Distortions such as a road would mean that the hexagon is no longer a viable polygon with which to describe market areas. If squares, or more probably rectangles, better describe market areas, the basic principles do not change. However, this criticism does not disprove the basic tenet of the theory that cities disperse and specialize according to the level of goods and services they provide. This is, of course, only part of the story. The even more critical part is the daily trip to work and the disassociation of place of residence from place of work. In other words, CPT provides only a partial explanation of the growth of service centers and central places.

In short, the several strengths of the model are largely theoretical. CPT represents a purely market explanation for the emergence, location, and size of cities. It emphasizes the importance of transportation costs in the decision making of both the firm and the customers it serves, and as such, explains why people prefer to shop at the stores closest to them, all other things being equal. It serves as the basis for several marketing research models used in determining the optimal location for new stores and in mapping the geography of trade and service activities of rural and urban trade centers.

Central Place Theory and Competitive Behavior

We have already looked at competitive behavior as it influences location decisions. For example, Weber's analysis (1909) argued that the competitive firm will choose a location so as to minimize costs. His models assumed that there was significant competition, that the entry of new firms would bring prices down to the costs of production, and that the seller would choose a central location with buyers coming to and delivering from that location. Such analyses were also influenced by von Thünen's work (1826), which also assumed a central location for the firm.

This is in contrast with the so-called market area analysis, which conceives of buyers scattered over an area. Each seller then has some monopoly power stemming from his or her choice of location. The scattering of firms becomes a part of the competitive strategy. In short, until the works of Lösch and Christaller, the assumption was generally that of pure competition, ignoring the possibility of at least some monopoly power. This assumption is becoming less and less tenable in today's society. The following analysis summarizes contributions from Greenhut's work (1963) in the analysis of monopoly power and monopoly behavior in a spatial setting.

Greenhut began by discussing the profit-maximizing, nondiscriminatory FOB mill price of the spatial monopolist. His assumptions were as follows:

1. A straight-line demand curve forming an isosceles triangle,
2. Absence of price discrimination under a FOB pricing system,
3. Costs of production equal to zero,
4. The monopolist as a profit maximizer, and
5. Two sets of buyers with identical demands: one set located at the production point where the net mill price is paid without any transportation, the other at some distance, paying the net mill price plus transportation.

Variables involved:

- m = net mill price
- b = y = price intercept
- K = freight cost per unit to the most distant buyer
- P = monopoly price in the absence of freight = b/2 when costs are zero and b/2 + MC/2 when costs are positive
- ox = bx = 1/2 0b and p = b/2, where elasticity = 1

In other words, the nondiscriminating monopolist will absorb half the freight to the distant buyer by lowering price by Y/2. If the monopolist is nondiscriminating, the price to the near buyer will be lowered by the same amount.

If marginal cost is greater than zero, MC > 0 as in Figure 3-6, we derive the following:

- BC = 1/2 BD since MR cuts the middle of any straight line drawn perpendicular to the 0Y axis
- YDB and PDC are similar triangles, therefore,
- CP = 1/2 YB, therefore,
- AP = AC + 1/2 YB
- YB = 0Y – 0B
- AP = AC + 1/2(0Y – 0B)
- AP = AC + 1/2 0Y – 1/2 0B
- 0B = AC, so AP = AC + 1/2 0Y – 1/2 AC
- AP = 1/2 0Y + 1/2 0B

For the nondiscriminating case, the following relations exist:

- b – m = the number of units purchased by the buyers located in the seller's town (because of the isosceles triangle)
- m + K = the price per unit to distant buyers
- b – (m + K) = the number of units purchased by distant buyers

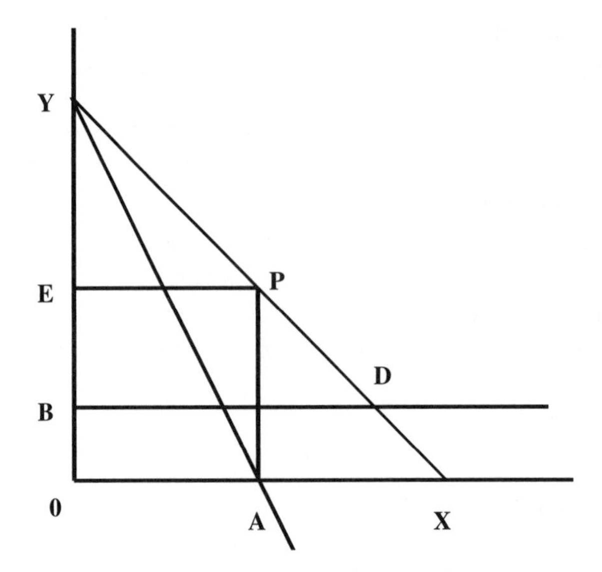

FIGURE 3-6. Monopoly demand curve when MC > 0.

Relationships that follow from above:

- $2b - 2m - K$ = the total number of units purchased
- Total Revenue (R) = $m(2b - 2m - K)$
- $dR/dm = 2b - 4m - K = 0$ for maximum revenue with no costs
- $m = b/2 - K/4$
- but, $b/2 = P$, so $m = P - K/4$

This result can be interpreted as saying that the producer reduces his or her mill price by K/4, that is, by one quarter the cost to his or her distant buyers to maximize profits.

The following holds if a third group is located the same distance as the second but in the opposite direction:

From the original set of relationships,

- $3b - 3m - K$ is the total amount purchased
- $R = m(3b - 3m - 2K)$
- $dR/dm = 3b - 6m - 2K = 0$
- $m = b/2 - K/3 = P - K/3$

We can continue for any number of buyers located on the market periphery like spokes of a wheel.

One interesting case is that of four spokes of a wheel with customers at the center: The freight absorption is then 2K/5 from $dR/dm = 5b - 10m - 4K = 0$.

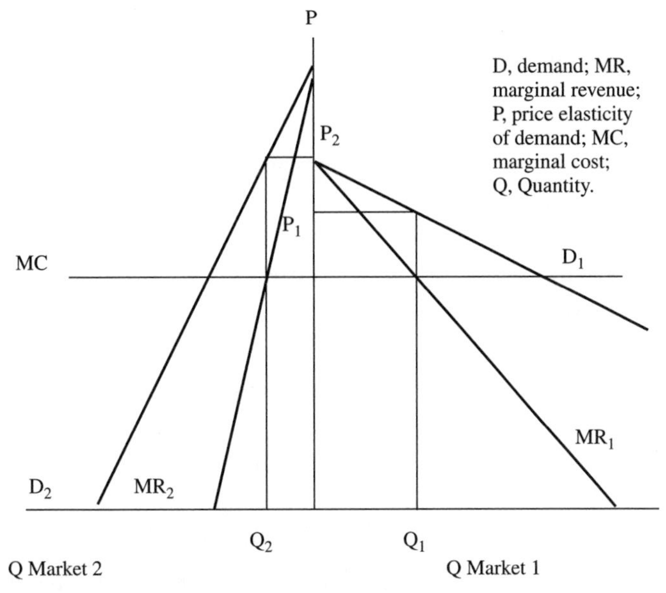

FIGURE 3-7. Two markets served by a discriminating monopolist. D, demand; MR, marginal revenue; P, price elasticity of demand; MC, marginal cost; Q, quantity.

If one buyer is located at the seller's factory and the other buyers are assumed to be evenly scattered on a straight line that runs from the seller's factory, the freight absorbed is, as in the initial case: K/4 from $dR/dm = 2b - 4m - K = 0$; for the multiplier applicable to m is always twice the multiplier applicable to b, which, in turn, is always twice the multiplier prefixed to K.

Are there any examples of this type of discrimination? While not a perfect example, let's say a person living in Duluth, Minnesota, purchases furniture in the Twin Cities (Minneapolis and St. Paul), 150 miles away. Most Twin Cities furniture stores provide free delivery to the buyer's Duluth home. In other words, furniture stores absorb freight to the most distant buyer. If the true price is the FOB price at the store plus transportation, this represents a classic case of freight absorption for distant buyers.

What about the discriminating monopolist? The classic argument is that the discriminating monopolist will discriminate against buyers with the more inelastic demand in favor of buyers with an elastic demand.

Looking at Figure 3-7, the following assumptions hold:

1. There are two separate markets, indicated by the demand (D) and marginal revenue (MR) curves.
2. Each of the two markets is separate from the other in that one market cannot buy the product at a lower price and then sell the product to the other market at a higher price.

3. The two market demand curves exhibit differing elasticities—the price elasticity of demand (P) in market 1 is higher than in market 2 at all comparable points.
4. The marginal cost (MC) is the same for the two markets. The firm maximizes its profits when it equates marginal cost with marginal revenue in each of the separate markets.

Under these assumptions, the common marginal cost will lead to two different prices. The price elasticity of demand is greater at all comparable points in market 1 than in market 2. This means, of course, that the consumers in market 1 are more flexible than the consumers in market 2. Perhaps this flexibility stems from the fact that consumers in market 1 live some distance from the place of business. This flexibility earns the consumers in market 1 a lower price (P_1), compared to the price in the less elastic (flexible) market 2 (P_2).

There are really three degrees of price discrimination, ending with perfect discrimination, whereby every buyer pays the price he or she is willing and able to pay; that is, the seller charges the price the market is able to bear. Under these conditions, the market is more efficient, but the monopolist gains the maximum possible profit.

An interesting example of such discrimination occurred in the airline industry. While waiting at an airport for a plane that was late, a colleague took a very informal, nonscientific survey of the other passengers in the waiting area. He counted only the people who were flying coach on flights originating in Minneapolis and terminating in New York. He tallied twenty responses. No two passengers paid the same price for their ticket—perfect price discrimination.

The question that needs to be asked for both types of monopoly theories: Is this bad? From a market efficiency point of view (virtually the only point of view accepted by mainstream economists), both models represent an improvement. However, if there is concern about monopoly pricing and monopoly profits, both models would be troublesome.

Service Centers in Rural Areas

Some activities, like building and maintaining active and viable centers for the delivery of essential services in rural areas (that is, areas outside our presently defined MSAs), go well beyond the explanatory powers of CPT. Such activities redefine commonly held notions about the structure of rural communities and the role of services, particularly those initiated, regulated, or provided by local governments. They depend, however, on the fundamentals of CPT with local understanding and enrichment through public discourse. We start the discussion of this process with the mapping of trade and services areas.

Mapping Trade and Service Areas

We delineate trade and service areas for the full range of goods and services purchased by households and businesses, with varying sizes for the areas depending upon the range of the good or service, local competition, and the fundamentals of CPT. Berry and Pred (1961) list over 450 authors credited with one or more central place studies. The bibliography covers sixteen topical areas, starting with statements of CPT and including, among other topics, rural neighborhoods and communities, medical service areas, planned shopping centers, consumer shopping and travel habits, and urban business structure and urban land use theory. Anding (1990) and Borchert and Yeager (1968) delineate systems of trade areas for the Upper Midwest Region. These include (1) complete shopping trade areas, (2) secondary wholesale-retail trade areas, (3) primary wholesale-retail trade areas, and (4) metropolitan trade and service areas. Morrill (1970) provides a variety of real-world applications of CPT and its extensions from local to global environments in accounting for the spatial organization of society.

We look to Reilly's Law of Retail Gravitation for an early application of central place concepts. This application is a distance-based "gravity" model of local shopping behavior. We calculate a trade area boundary between two trade centers using Equation 3-1 (Stone, 1997):

$$TAB_{ab} = D_{ab}/([1 + (POP_a/POP_b)^{1/2}] \qquad (3\text{-}1)$$

where TAB_{ab} is the trade area boundary between trade center a and trade center b, D_{ab} is the distance between the two trade centers, POP_a is the population of trade center a, and POP_b is the population of trade center b. This is equivalent to the statement that the ratio of the two distances to their common boundary is equal to the square root of the two population ratios, that is,

$$\frac{D_a}{D_b} = \left(\frac{POP_a}{POP_b}\right) \qquad (3\text{-}2)$$

Using this distance-based model, we calculate the trade area boundary between two trade centers—populations of 250 and 25,000, respectively, and 60 miles apart in Equation 3-1, as follows:

$$TAB_{ab} = 60/[1 + (250/25000)^{1/2}] = 55 \text{ miles from trade center b} \qquad (3\text{-}3)$$

We use a variation of Reilly's law in the delineation of the air node regions mentioned earlier. Rather than retail trade areas, we focus on local labor markets as the central organizing rationale in area and regional delineation. We seek information on building and maintaining the critical linkages between rural and metropolitan area organizations, especially those that provide rural residents quick and easy access to metropolitan area services for both businesses and households. The new business and personal networks of communication and information transfer and the new forms of specialization in service delivery emerging from the active and effective use of these networks already support new and different ways of bundling and packaging rural services for rural residents. The studies of individual behavior settings and alternative service delivery options in rural communities, for example, provide new insights into the most effective ways of combining public and private resources and capabilities to meet these new challenges of community survival and progress in rural areas.

Identifying Individual Behavior Settings

The writings of Roger Barker (1968, 1978) and Karl Fox (1985, 1994) present a unique approach to community surveys that goes well beyond the mapping of trade and service areas. In his earlier writings Barker identified, described, and catalogued individual behavior settings as basic elements of community organization. He described a behavior setting as "a place where most of the inhabitants can satisfy a number of personal motives, where they can achieve multiple satisfactions. In other words, a behavior setting contains opportunities. . . . The unity of a behavior setting does not arise from similarity of the motives of the occupants. [A] setting exists only when it provides its occupants with the particular psychological conditions their own unique natures require. Heterogeneity in the personal motives of the individual inhabitants of a setting contributes to the stability of the setting" (Barker, 1968, p. 219).

In Midwest (the code name for the small Kansas town Barker studied in his early years), Barker identified 884 nonhousehold behavior settings, with the largest number being school sponsored activities, followed closely by volunteer associations. In most cases businesses were small and consisted of only a single behavior setting. Schools were much more complicated. Each class in an elementary school was a separate behavior setting. A high school was even more complicated, with each of its varied activities, academic and extracurricular, being separate behavior settings. A particular environment elicits a unique behavior setting, hence the expression "ecobehavioral approach" that appeared in the literature describing Barker's work.

Barker's classification of behavior settings identifies each participant, if gainfully employed, by occupation, following the *Dictionary of Occupational Titles*. The classification also recognizes levels of general educational development and vocational preparation. In addition, the private-enterprise categories permit sorting communities according to their status in trade center hierarchies as described by Borchert and Yeager (1968).

Service Delivery Options

Osborne and Gaebler (1993) introduced a number of criteria for choosing the best alternatives in service delivery and then monitoring and evaluating their performance. The criteria are as follows: service specificity, availability of producers, efficiency and effectiveness, scale of the services, relating benefits and costs, responsiveness to consumers, susceptibility to fraud, economic equity, equity for minorities, responsiveness to government direction, and size of government. The authors also list the qualities desired in service producers, with each sector playing to its particular strengths. They list the tasks best suited to each sector, like policy management for the public sector, economic tasks for the private sector, and social tasks for the third sector—not-for-profit organizations.

The authors note what each sector is best in providing. The public sector is best in regulation, ensuring quality, preventing discrimination and exploitation, ensuring continuity and stability of services, and ensuring social cohesion, besides policy management. The private sector is almost the opposite, being best at performing complex tasks, replicating the successes of other organizations, delivering services that require rapid adjustment to change, delivering services to very diverse populations, and delivering services that become obsolete quickly. The third sector is best at tasks that generate little or no profit margin, require compassion and commitment to other individuals, involve a comprehensive, holistic approach, require extensive trust on the part of customers or clients, need volunteer labor, and demand hands-on personal attention.

An example of a major effort to provide the quantitative bases for improving the delivery of health services in rural area is the report entitled *Rural Health in the Northwest Area,* published by the Institute for Health Services Research, School of Public Health, University of Minnesota. This report is directed primarily at acute medical problems in an eight-state area extending from Minnesota and Iowa in the Midwest to Washington in the Pacific Northwest. Neither mental health services nor geriatric services are covered, partly because of the lack of data—a situation that will become even worse as both types of services move more and more to community-based, rather than institutional, care.

Central Place Labor Markets

Central place labor markets critically affect the economic performance of the central place as well as the larger region served by its largest central place. The spatial and functional organization of these markets is an important key to predicting likely future changes in a region's industry structure as well as its productivity and competitive position in domestic and international markets.

Alternative Market Perspectives

We refer, first, to the division of labor in a regional economy and the local economic conditions that contribute to its differentiation from one part of a region to another. This correlates directly with the division of production, the concentration of technology-intensive and high-order services in the large metropolitan centers, and the dispersion of agriculture and other natural-resource-based activities to the region's periphery (Schaeffer and Mack, 1996).

New International Division of Labor

The new international division of labor (NIDL) builds on this relationship. The NIDL refers to (1) the manufacturing industries in developing countries that produce for world markets and (2) the increased subdivision of the manufacturing process into smaller and smaller steps. It is viewed by some as a new form of expansion and accumulation of capital that began in the 1960s from the emergence of a single market for labor and a world market for industrial sites.

An alternative approach to the study of labor markets takes into account the efforts of innovative entrepreneurs successfully engaging in the replacement of existing technology-intensive products with updated or entirely new versions of the original products. The alternative perspective also addresses the special role of product exports and producer services in the local division of labor and the regional organization of local labor markets.

By labeling the focus of this critique the international division of labor, we can ignore, on theoretical grounds, the individual elements that make up the larger system. This occurs at the risk of emphasizing only the quantitative outcomes of local labor market activities without an understanding of the underlying economic processes that contribute to these outcomes. Meanwhile, the reality of labor markets and the changing occupational distributions of individual industries, along with changes in industry composition in these markets, are ignored.

Finally, the use of highly aggregated numbers on industry production and related labor skills leaves the question unanswered as to how outcomes of local labor market performance yield an international division of labor. For this, we seek information on the functioning of local labor markets, individually and as interacting elements in a regional production system.

Wage Rate Differences

We refer to the commuting area—the daily journey to work of an urban-centered workforce—as the labor market area (LMA). It is the primary building block in a "bottom-up" and central place approach to regional economic organization. LMAs are also based on the principle of proximity in their delineation and definition. New research findings confirm earlier insights indicating that the most strategic relationships among firms in network systems serving global markets are local because of the importance of timeliness and face-to-face communication for rapid product development (Saxenian, 1994). The economic activities within a region account for a corresponding differentiation of their role in regional economic organization.

Concentration of high-order producer services and infrastructure—communications, transportation, energy systems, along with the region's major educational and research institutions and technology-intensive manufacturing—characterizes the economy of the metropolitan core area, which is usually the largest and most densely populated labor market area in an economic region. Beyond the core area lie other LMAs, including other highly urbanized metropolitan LMAs, nonmetropolitan LMAs serving as multicounty shopping and service centers, and the most sparsely populated LMAs of small towns and open-country settlement. Rural and nonmetropolitan LMAs are the most numerous in each economic region, but are not highest in total population and economic activity. They include the overspill areas bordering the core area as well as the most distant, natural-resource-based peripheral areas.

An excess of local jobs over resident jobholders identifies the central place of an LMA. This definition of a central place focuses on the critical resource of most regional economies, that is, their workers of many trades and skills supporting diverse and dynamic local labor markets (Saxenian, 1994). The in-commuting of nonresident jobholders overcomes the deficit in resident jobholders for central places. This accounts for the varying concentrations of economic activity, depending on the types and sizes of LMAs and their location relative to metropolitan core areas. This also accounts for the types and sizes of businesses in the various LMAs (Reynolds and Maki, 1990a,b).

Daily Commuting Area and Local Labor Markets

Local labor markets, when differentiated from each other by the drawing of boundaries between them, form LMAs. A collection of LMAs forms an economic region. Each region has a center—a metropolitan core area—and a periphery. The metropolitan core area is also an air node in the U.S. air transportation network of air nodes and connecting cities within each region. A trading region is defined by the proximity of its contiguous labor market, or daily commuting areas to the metropolitan core. The LMAs differ from one another by size, location, and economic base as represented by their export-producing industries.

We present a market economy perspective in a proximity-based regional delineation based on the 29 U.S. air node centers, excluding Alaska and Hawaii, and

TABLE 3-1 Total population and land area in specified air node regions, by region population rank, United States, 1990

				Total Population					Land Area (sq. mi.)			Pop. Density	
					Metropolitan (MSAs)								
Rank	Air Node Region	Core State	No.	Region Total (thou.)	Total (thou.)	Core* (thou.)	Other (thou.)	Non-metro (thou.)	Total (thou.)	MSAs (thou.)	Nonmet. (thou.)	MSAs (sq.mi.)	Nonmet. (sq.mi.)
1	Los Angeles	CA	3	22,727	22,025	8,863	13,162	702	905	490	416	45	2
2	Chicago	IL	20	16,185	13,581	7,411	6,170	2,604	296	104	191	130	14
3	New York	NY-NJ	12	15,020	14,855	8,547	6,308	165	33	29	5	517	36
4	Houston	TX	8	12,741	10,393	3,322	7,071	2,348	537	163	375	64	6
5	San Francisco	CA	4	12,018	11,040	1,604	9,436	978	436	200	237	55	4
6	Atlanta	GA	24	11,077	7,211	2,960	4,251	3,866	445	120	325	60	12
7	Dallas	TX	7	11,071	8,003	2,676	5,327	3,068	957	140	816	57	4
8	Pittsburgh	PA	19	10,216	7,815	2,395	5,420	2,400	247	84	163	93	15
9	Detroit	MI	23	10,017	8,422	4,267	4,155	1,595	179	70	109	121	15
10	Boston	MA-NH	16	9,920	8,678	5,455	3,223	1,242	235	78	157	111	8
11	Denver	CO	2	8,287	5,724	1,623	4,101	2,563	2,023	178	1,845	32	1
12	Minneapolis-St. Paul	MN-WI	10	7,792	4,122	2,539	1,583	3,670	1,041	134	907	31	4
13	Seattle	WA	1	7,775	5,875	2,033	3,842	1,900	708	121	587	49	3
14	Kansas City	MO-KS	6	7,566	4,974	1,583	3,391	2,592	515	102	413	49	6
15	Raleigh	NC	28	7,299	4,934	858	4,076	2,365	222	77	145	64	16
16	Orlando	FL	26	6,835	6,205	1,225	4,980	630	140	73	67	85	9
17	Cincinnati	OH-KY-IN	21	6,747	4,813	1,526	3,287	1,935	184	59	125	81	16
18	Newark	NJ	15	6,411	6,002	1,916	4,086	409	50	38	12	160	34
19	Syracuse	NY	13	6,348	5,016	742	4,274	1,332	182	78	104	64	13
20	Washington, D.C.	MD-VA-WVA	18	6,019	5,370	4,223	1,147	650	94	48	46	112	14

21	Memphis	TN-AK-MS	11	5,996	2,392	1,007	1,385	3,604	434	47	387	51	9
22	Miami	FL	25	5,425	5,193	1,937	3,256	231	58	41	17	127	14
23	Charlotte	NC-SC	27	5,401	3,911	1,162	2,749	1,490	156	67	89	58	17
24	Philadelphia	PA-NJ-DE-MD	17	5,380	5,380	4,922	458	0	21	21	0	258	n.a.
25	Baltimore	MD	14	5,211	4,478	2,382	2,096	733	65	33	32	134	23
26	Dayton	OH	22	4,810	3,689	951	2,738	1,121	88	41	47	90	24
27	Nashville	TN	29	4,756	2,908	985	1,923	1,847	199	60	139	48	13
28	St. Louis	MO-IL	9	4,751	2,794	2,492	302	1,957	232	33	199	84	10
29	Salt Lake City	UT	5	3,252	1,709	1,072	637	1,543	1,295	32	1,263	53	1
	Total or average			247,053	197,512	82,678	114,834	49,540	11,977	2,761	9,218	72	5

Sources: U.S. Bureau of the Census, Statistical Abstract of the United States, 1995, Table 43; Reynolds and Maki, 1990a.
*MSAs, PMSAs, and NECMA (Boston).

the daily commuting areas using the proximity criterion (that is, distance of the LMA to its closest metropolitan core area).

Table 3-1 lists the 29 air nodes in the U.S. air transportation system for the contiguous 48 states. Also presented is the total population of the corresponding proximity-based economic region and its subregions. These include the core metropolitan statistical area (MSA), other MSAs, and nonmetropolitan areas.

Air node region rank is given by the metropolitan core (that is, commuting) area population in 1990. The MSA is generally smaller than the corresponding commuting area, although it is the largest MSA in each region. Other MSAs and nonmetropolitan areas account for the remainder of the regional population. Thus, the Minneapolis-St. Paul MSA, with a total population of 2,539,000 in 1990, served as the core MSA for the proximity-based air node region, with a total population of 7,792,000. The Detroit region, which is a second hub serviced by Northwest Airlines, was about 2.2 million larger. It ranked ninth in total population. The two air node regions belong to the two larger regional groupings, namely the West/Plains and North mega-regions, respectively. West/Plains has the largest land area but the lowest population density, with the Minneapolis-St. Paul region being the lowest of the low—with an average of only 31 persons per square mile for its MSA. The North mega-region has the largest total population for its core MSA and the highest population densities—with as many as 517 persons per square mile.

Daily Commuting Area

The daily commuting area of a resident population represents the geographic boundaries of the local labor market. The nonmetropolitan LMAs are marked by the lack of even one place with a population of more than 50,000. While a majority of counties in the United States are classified as nonmetropolitan, 252 of the 382 LMAs are metropolitan-centered. They contain one or more of the 184 MSAs, including individual counties or groups of counties, each with one or more places with a population of at least 50,000. The 29 air node regions may form even larger regions, such as the West/Plains region (numbers 1 through 8 and 10), the North region (numbers 9 and 12 through 23, and the South region (numbers 11 and 23 through 29), for broad interregional comparisons.

Table 3-2 ranks the 29 core MSAs by their 1980–94 population growth rates, grouped by mega-region. Population growth in the Minneapolis-St. Paul MSA ranked well above the Detroit MSA and above the average of core and noncore MSAs. Three of the 29 core MSAs, including Detroit in the North mega-region, declined in total population during the 1980–90 period, but not in the 1990–94 period. Three of the fastest growing core MSAs in the 1980–94 period (all in the South mega-region) Atlanta, Raleigh-Durham, and Orlando—were among the

TABLE 3-2 Total population change in core MSAs and ranking (1980–94 change), United States, 1980–94

		MSA	Region	Total Population			Annual Change		
Rank	Core MSA	Code	Code	1980 (thou.)	1990 (thou.)	1994 (thou.)	1980–94 (pct.)	1980–90 (pct.)	1990–94 (pct.)
Above-average annual growth:									
1	Orlando	5960	26	805	1,225	1,361	3.82	4.29	2.67
2	Atlanta	520	24	2,233	2,960	3,331	2.90	2.86	3.00
3	Raleigh	6640	28	665	858	965	2.70	2.58	2.98
4	Dallas	1920	7	2,055	2,676	2,898	2.49	2.68	2.01
5	Seattle	7600	1	1,652	2,033	2,180	2.00	2.10	1.76
6	Houston	3360	8	2,753	3,322	3,653	2.04	1.90	2.40
7	Charlotte	1520	27	971	1,162	1,260	1.88	1.81	2.04
8	Salt Lake City	7160	5	910	1,072	1,178	1.86	1.65	2.39
9	Washington	8840	18	3,478	4,223	4,474	1.82	1.96	1.45
10	Nashville	5360	29	851	985	1,070	1.65	1.47	2.09
11	Miami	5000	25	1,626	1,937	2,025	1.58	1.77	1.12
12	Denver	2080	2	1,429	1,623	1,796	1.65	1.28	2.56
13	Los Angeles	4480	3	7,477	8,863	9,150	1.45	1.72	0.80
14	Minneapolis-St. Paul	5120	10	2,198	2,539	2,688	1.45	1.45	1.44
	Total or average			29,103	35,478	38,029	1.93	2.00	1.75
Below-average annual growth:									
15	Kansas City	3760	6	1,449	1,583	1,647	0.92	0.89	1.00
16	Baltimore	720	14	2,199	2,382	2,458	0.80	0.80	0.79
17	Memphis	4920	11	939	1,007	1,056	0.84	0.70	1.19
18	San Francisco	7360	4	1,489	1,604	1,646	0.72	0.75	0.65
19	Cincinnati	1640	21	1,468	1,526	1,581	0.53	0.39	0.89
20	St. Louis	7040	9	2,414	2,492	2,536	0.35	0.32	0.44
21	Chicago	1600	20	7,246	7,411	7,668	0.41	0.23	0.86
22	New York	5600	12	8,275	8,547	8,584	0.26	0.32	0.11
23	Syracuse	8160	13	723	742	754	0.30	0.26	0.40
24	Philadelphia	6160	17	4,781	4,781	4,949	0.25	0.00	0.87
25	Boston	1123	16	3,149	3,228	3,260	0.25	0.25	0.25
26	Dayton	2000	22	942	951	956	0.11	0.10	0.13
27	Newark	5640	15	1,964	1,916	1,934	–0.11	–0.25	0.23
28	Detroit	2160	23	4,388	4,267	4,307	–0.13	–0.28	0.23
29	Pittsburgh	6280	19	2,571	2,395	2,402	–0.48	–0.71	0.07
	Total or average			43,997	44,832	45,738	0.28	0.19	0.50
	Total or average, 29 MSAs			73,100	80,310	83,767	0.98	0.95	1.06

Source: U.S. Bureau of the Census, Statistical Abstract of the United States, 1995. Table 43.

top three or four in both the 1980–90 and 1990–94 periods. Dallas dropped from third to ninth during this period. Four of the five largest core MSAs in 1990—Los Angeles-Long Beach, New York, Chicago, Philadelphia, and Detroit—were well below average in population growth during one or both of the two periods. Four of the larger core MSAs—Washington, Houston, Atlanta, and Dallas—were among the fastest growing. Neither prior population size nor density predicted likely population growth. The core MSAs, in total, grew less rapidly than the noncore MSAs in the 1980–90 period but more rapidly in the 1990–94 period.

The 29 air node regions and their core areas and the three mega-regions illustrate the use of contiguity and proximity criteria in regional delineation. Moreover, each of the regions exercises some comparative advantage due to history, climate, amenities, and natural resources, but most of all, its human resources. Use of air node regions, including the identification and delineation of their core areas as important sources of many critical regional resources, is an outcome, in part, of the widely acknowledged inadequacies of administrative regions for urban regional analysis and study. It also relates to the diverse effects of global competition and technology-intensive industries—which also are air-transport-dependent—on all levels of business, government, and household interactions, from the neighborhood and municipality to the mega-region and beyond.

Let us take, for example, the use of individual counties as building blocks for the labor market areas in the Minneapolis-St. Paul air node region. Of the 281 counties in the Minneapolis-St. Paul region, based on the 1980 commuting area delineations (Tolbert II and Killian, 1987), 32 are included in thirteen MSAs. Thirteen of the counties are in the Minneapolis-St. Paul MSA and also are part of the sixteen-county Minneapolis-St. Paul LMA. An additional 94 non-MSA counties are part of the remaining twelve LMAs that include the MSA counties outside the metropolitan core area. Even a dominantly rural region like Minneapolis-St. Paul, which has 155 counties in thirteen entirely rural LMAs, forms a highly integrated trading region, with much internalization of trade between metropolitan and nonmetropolitan areas.

The metropolitan core area is of critical importance to the rural areas with an expanding industrial base because of its market and nonmarket linkages to core area producer services and transportation infrastructure (Glasmeier, 1993). Also important are the secondary core areas. These include the endogenous growth centers, characterized by their increasingly diverse local labor markets (Hansen, 1993). The commuting areas of both primary and secondary core areas also serve as the functional economic community for solving areawide problems. These problems can be approached by strengthening and extending the networks of civic engagement that facilitate the flows of information about technological developments, employment, entrepreneurial opportunities, and related factors affecting the competitive position of export-producing businesses in their respective markets.

A single region is a mixture of metropolitan and nonmetropolitan, urban and rural areas. A rural area is marked by the lack of even one place with a population greater than 50,000. A majority of counties in the United States are still classified as nonmetropolitan. Each region has a center—a metropolitan core area—and a periphery. Table 3-3 lists the 26 LMAs, identified by their trade centers, included in the Minneapolis-St. Paul Economic Region. Of the 281 counties in the region, 32 are core counties, 31 are not core counties but belong to an MSA as defined by the U.S. Bureau of the Census, and 218 are nonmetropolitan counties. We illustrate the meaning and use of the several measures of interregional trade with findings from the University of Minnesota IMPLAN regional database and modeling system.

High-order, high-margin services and technology-intensive manufacturing, for example, are increasingly important parts of the emerging industrial complexes in metropolitan areas. They concentrate in these areas because of their active and diverse local labor markets. Commodity-producing, low-margin industries, on the other hand, concentrate in rural areas because of low site costs as well as the local availability of highly productive workers. The urban periphery, while the place of business for largely commodity-producing industries, also has an expanding service-providing sector because of a growing commuter population that initially replaced farming as the local economic base.

Table 3-4 presents estimates of gross out-shipments, or exports, of local industry production for the core area and six clusters of LMAs in the Minneapolis-St. Paul Economic Region. In this table, North refers to the 74 counties in nine LMAs, while South refers to the remaining 191 counties in seventeen LMAs. The 25 LMAs aggregate to the following: Northeast combines Duluth-Superior, Ashland, and Houghton-Marquette; North Central combines Bemidji, Alexandria, and St. Cloud; Northwest combines Bismarck, Grand Forks, and Fargo-Moorhead; Southeast combines Eau Claire, Wausau, Winona, La Crosse, Rochester, and Waterloo; South Central combines Willmar, Mankato, Worthington, Spencer, and Mankato; and Southwest combines Aberdeen, Sioux Falls, Sioux City, and Norfolk. Inclusion of the core area of sixteen counties provides comparison of its external trade with the external trade of the more rural areas of the region. Even with high levels of industry aggregation, subregional differences in export composition emerge in the same industry comparisons. These findings, aggregated from more than 500 individual sectors, illustrate sharp differences in commodity exports and, hence, the economic base and its occupational requirements among the primary and secondary core areas and the dominantly rural areas.

TABLE 3-3 County status and population, by LMA, Minneapolis-St. Paul Region

Labor Market Area	Counties				Area (sq.mi.)	Population (1991)	
	Total (no.)	Core (no.)	MSA (no.)	Rural (no.)	Total (no.)	Total (thou.)	Density (no.)
Minneapolis-St. Paul, MN-WI	16	2	11	3	7,830	2,624.6	335
Northeast:							
Duluth-Superior, MN-WI	11	2	2	7	23,902	396.0	17
Ashland, WI	9	1	0	8	9,355	179.7	19
Houghton-Marquette, MI-WI	13	2	0	11	14,579	256.8	18
Total, Northeast	33	5	2	26	47,836	832.5	17
North Central:							
Bemidji, MN	7	1	0	6	9,304	159.8	17
Alexandria, MN	7	1	0	6	5,869	144.4	25
St. Cloud, MN	4	1	3	0	3,315	225.0	68
Total, North Central	18	3	3	12	18,488	529.2	29
Northwest:							
Fargo-Moorhead, ND-MN	13	2	2	9	15,175	238.1	16
Grand Forks, ND-MN	20	1	1	18	21,984	243.6	11
Minot-Bismark, ND	13	2	2	9	18,790	197.4	11
Total, Northwest	46	5	5	36	55,949	679.1	12
Southeast:							
La Crosse, WI-MN	5	1	1	3	3,500	202.4	58
Winona, MN-WI	4	1	0	3	3,032	103.8	34
Eau Claire, WI	4	1	2	1	2,732	182.6	67
Wausau, WI	4	1	1	2	4,619	195.4	42
Iron Mountain, MI-WI	8	1	0	7	7,878	170.0	22
Total, Southeast	25	5	4	16	21,761	854.2	39
South Central:							
Waterloo, IA	13	1	2	10	7,321	346.2	47
Mason City-Ft Dodge, IA-MN	17	2	0	15	9,550	300.3	31
Mankato, MN	8	1	0	7	4,204	246.8	59
Rochester, MN	6	1	1	4	3,950	244.2	62
Total, South Central	44	5	3	36	25,025	1,137.5	45
Southwest:							
Sioux City, IA-SD-NE	9	1	2	6	5,025	197.9	39
Norfolk, NE	10	1	0	4	6,286	116.9	19
Sioux Falls, SD	22	1	1	20	14,927	312.9	21
Aberdeen, SD-ND	26	1	0	25	32,868	145.3	4
Spencer, IA	9	1	0	8	5,211	138.2	27
Worthington, MN	8	1	0	7	4,854	108.8	22
Willmar, MN-SD	15	1	0	14	11,483	190.3	17
Total, Southwest	99	7	3	89	80,654	1,210.3	15
Total, all areas	281	32	31	218	257,543	7,867.4	31

Source: University of Minnesota IMPLAN regional modeling system.

TABLE 3-4 Total exports of industry output, by subregion, Minneapolis-St. Paul Economic Region, 1990

	Mnpls.-St. Paul	North Region LMAs			South Region LMAs		
Industry	Core Area (mil.$)	East (mil.$)	Central (mil.$)	West (mil.$)	East (mil.$)	Central (mil.$)	West (mil.$)
Agriculture	898	648	1,071	1,902	1,213	1,986	2,436
Agricultural services, forestry, fisheries	130	76	35	16	129	61	67
Mining	126	2,223	39	381	80	197	101
Construction	228	21	55	53	47	59	138
Manufacturing	27,596	4,555	2,621	2,349	7,967	11,926	10,950
Trans., comm., utilities	2,088	473	447	601	164	141	357
Wholesale trade	2,090	86	71	210	167	226	282
Retail trade	743	91	151	78	124	37	157
Finance, insurance, real estate	2,057	161	233	97	93	333	387
Private services	3,245	382	325	530	433	1,159	1,007
Government	108	5	24	33	3	5	7
Total	39,309	8,721	5,071	6,250	10,421	16,130	15,889

Source: University of Minnesota IMPLAN System.

Table 3-5 shows a remarkable similarity, largely, in percentage distributions of the major industry groups among the seven subregions. Agriculture and manufacturing together, except for mining in the northeast, emerge as the dominant segment of each subregion's economic base. The local economic bases of individual communities and labor market areas differ greatly, of course, because of specialization in particular types of manufacturing and services.

The distributions of gross industry output and employment—across all industries, not simply export-producing ones—start showing the variations in economic activity when we consolidate the six subregions and compare them with the three subareas of the metropolitan core area. The core LMA, for example, accounts for 40 percent of the region's total industry output.

The net exports of each area—that is, gross industry out-shipments less gross industry in-shipments—show even greater area-to-area differences. Construction and the service-providing industry groups, for example, consistently show net in-shipments for the LMA clusters outside the core area. Construction is the most highly import-dependent industry group. However, the aggregation of many individual industries into the broader categories, like manufacturing, obscures the sharply contrasting composition of exports of the core area and its surrounding LMAs. The additional data would more clearly identify opportunities for the internalization of inter-area trade.

TABLE 3-5 Total exports of industry output, by subregion, Minneapolis-St. Paul Economic Region, 1990

	Mnpls.-St. Paul	North Region LMAs			South Region LMAs		
Industry	Core Area (pct.)	East (pct.)	Central (pct.)	West (pct.)	East (pct.)	Central (pct.)	West (pct.)
Agriculture	2.3	7.4	21.1	30.4	11.6	12.3	15.3
Agriculture services, forestry, fisheries	0.3	0.9	0.7	0.3	1.2	0.4	0.4
Mining	0.3	25.5	0.8	6.1	0.8	1.2	0.6
Construction	0.6	0.2	1.1	0.8	0.5	0.4	0.9
Manufacturing	70.2	52.2	51.7	37.6	76.5	73.9	68.9
Transportation, communications, utilities	5.3	5.4	8.8	9.6	1.6	0.9	2.2
Wholesale trade	5.3	1.0	1.4	3.4	1.6	1.4	1.8
Retail trade	1.9	1.0	3.0	1.2	1.2	0.2	1.0
Finance, insurance, real estate	5.2	1.8	4.6	1.6	0.9	2.1	2.4
Private services	8.3	4.4	6.4	8.5	4.2	7.2	6.3
Government	0.3	0.1	0.5	0.5	0.0	0.0	0.0
Total	100.0	100.0	100.0	100.0	100.0	100.0	100.0

Source: University of Minnesota IMPLAN System.

The dominant basic industry groups are also the dominant import-dependent industry groups, except for construction. Export growth means import growth because of the corresponding increases in demand for the imported intermediate production inputs. As an area grows and diversifies, however, import replacement occurs for both intermediate inputs and final purchases. Imported finished goods and services dominate total imports in the periphery of an economic region, while imported intermediate goods and services are dominant in its core area. Again, the import dependencies of the core area contrast sharply with those of the periphery, which, of course, is a measure of the opportunities for internalizing the export trade of individual LMAs.

Workforce Characteristics

The competition for land and space in an economic region differs from place to place because of proximity to a metropolitan core area and the development of its location-dependent activities. These activities account for the changing place-to-place geography of production in a region and the changing local division of labor. Their type and intensity are among the high-order determinants of the economic viability of any regional system of local labor markets.

The regional database for assessing rural-metropolitan linkages includes a common 1990 data set of industry and commodity exports and imports, as well as industry and commodity sales and industry employment and value added, for all U.S. counties and for each of the more than 500 producing sectors possible in a single county. Supplemental data also show the sharp gradient in land values per acre, or per square foot, as land values are more commonly measured in the central city. These values range from less than $.05 per square foot for strictly agricultural land in the urban periphery to $.50 per square foot for five-acre residential lots, $5 per square foot for prime residential lots, and $50 per square foot for commercial parcels in established metropolitan area suburbs.

Tracking Local Market Changes

An end in view of our examination of the theory and reality of local markets and central places is the building of functional economic communities. The continuing changes in the income and geographic distribution of the nation's population, coupled with its ever-improving access to the latest technology advances, is radically changing the patterns of urbanization and market organization of urban regional activity. Accurate tracking of local market changes requires a variety of monthly, quarterly, and annual economic indicators. These include various measures of employment, earnings, personal income, population, and economic activity. Critical local market issues, such as the activities of large service providers and the various forms of spatial mismatch between place of work and place of residence, require still other measures of economic performance. We start with the national economic indicators that parallel our discussion of macroeconomics in Chapter 1 and the most closely related local economic indicators.

Economic Indicators

Most commonly cited national economic indicators are the federal funds rate and the yield on 30-year treasury bonds, along with measures of capacity utilization, consumer confidence, consumer prices, gross domestic product, payroll employment, unemployment, and an index of leading economic indicators. The last six or seven of these nine indicators have urban regional counterparts.

The *Wall Street Journal Classroom Edition* (http://info.wsj.com/classroom/Indicators, available to *Journal* subscribers) describes each of the national indicators and their data sources and uses in tracking the nation's economy. The *federal funds rate,* for example, is what the Federal Reserve bank charges other banks for overnight loans. It not only affects the cost of money for both businesses and households; it also shows the likely direction of federal monetary policy and, subsequently, of the national economy. The *yield on 30-year Treasury bonds* also shows how much it costs to borrow money, but in the long term. This measure represents the combined value of the trading price of the Treasury bond and the interest paid to its holder. Both rates are quoted daily.

Capacity utilization shows the percentage of the nation's industrial capital stock actually used in producing the nation's gross product. It anticipates employment trends. Lower capital utilization eventually triggers a slowing down of employment growth, including overtime. The St. Louis Federal Reserve Bank, among others, releases this series of 75 different capital utilization indexes each month.

Consumer confidence has two separate measures each month—one from the Conference Board based on a mail survey of 5,000 households by National Family Opinion, Incorporated, of Toledo, Ohio, and the other from the University of Michigan, based on its telephone survey of 500 households. The mail survey asks respondents to look six months ahead, focusing on job availability. The telephone survey asks respondents to look one year ahead, focusing on family finances and overall business conditions. This is a measure used with other indicators, like consumer prices and payroll employment, to predict changes in consumer spending.

The *consumer price index (CPI),* the most commonly watched inflation indicator, is released monthly by the U.S. Bureau of Labor Statistics. It is available for selected MSAs. The CPI is used in many union (and nonunion) contracts to automatically adjust workers' wages to inflation. It is based on compilations of prices charged for items sold to consumers living in urban areas, who account for 80 percent of the nation's population. The 364 individual items are weighted according to their importance in the market basket of goods and services purchased by consumers.

The *gross domestic product,* defined as the value of goods and services produced in the U.S. economy and reported quarterly by the U.S. Department of Commerce, is the broadest measure of economic activity. It is available for individual states and MSAs. It represents the nation's total final product or total factor income. As final product, it is the sum of the personal consumption expenditures of households in the United States; the capital expenditures of businesses; the purchases of local, state, and federal governments and the compensation of their employees; and net exports of goods and services. As factor income, it is the sum of private sector employee compensation, proprietors' income, indirect taxes, and other property income, including capital consumption allowance, or depreciation.

Payroll employment data, released monthly by the U.S. Bureau of Labor Statistics, come from an establishment survey that covers all wage and salary workers on company payrolls. It excludes farm workers, self-employed workers, unpaid volunteer workers, and domestic workers. The U.S. Department of Commerce reports the corresponding employment estimates annually, including state, local, and federal government, for every county in the United States, with roughly a two-year lag; that is, the 1995 estimates were available by early 1998.

Payroll employment is an important measure of business expansion and contraction and related changes in job creation.

The *unemployment rate,* released monthly by the U.S. Bureau of Labor Statistics and also available for selected MSAs, is based on a survey of 60,000 households. The civilian unemployment rate measures the fraction of people over age sixteen who are looking for work but cannot find a job.

The *Index of Leading Economic Indicators,* released monthly by the U.S. Department of Commerce (and published by the Conference Board since January 1996), predicts swings in the national business cycle by three months or more. The leading index has eleven components, each weighted equally:

- Average workweek of production workers
- Average weekly initial state-unemployment insurance claims
- Manufacturers' new orders of consumer goods and materials industries
- Contracts and orders for plants and equipment
- Building permits
- Change in manufacturers' unfilled orders for durable-goods industries
- Change in sensitive-materials prices
- Stock prices of 500 common stocks
- Money supply (M2)
- Index of consumer expectations

Three consecutive monthly changes in the composite index have signaled, in past years, a possible recession or economic recovery. The index has become less reliable in recent years as a predictor of business cycle changes.

We find several approaches to the preparation and use of urban regional indicators. Data Resources, for example, prepared a series of leading indicators for the *Boston Globe.* Six sectors represented New England with a total of ten leading indicator series, all seasonally adjusted and weighted equally:

- Employment and unemployment: average weekly hours of production workers (moving average), New England manufacturing; inverse of layoff rate in New England employment (moving average)
- Consumption and distribution: percent of companies in the Boston area reporting more orders received (moving average); percent of companies reporting slower deliveries, United States
- Fixed capital investment: new building permits, New England (moving average)
- Inventory investment: percent of companies reporting higher inventories (moving average), New England
- Prices, costs, and profits: index of stock prices (moving average), United States; percent change in prices of raw materials, United States (moving average)

- Money and credit: money supply, United States; change in consumer installment credit outstanding, New England

Michigan metropolitan area indicators were developed by Kozlowski and Associates for eleven standard MSAs in Michigan under the auspices of the W.E. Upjohn Institute for Employment Research (as reported by Kozlowski and Hansen, 1998). Kozlowski later proposed an index of leading indicators based on a composite of four local quarterly indicators: average workweek of production workers in local manufacturing, average weekly initial claims for unemployment insurance (inverted), constant dollar value of total deposits at local commercial banks, and number of private housing units authorized by building permit.

In their most recent research, Kozlowski and Hansen (1998) evaluated the performance of time series generated from surveys of purchasing managers in regional markets in periods of recession and recovery in the late 1980s and early 1990s. The regional time series are correlated with the National Association of Purchasing Managers composite monthly index (NPMI), which is based on monthly surveys of purchasing managers in U.S. manufacturing firms. Its origins date back to the 1930s. Kozlowski and Hansen found that the purchasing managers monthly index contributes "significant information about growth in manufacturing output," although the contribution seems to be "quantitatively marginal." Prompt release of its data each month "appears to be its main advantage." The NPMI and its regional counterparts have some obvious flaws. The survey questions elicit qualitative responses. The seemingly unrelated selection and weighting of components result in composite indexes that are not uniform around the country. Furthermore, survey respondents interpret questions differently. Comparable quantitative data are, therefore, not available from the regional surveys.

Large Service Providers

One criticism of CPT is more fundamental, even from the service center or central place perspective. If the level of competition described in the model does not exist, and just as importantly, if transportation and search costs are not major factors in the customer's desire to buy from a particular firm, the model does not apply. The CPT conditions are probably not met for many manufacturing enterprises, for example. When factors such as availability of labor, taxes, and government services provided at a particular location, or the amenities a particular region holds are the most important to firm location, the hierarchy of central places does not make sense. Further, once a manufacturing firm does locate in a region, external economies lead to agglomerations of suppliers and population serving industries that are not easily explained by CPT. Nor are the conditions met for a dominant service provider, such as an international airlines company with an 80 percent share of the air transportation market of a major air node. Market factors, like restricted site capacity and predatory pricing, determine the size and attributes of the large provider's local service area and its market performance.

Restricted Site Capacity

In most local markets, a large service company is often the dominant service provider. This is particularly true of the U.S. airlines with 70 to 80 percent or greater market share at their respective air nodes. The major airlines maintain dominance because of restricted site capacity and control of a large majority of airport slots for parking and servicing aircraft. Airport dominance also allows for the effective use of price discrimination in filling all remaining seats just prior to flight departure. The market segment with price-inelastic demand, typically the business traveler, pays the full price. The market segment with price-elastic demand—the leisure traveler—gets the cheap seats. Eventually, an airline dominant at one air node finds ways of gaining dominance at another air node.

From the airline industry perspective, air node dominance brings stability as well as profitability. The industry is no longer at the mercy of a boom-and-bust cycle, unable to acquire the financial resources for long-term investment in new equipment and new markets.

Predatory Pricing

The dominant airline maintains control of the pricing of air transportation, not only by a large majority control of available slots, but also by the threat and even the actual exercise of predatory pricing. The dominant airline, of course, is in a position to sustain losses from price competition longer than a newly established airline vying for a larger market share at the expense of the dominant airline.

Place of Work and Place of Residence

Another set of emerging issues in the larger central places of a growing metropolitan-centered region is the spatial mismatch between place of work and place of residence. The issues involve more than concentrated poverty in the central cities and concentrated resources in the surrounding suburbs that together generate adverse consequences for the quality of education in the central cities. They include the lack of public discourse and meaningful communication among community leaders and interest groups about these issues facing the metropolitan region. They include also the gradual encroachment of crime and civil disorder into the downtown districts of central cities that house the new engines of economic growth of once agriculture-dependent trading regions.

In spite of its well-developed metropolitan core area, the Minneapolis-St. Paul Economic Region, for example, still ranks below the top half of the 25 economic regions in competitive position. Its core area faces the strong likelihood of even more concentrated poverty, declining residential property values, an expanding core of crime, and an increasingly dysfunctional public education system. Rural labor market areas bound the core area. Some of these experience rapid population growth because of the outward migration of metropolitan area residents and jobs. More distant rural areas experience generally declining population and workforce, although they still produce the region's primary exports to domestic and international markets. Thus, the extended core area and other metropolitan areas of the region outside the core area account for much of the region's recent economic growth.

Among policy choices is the maintenance and renewal of the region's core competencies. State and local governments have the primary responsibility for the physical and economic infrastructure of globally competitive businesses and industries. They share with the federal government responsibility for maintaining each of the 29 U.S. air transportation nodes and the interstate surface transportation that link together every metropolitan area into a global communications and transportation network. These governments have accountability for public education and a large number of the region's health care delivery systems. They also provide the long-term industry and occupation forecasts for employment counseling and education curriculum planning.

An essential part of making sound policy choices is a sound sense of the interconnectedness of export-producing activities—locally and regionally as well as nationally and internationally. Much of the transportation infrastructure is based on farm-to-market roads and farm commodity shipments by truck, rail, and barge. Yet, these shipments account for only a small part of the total value of all commodity traffic, and even a smaller part of all transportation expenditures.

Alternative market arrangements, like congestion pricing in the metropolitan areas to more efficiently allocate the use of highways during peak travel periods, are also lacking. Even with such arrangements, the more fundamental questions remain: What is the full cost of urban sprawl, and what are the available effective means for containing this sprawl? What is the role of transportation in adding to or reducing urban sprawl with all its consequences? On top of all this, planning for the future is still guided by forecasting of the past.

Another set of policy choices pertains to the regulation of business enterprises engaged in interregional and international trade. The costs of business compliance with new federal regulations are constraining the capacity of small businesses to successfully compete in domestic and global markets (Galloway and Anderson, 1992; Lipke, 1993). The addition of new health care costs makes the future success of many businesses even more questionable. Finally, many state legislatures may have an antibusiness legacy that further discourages new business enterprise, much to the dismay of industrial development agencies of cities that may be encouraging new business development.

A third set of policy choices addresses the information-producing and information-transfer functions of state governments and postsecondary education institutions. Until credible new ways are found to monitor the efficacy of existing educational programs while anticipating future skills requirements of local export-producing businesses and industries that are credible, postsecondary education will lag in preparing students for emerging future job opportunities rather than for jobs of the past.

Whether or not the three sets of policy choices receive the attention they deserve may prove irrelevant, given the pressing issues of day-to-day travel to work from the surrounding suburbs to downtown office towers or living in crime-ridden central city neighborhoods of concentrated poverty. In such a scenario, what options remain for central cities not only to survive the consequences of neglect of these central issues of our time but also to actually achieve a truly remarkable turnaround toward a new vision of the central city as the modern-day Athens of their extended economic region?

Notes

1. Samuelson warns the reader: "Here's what you should have learned in college but almost certainly didn't." His idea is to explain the "stark contrasts" in economic performance between countries in terms of microeconomics, that is, the study of individual markets and the role of incentives, often small, that ultimately have large consequences (*Newsweek,* April 10, 1998, p. 47).

2. For the discussion of Athens and the world cities that follow, we refer to Jim Antonious, *Cities Then and Now* (New York: Macmillan, 1994).

4

Exports and Economic Base

Exports define the economic base of a local area. Exports are the goods and services produced locally but purchased by people and businesses from outside the locality. They are measured by the payments. Any payment for a good or service in one locality that originates in another locality is considered to be an export. For example, if a tourist from New York City eats in a restaurant in Des Moines, Iowa, the meal is considered to be an export from Des Moines to New York, even though the person spending the money is physically located in a Des Moines restaurant.

Urban regional economics, with its focus on the spatial dimensions of regional activity, provides a special understanding of the linkage between a region's economic base and its economic performance. For example, measures of economic well-being, like personal income per person, vary from one part of a region to another. More likely than not, the personal income measures are high in the metropolitan core areas and low in the rural periphery. They correlate closely with levels of investment per worker. Where perceived profits are high (with little risk of failure), investment and production, jobs, and people follow. Such activity probably involves a location decision. Other measures of personal well-being may differ from the economic ones, with high-amenity rural areas ranking higher than some metropolitan area neighborhoods.

Theorizing about the location of investment and production has a long history in scholarly literature. This has been enriched greatly by the gradual inclusion of all economic units—agricultural and nonagricultural firms, rural and urban households, and public facilities. Economic Base Theory asserts that the economic structure of a locality consists of two major categories: *basic activities,* which result in a product or a service that is sold outside of the defined locality, and *nonbasic activities,* whose goods and services are consumed at home. The location paradigms discussed in Chapter 2 are now widely accepted and applied in regional science.

We begin with an analysis of the reason for new businesses and related investments that trigger more trade between regions. Economies from scale and comparative advantage will be shown to provide additional motivation for trade.

The discussion then turns to exports and the economic base of a region as a consequence of the many different, independent location decisions for an area. These decisions result in gains in population and growth and in the region's diversity of economic activity, facilitated by an active, ever-changing local labor market. Other local markets emerge to serve area residents, like the farmer's elevator for buying and selling locally produced grains or the realtor trading in local real estate. Still other markets emerge, in which businesses trade among themselves or provide goods and services to local people for their everyday living. Access to active markets outside the locality also improves, thus bringing about an even greater diversity and a larger scale of local economic activity because of the growing inter-area trade in goods and services.

How Regions Grow

Economists have had many and long debates about the fact that some regions grow faster than others, and we have had discussions about concepts that are most accurate or convenient for explaining these differential rates of regional growth. Of course, discussion must start with the people who found new businesses or expand existing ones, and with the local environments in which these activities are most likely to succeed. David Birch, along with Paul Reynolds and others, point to new business foundings as a source of economic growth through new job creation (Birch, 1981; Reynolds, Miller, and Maki, 1994a,b). New businesses involve investment in people and places. Eventually, a few of these businesses generate local surpluses of goods and services that ultimately find buyers outside their locality. They become involved in exporting their products to different domestic and international destinations. Albert Hirschman viewed this process and called it *linkage*. Hirschman's contribution in *Strategy of Economic Development* (1958/1978) gave us new insight regarding content and concepts that relate particularly to the role of investment in regional growth. The contents and conclusion of his analysis stood in contrast to traditional neoclassical theory. Essential to the viability of either approach is interregional trade. Both investment and trade are critical for regional growth, whether from the perspective of the individual entrepreneur founding a new business or the academician expounding a new strategy of economic development.

Investment and Linkage

The story of business foundings, presented later in this chapter, has much to say about investment and linkage at a local level of economic activity. We start, however, with an overview of economic activity in many localities and how these activities are linked, one to another. Hirschman is known for introducing the concept of linkage in the more general economic analysis.[1] In the 1950s he was occupied by the processes of industrialization in developing countries, especially in Latin America. Hirschman had a critical attitude toward the Harrod-Domar models, in which one fundamental assumption was that economic growth depends on the capital-output ratio and that the amount of capital was decided by the attitude toward savings and foreign direct investment.

Concept of Linkage

In Hirschman's theory of economic development, the level of investment plays an important role. This has important implications: for economic theory, the formulation of the concept of economic linkages; for investment, the dynamic consideration based on the effect of linkages.

The concept of linkages developed as a counterconcept to the assumptions used in the 1950s, and still used by many today. We assume, for example, that factors of production are evenly distributed in the economic sense, and that natural resources can be used anywhere, as long as we take transportation costs into account. We may even assume that entrepreneurship is evenly distributed over space and is mobile so as to receive the highest reward for its services. Hirschman's main thesis was that "development depends not so much on finding optimal combinations for given resources and sectors of production as on calling forth and enlisting for development purposes resources and abilities that are hidden, scattered, or badly utilized" (Hirschman, 1958). This is the theoretical point of departure for the book as it addresses the strategy of unbalanced growth.

Hirschman put forth the concept of linkage in 1958 with a dynamic point of departure, originally as an aid to analyzing processes of development and change. His point of departure is a sector in a region/country (or a *cluster,* as he calls it) that has a high import share. Such a sector may want to increase its production domestically or in a region, probably for domestically produced capital goods.

Another type of stimulus in economic development that Hirschman identified is investment that changes the competitive situation. He called this dynamic forward linkages: "the existence of a given product line A, which is a final demand good or is used as an input in line B, acts as stimulus to the establishment C which can also use A as an input" (Hirschman, 1992). Forward linkage is what we get when we create activities to buy goods from the sector that we start with. The pressure to create such investment processes that contribute to forward linkages comes, in large measure, from existing producers who want to increase the size and scope of their product markets.

Hirschman saw the advantage of input-output analysis in demonstrating the existence of forward and backward linkages in the development processes. His line of thought was that an input-output table for a country in an early phase of its industrializing process will have many empty cells. As time passes, more and more cells will be filled in as the effects of forward and backward linkages spread in the economy. Input-output analysis was developed in the 1950s and, as data were developed, could be used to describe dynamic forces in the industrializing processes in both developed and undeveloped countries.

Hirschman suggested, however, that using input-output analysis to measure forward and backward linkages was not always a good idea. "Input-output analysis is by nature synchronic, whereas linkage effects need time to unfold. In a country setting out to industrialize, existing input-output tables cannot reveal which additional industrial branches are likely to be created in the wake of industrial investment in a given product line. The input-output framework is even less suited to tracing backward linkage effects towards the machinery and equipment industries" (Hirschman, 1992). He continues by saying that input-output analysis is a meaningful tool when we know that the change is in a sector that already is in the input-output table. His analysis and formulation of concepts can be compared to Rostow's Leading Sectors (Rostow, 1971) and Perroux's Propulsive Industry.

Industrialization Processes and Linkage

Hirschman found it important to differentiate between countries that are starting to industrialize and those that are at a later stage of industrialization. Industrialized countries have a more balanced development in the production of finished goods, investment goods, and raw materials. In such countries, the growth process will to a high degree be an interaction between forward and backward linkages.

In countries that started their industrial development in the twentieth century, Hirschman saw two important factors that generated growth in the economy: export of raw materials from the country and import substitution. The dynamics of the backward linkages are related to a process in which import substitution is taking place. In the time period analyzed (1945–60), import substitution was an important and interesting phenomenon. Many merchants imported goods, but after some time they started to produce these goods themselves. This development was supported by the authorities because of the creation of new employment, the creation of value added, and the effect on the trade balance.

Hirschman extended the use of the linkage concept, and his point of departure was that forward and backward linkages identify the processes of investment that are important in an economy in the process of development. The forward and backward linkages can indicate how activities can spread from a cluster of producers producing more or less the same good. It is also important to say something about how such a process can start. He reasoned that for many peripheral regions, an increase in income first comes from an increase in the export of agricultural products, mining, or other resource-based activities. This increases the demand for consumption goods in such a way that it becomes attractive to start production of these goods. This process happens by stages. The degree to which import substitution actually takes place depends on several factors, among them more income for consumption. It is interesting to see how many of these points of view and concepts that Hirschman discussed show up years later in Michael Porter's analysis of competition.

Trade and Linkage

Linkage is, of course, central to interregional trade. The trade-related linkages are many and varied. They include the prior investigation of destination markets and the participants in these markets, their preferences, their financial situation, the actual shipments, the follow-up, and so on. Large numbers of business-services providers exist in the large metropolitan labor markets to serve export-producing businesses in the much larger metropolitan-centered region served by these businesses. The types and importance of these businesses is documented with the use of inter-industry transaction tables, presented later in this chapter.

The Reason for Trade

The obvious reason for trade is the comparative advantage that a region enjoys in the production of particular products that can be traded for other products produced elsewhere and for which active markets exist in the destination regions.

Influence of Economies from Scale

If there were no economies from scale, there would be less reason for trade. Assume a plane with an even scattering of self-sufficient farms. Each farm produces all products for its own needs. Assume further that the techniques and costs of production are the same for all farms. Is there any reason for trade? No, because all produce at the same costs.

For example, if farmer A produces a pair of shoes for $20 and wants to sell these shoes to farmer B, who could also produce a pair of shoes for $20, what would be the gain to farmer B? With no gain, it can be shown that there is a loss in such a transaction. That loss is the transportation costs incurred when the shoes are shipped from farmer A to farmer B. If farmer A absorbs the transportation costs, farmer A loses. If farmer B incurs these costs, he or she is paying more than $20 for a pair of shoes that could be produced locally for $20, again a loss.

Now assume there are economies from scale. Farmer A could produce shoes for several families at a per unit cost of less than $20 per pair. The per unit costs fall because of gains that accrue to a producer when larger quantities are produced. Let's say the unit costs for producing 50 pairs of shoes is $15 per pair. Let's also say that the producer has 50 other families living within a circle surrounding the producer, where the transportation costs are below $5. Now farmer A can offer shoes at a delivered price below what the other farms could realize by producing their own shoes.

It is also true, of course, that the other farmers would have to be able to produce products at reduced per unit costs in order to be able to pay for farmer A's shoes. Each farm will begin to specialize in producing products for which they have comparative advantages and trading for products for which their neighbors have comparative advantages. The result is a pattern of cities as described in the central place model of Chapter 3.

We have used the term *comparative advantage* several times in this book so far. We now turn our attention to this concept.

Principle of Comparative Advantage

The assumptions of the previous discussion will now be altered somewhat. First, let's divide the region into two parts, the east and the west. Let's assume that the labor force is equally skilled in both regions. However, for other reasons, labor in the west can produce more shoes in an hour than can similar labor in the east. A laborer in the east can produce more apples in an hour than can the same laborer in the west. These differences could be due to any number of factors, such as better weather and soil conditions for producing apples, or previously developed agglomeration economies.

If we go no further than this, the answer is simple. Under the conditions of absolute advantage described above, the west would trade shoes for the east's apples. This trade takes place under the principle of absolute advantage, which should make a lot of sense. If I am better at doing one thing and you are better at doing another, I should do what I do best while you do what you do best, and then we should trade. In this way, total production of both items would be larger than if we each produced both, there is more to go around, and through trade, we would both be better off.

The trouble with the principle of absolute advantage is that it provides no basis for trade if one region is better at producing all products than the other. What if, for example, the west can produce more shoes and more apples of equal quality with one day's labor than can the east? In a brilliant work by David Ricardo (1817), it is illustrated how even in this case there is a basis for trade. We illustrate such a case again with two regions—east and west, each producing both shoes and apples. Suppose the west could produce 10 pairs of shoes or 40 bushels of apples and the east could produce 5 pairs of shoes or 3 bushels of apples of equal quality with one day of labor. Under the principle of absolute advantage, there would be no reason for trade.

To understand the principle of comparative advantage, we must appreciate what is going on in the workplaces of each region. First, realize that there is more than one unit of labor in the two regions. If laborers worked for two days in the west, they could produce 20 pairs of shoes or 80 bushels of apples; three days' labor could produce 30 pairs of shoes or 120 bushels of apples, and so on. This is the production possibility for the region. However, a 30-to-120 ratio is equivalent to a 20-to-80 ratio or to a 10-to-40 ratio. All reduce to the same rate of trade-off between the two products.

To find the comparative advantage, multiply the east's 5-to-3 ratio by 2. We now have two days' labor producing 10 pairs of shoes and 6 bushels of apples, which, remember, reduce to the 5-to-3 ratio with which we began our discussion. Where is the advantage?

The east can get 6 bushels of apples at home by moving two days' labor out of shoe production and into apple production. By doing so, it sacrifices 10 pairs of shoes. If it could get any more than these 6 bushels for 10 pairs of shoes through trade, it would be better off. The west, for example, can gain 10 pairs of shoes by moving a day's labor out of apple production. Of course, when it moves that day's labor, it sacrifices 40 bushels of apples. If the west could get 10 pairs of shoes for anything less than 40 bushels of apples, it would be better off.

Let's say that the east produces 10 pairs of shoes and trades these shoes to the east for 12 bushels of apples. Clearly the east is better off because it receives 12 bushels of apples for 10 pairs of shoes through trade, whereas it would have received only 6 bushels of apples by sacrificing 10 pairs of shoes at home. The west is also better off through trade. It would have sacrificed 40 bushels of apples to gain 10 pairs of shoes in the absence of trade. It had to sacrifice only 12 bushels of apples for 10 pairs of shoes through trade.

The combination of economies from scale and absolute or comparative advantage represents the necessary and sufficient conditions for trade to take place. Once trade takes place, whether through the principles of central place theory or through the export base model to be discussed next, a region becomes identified by the package of products it produces. This package represents the economic structure of the region, as identified earlier in this chapter, and pretty much charts the future of the region. We turn next to the principle of economic base.

The Economic Base Model

Bringing together the individual business decision to locate, invest, and export, and the emergence of an active and diverse local labor market in a single paradigm of the local economic base is a central contribution of urban regional economics. This task is probably best approached from the perspective of the economic base model, starting with its historical antecedents and then moving on to a version that relates also to an emerging local labor market. Such a model also joins the export-producing and resident-based activities into a functional economic community. Discussion of exports and the economic base often occurs without reference to its geography, that is, the area in which the export-producing industries reside. The reference area becomes simply an afterthought once the question arises. In this discussion, however, the labor market area serves as a basic building block in the delineation of an economic region and a functional economic area. The size and scope of the economic base changes over time, however, as a result of the comparative advantage of the given labor market area relative to other labor market areas, including those nearby as well as more distant ones.

Historical Antecedents

Economic base is demand oriented, predicated on an earlier theoretical construct that was developed by John Maynard Keynes. The most fundamental form of the Keynesian model is the assertion that regional income is a function of the level of household consumption demand and investment, or

$$Y = C + I \tag{4-1}$$

where Y represents the level of regional income, C represents the level of household consumption, and I represents the level of investment in physical capital. The model further asserts that consumption is, itself, a function of the level of income, or

$$C = a + bY, \tag{4-2}$$

where C and Y are defined as before, a is the level of autonomous spending (spending that does not depend on income), and b is the so-called marginal propensity to consume, or the fraction of any additional dollar that is spent on consumption. By the way, $1 - b$ then becomes the marginal propensity to save, or the fraction of any additional dollar that is saved.

Substituting Equation 4-2 into Equation 4-1 gives us the following:

$$Y = a + bY + I \tag{4-3}$$

Subtracting bY from both sides:

$$Y - bY = a + I \tag{4-4}$$

Factoring the Y:

$$Y(1 - b) = a + I \tag{4-5}$$

Solving for Y:

$$Y = a[1/(1 - b)] + I[1/(1 - b)] \tag{4-6}$$

The term $1/(1 - b)$ represents the Keynesian multiplier. Any change in autonomous (non–income-related) spending leads to a $1/(1 - b)$ multiplied change in the region's income level. Savings $(1 - b)$ represents the leakage from the income stream that keeps the value of the multiplier from being infinite.

If we add exports to the spending stream, imports become the leakage. If we add government spending to the spending stream, taxes become the leakage. Thus, if the marginal propensity to import were given as

$$M = C + dY \tag{4-7}$$

where M represents the level of imports into a region, d represents the marginal propensity to import, and Y is as before, it can be shown that the expanded multiplier is

$$1/(1 - b + d)$$

In other words, the expanded multiplier is less than the simple multiplier by the additional leakage, d. In similar terms, the multiplier is lower yet to the extent of government taxation, and can be shown to be

$$1/(1 - b + bt + d - dt)$$

where t is the marginal tax, and bt and dt in the final multiplier allude to the fact that taxation is a function of income, but the propensities to spend and import are functions of disposable income.

This multiplier is used in conjunction with any form of autonomous spending. If we are speaking of export base logic, the level of a region's income is determined by taking this multiplier times the exogenous (external) level of spending, or exports. Such a model recognizes that (1) localities are more open (depend more on trade) than national economies, (2) internal levels of consumption depend very much on income earned from external sources, and (3) simple levels of government activity in terms of spending and taxation are less important to the economic development of subnational localities than national economies because of a locality's greater dependence on imports. The higher import levels result in significant leaks from the effects of any change in government activity.

The economic base multiplier is similar in concept to the Keynesian multiplier. Total activity in the region is seen to be dependent on two forms of activity, basic and nonbasic:

$$T = B + NB \tag{4-8}$$

where T represents total activity in the locality. Activity can be measured in terms of employment, income, value added, or industrial output. B represents activity oriented towards exports, and NB represents nonbasic activity that is local in its orientation.

Nonbasic activity can then be identified as follows:

$$NB = f + gT \tag{4-9}$$

where NB and T are as before, f represents any autonomous local activity, and g represents the proportion (in percentage terms) of total activity that is nonbasic.

Equation 4-8 can be manipulated in the same way as was Equation 4-3 from the Keynesian model as

$$T = [1/(1 - g)]f + [1/(1 - g)]B \tag{4-10}$$

The naïve model of economic base assumes f to be equal to zero so that total activity depends entirely on the level of basic activity. Also, since $g = NB/T$ and $(1 - g) = B/T$, it follows that $1/(1 - g) = T/B$. In other words, the export base multiplier is equal to the total level of local activity over the level of basic activity.

Delineating the Economic Base

The first step in deciding how to restructure a community is to delineate its economic base. Without the economic base, the community would not exist. It provides the means of trade with the rest of the world and the income payments for maintaining the industry production processes. However, use of different delineation criteria complicates its measurement. These criteria include the basic-to-nonbasic ratio, the area income multiplier, transfer payments and economic base, and some form of social accounting.

Basic-to-nonbasic ratio. Homer Hoyt, an economist with the U.S. Federal Housing Administration, formulated the essential outline of the economic base idea more than 65 years ago. His initial formulation of the economic base ratio was very simple: The ratio of basic to nonbasic employment in any city is 1 to 1, making the base multiplier equal to 2. Hoyt's intent was to use the basic-to-nonbasic ratio to predict employment and population growth in a city. Given well-founded forecasts of export-producing employment, a reliable estimate of the export base multiplier, and reliable estimates of the employment-population multiplier, it is possible to construct both employment and population forecasts. Population forecasts can then be used for determining housing needs as well as for other metropolitan planning purposes.

Income multiplier. More than 35 years ago, Charles Tiebout (1962) published a series of articles critical of the economic base concept. He argued that it overemphasized employment as the measure of regional activity. He proposed, instead, the use of income as a measure of the economic base. He proposed, further, to identify the income affected by exogenous determinants of change as basic and the endogenously affected income as service or residentiary.

In 1962, the Committee for Economic Development published Supplementary Paper No. 16—The Community Base Study, by Charles Tiebout. It joined the economic base idea with the mainstream of current economic thought. He introduced the use of the Keynesian multiplier, described above, as a model for the construction of the community base multiplier. He also provided a rationale for the implementation of a community base study.

Transfer payments and economic base. An occasional extension of the income measure of economic base is the use of transfer payments, such as Social Security, and commuter spending to represent "new dollars" coming into an area for the purchase of goods and services, much like the new dollars earned from local exports of goods and services. In a Missouri study, the social security income is included with worker earnings from so-called "basic" industries, namely farming, agricultural services, forestry, fisheries, mining, manufacturing, federal government, and commuter spending, along with some fraction of worker earnings from the remaining industries in an area.[2] This study shows social security payments to area residents accounting for 28.3 percent, while commuter earnings total 7.5 percent, of the local economic base. Use of this definition of an area's economic base raises the immediate problem of excluding some income payments, but then double-counting others, as we show later in Chapters 6 and 7.

One problem with the inclusion of social security payments and commuter earnings under the rationale of exogenous income is the obvious omission of other monetary inflows, like building and equipment loans from nonresident financial institutions. A more serious flaw is the double role assumed by transfer payments, with the influence on (1) personal consumption expenditures that help "drive" the area economy and (2) size and composition of the economic base. Possibly none of the commuter earnings are spent in the commuter's place of work or destination area. The exogenous income flows in the economic base model derive from local exports. The economic base multipliers derive from the backward linkages of the export-producing businesses to their input supply sources in the area. These are added indirect effects of spending and respending the income flows attributed to local exports. The social security payments relate to the much earlier income flows from local workers to the federal government and not to current exports of goods and services or income transfers. Neither social security payments nor commuter earnings have backward linkages to local input suppliers of intermediate goods and services, like the local export-producing businesses, which in reality are the area's economic base. Their backward linkages to final goods and services are taken into account in the indirect effects of the household spending cited in Chapter 7.

Social accounting matrix. A growing number of countries, including those in the Organization for European Economic Development that sponsored Sir Richard Stone's early studies (1961), have adopted the national income and product accounts and modeling systems he proposed. The accounting systems followed the principles and procedures he established during a quarter century of research and reporting. The social accounting matrix shows the origins and destinations of income and product flows in a given economy. It can be of great use in representing the actual winners and losers in the many different forms of income redistribution resulting from various legislative and market decisions. The social accounting matrix is at the heart of the third generation of economic base modeling. (Various forms and uses of the social accounting matrix are described in Chapter 7.)

The most recent uses of economic base modeling rest upon the theoretical and empirical advances of a half century or more of economic research, particularly the improvements in data collection, processing, and analysis over the last two decades. Using base models rests, finally, on access to data and constructs for estimating income flows in a regional economy. Income flows that bring in the new dollars for buying locally produced goods and services, as well as imports, now define the dimensions and determinants of the community economic base.

Economic Base Structures and Processes

High-order, high-margin services and technology-intensive manufacturing are increasingly important parts of the emerging, export-oriented industrial complexes in metropolitan labor market areas. They concentrate in these areas because of the presence of active and diverse local labor markets. Commodity-producing, low-margin industries, on the other hand, concentrate in rural areas because of low site costs as well as local availability of highly productive workers. The urban periphery, while the place of business for largely commodity-producing industries, also has an expanding services-producing sector because of its growing commuter population. This sector has replaced farming as the local economic base.

Growing and Declining Sectors

The statistical series presented here illustrates important differences among U.S. regions in industry performance. These differences can be attributed to corresponding differences in each region's core competencies. We view these differences from the vantage of economic base theory, strengthened in part by a generous use of interregional trade theory and regional income and product accounts.

Table 4-1 presents the summary results of this effort. An *excess earnings* value for each industry group in a given labor market area (LMA) measures the industry sector's importance to the area's economic base. Excess earnings represent labor earnings in excess of what would be expected for each area industry from the U.S. distribution of labor earnings for those same industries. Manufacturing, for example, declined in relative importance, from 57.6 to 45.6 percent of the excess of earnings for all industries in the 29 metropolitan core LMAs from 1974 to 1986. It increased in relative importance in the remaining LMAs, from 49.5 to 52.1 percent during the same period. The relative annual change columns show that manufacturing declined in relative importance, largely representing the lagging growth or lack of growth in nondurable manufacturing. All are relative (to the United States) differences. Because of the overall rapid growth of labor earnings in human services (health care, education, and social services), relative positive differences were less in 1986 than 1974. Manufacturing, on the other hand, declined in relative importance during this period. It is an industry that is highly differentiated by its product and place specialization and its high propensity for importing and exporting among trading regions. The large increase in business services is in part attributed to the shift of business-services activities within the manufacturing sector to an independent business-services sector. A similar phenomenon occurred many years earlier in agriculture. The shift to services is thus positive for business services but negative in human services in the core LMAs.

Two of the largest shifts in the economic bases of LMAs are indicated by the percentage changes in agriculture and business services. Agriculture declined in relative excess earnings in the noncore LMAs, from 28.9 percent in 1974 to 15.2 percent in 1986. Business services increased in the core LMAs, from 12.6 percent in 1974 to 29.8 percent in 1986. Core LMAs, like Minneapolis-St. Paul,

TABLE 4-1 Proportion of total excess earnings and annual rates of change for specified industry group, U.S., 1974–1986

	Metro LMAs		Nonmetro		All Regions		Relative Annual Change		
Industry Groups	1974 (pct.)	1986 (pct.)	1974 (pct.)	1986 (pct.)	1974 (pct.)	1986 (pct.)	Metro (pct.)	Nonmet (pct.)	Total (pct.)
Agricultural Products and Agriculture Services (SIC 01-09)	0.2	0	28.9	15.2	20.9	9.5	–10.3	–6.4	–6.4
Mining and Extractive Industries (SIC 10-14)	2.1	4.4	10.8	14.3	8.4	10.6	9.3	1.2	2
Construction (SIC 15-17)	1.7	1.2	1.7	3.1	1.7	2.4	–0.4	3.9	2.9
Manufacturing Nondurables (SIC 20-23, 26-31)	9.7	5.4	16.6	17.7	14.7	13.1	–2.4	–0.7	–1
Manufacturing Durables (SIC 24-25, 32-39)	47.9	40.2	32.9	34.4	37.1	36.6	1.1	–0.8	–0.1
Transportation, Communications, and Utilities (SIC 40-49)	4.1	3.7	2	2.9	2.6	3.2	1.5	2	1.7
Wholesale Trade (SIC 50-51)	14.9	12.4	1.3	1.2	5.1	5.4	1	–2	0.5
Retail Trade (SIC 52-59)	1.4	0.7	1.5	2	1.5	1.5	–3.6	1.2	0.1
Business Services (SIC 60-67, 73, 81, 86, 89)	12.6	29.8	1.4	3.6	4.5	13.4	10.2	7	9.6
Consumer Services (SIC 70, 72, 75, 76, 78, 79)	0.8	1.2	1.4	4.1	1.2	3	6.2	8	7.7
Human Services (SIC 80, 82, 83)	4.6	0.9	1.5	1.6	2.3	1.3	–10.4	–0.6	–4.6
Total	100	100	100	100	100	100	2.5	–1.2	0

Source: Reynolds and Maki, 1990, *Business Volatility and Economic Growth*. Prepared for U.S. Small Business Administration (Contract SBA 3067-OA-88).

are characterized by the increasing importance of business and related services to the area's economic base. The business-services sector is replacing many of the job losses in manufacturing and retail trade. The new jobs include well-paying positions in management and administration, and various professional and technical occupations.

We have said that a leading measure of the size and viability of a local economic base is the value of its exports, or out-shipments, to market destinations outside its own LMA. Another measure is its propensity to import goods and services from sources outside the area. The primary difficulty in the use of these two measures is the lack of any accurate monitoring of commodity or product flows from one area or region to another. We have, however, indirect measures of these two indices of actual shipments from a variety of data sources (including the U.S. transportation census) that show gross out-shipments and in-shipments for industries within each state.[3]

Each industry produces one or more commodities. Use of commodity rather than industry measures of exports and imports would reveal the balance of trade among individual commodity groups for each combined LMA. Measures of net commodity exports, for example, would show the excess of individual commodities produced within the area over its total imports. These measures would show the purchases of commodity imports by individual producing and consuming sectors and the extent to which each sector contributes to any trade deficits. The proportion of net exports accounted for by each locally produced commodity is also an alternative measure of an area's economic base.[4]

The dominant basic (exporting) industry groups are also the dominant import-dependent industry groups, except for construction. Export growth means import growth because of corresponding increases in demand for imported intermediate production inputs (Jacobs, 1984). As an area grows and diversifies, however, import replacement occurs for both intermediate inputs and final purchases. Imported finished goods and services dominate total imports in the periphery of an economic region, while imported intermediate goods and services dominate in its core area. Again, the import dependencies of core areas contrast sharply with those of the periphery, which is a measure of the opportunities for internalizing the export trade of individual LMAs. Dominantly agricultural areas share increasingly in the internalization of exports and imports within the larger economic region.

Vulnerable Regional Economies

One region's economic base differs from another for many reasons. One region may specialize in a particular product or commodity, such as farming. Such a region risks much by its extreme dependency on one industry. Another region may have a manufacturing industry as well as farming in its economic base. Its manufacturing may not compete, however, because of high costs. These costs are high because of high material input costs, like energy or labor. Costs may be high, also, because of low output per worker. Still another region may have a wide variety of manufacturing and service industries in its economic base but may still lack flexibility in adjusting to changing market conditions. Thus, multiple criteria must apply in the assessment of a region's economic base. Among the criteria are the risks and costs of production and the productivity and flexibility of resource use in production.

Extreme specialization of economic activity in a region usually incurs high *risks* for its trade and services sectors. They are dependent upon a stable and viable economic base for their business successes. Risk thus correlates with industry specialization for businesses that cater to local markets.

Risk also correlates with both early and late stages of a product cycle. Introduction of a new product incurs costs of investing in its development and marketing. Cost recovery depends on the market success of the new product.

Alternatively, an old product is in danger of losing its market to competing products on a quality or cost basis. Again, businesses dependent on the economic base created by the successful entry of a locally produced product that now faces new and stiff competition face an uncertain future. High-risk businesses face reduced prospects for business loans and favorable credit terms from suppliers.

Rural regions, removed from access to large consumer and producer markets, are singularly dependent on farming, fishing, or mining. They often lack prospects for diversifying their local economic base. In such cases, local credit sources soon disappear, while distant credit sources heavily discount high-risk loans.

Risk, when discounted heavily, adds to production costs. High wages and high regulatory and environmental requirements also add to the *costs* of doing business in a particular region.

Many rural regions offer cost advantages for export-producing businesses because of low site costs—land and buildings, labor, environmental regulations, and congestion. Transportation, business, and related services cost less in many rural areas than they do in metropolitan areas. Personal services and housing also cost less in rural than in metropolitan areas. For these regions, low costs are an advantage when accompanied also by low-cost access to large consumer and producer markets.

Productivity of resource use correlates with investment per worker, which is generally low in rural areas. It is much higher in metropolitan areas. Capital substitutes for labor, which increases overall factor productivity.

High labor productivity also correlates with investment in the *human agent*—education, on-the-job training, and lifelong learning. Attitudes, lifestyles, and social and political stability also make a difference. Convenient access to nearby educational institutions in rural areas, and to information sources in metropolitan areas, provide opportunities for rural residents to improve existing skills and develop new ones.

Despite an initial disadvantage in low investment per worker, residents in rural areas have opportunities to reduce these disadvantages by wisely directed investments in education and training for a changing job market. A strong work ethic and supportive attitudes and values, when coupled with new and superior skill development, make some rural areas superior places for new business formation, branch plant location, and expansion of existing businesses.

TABLE 4-2 Criteria for assessing a region's vulnerability to changing market conditions

Criterion	Core Area	Transitional Rural Area	Peripheral Rural Area
Risk	Low	Moderate	High
Cost			
Site	High	Moderate	Low
Transfer	High for commodities; low for information	Moderate	Low for commodities; high for information
Productivity	High	High in branch plants; moderate in small businesses generally	High in branch plants; low in small businesses generally
Flexibility	High	Moderate	Low

A fast-changing global economy makes *flexibility* in work practices and organization an essential precondition for a successful business enterprise. Flexibility of access to input sources and of procurement practices are also important. In both cases, the flexibility is a function of the learning processes in an organization and of the ability to gauge the likely outcomes of improved learning capacities.

Successful business organizations have the robustness or capacity for adaptation to changing market conditions and new production technologies. They are successful learning organizations.

Table 4-2 summarizes the four criteria for assessing the vulnerability of an area's economic base—risk, cost, productivity, and flexibility. The table relates these four criteria to the location of a rural area in the larger regional settlement system. Both the metropolitan core area and the rural area are parts of the larger system. The table also differentiates between site costs and transfer costs.

Site costs include labor costs as well as land and building costs, rent, taxes, pollution abatement, and other regulatory costs. Transfer costs include transportation charges and other marketing and transaction costs.

Different Ways of Thinking about Economic Growth

Much of past thinking about economic growth and development started with a traditional low-income rural agricultural economy losing its workforce to the higher-paying jobs in the new, modern industrial sector. The industrial sector owners, according to the theory, would plow their capital earnings back into more capital investment, creating more jobs and higher incomes. The resulting rural out-migration would increase until the traditional sector finally disappeared (Lewis, 1954).

Weisskoff (1985, p. 105), in a critique of the various theories of economic growth and development, claimed that the labor surplus theory "... fails to comprehend certain real factors that are always present and that are critical to understanding and anticipating real events." These omitted factors are as follows:

1. Importing of efficiently produced farm and manufactured products that drive traditional agricultural products and crafts out of domestic markets.
2. Historic role of internal commodity prices and the role of force in explaining the decline or rise of certain activities.
3. Openness of the economy as measured by imports, exports, and balance of trade.
4. Colonial tradition and its impact on the peasant economy through laws, history, and attitudes.
5. Population density as it affects the size, diversity, and availability of the local labor force.
6. Market size as it affects small producers in finding local markets for their products.
7. Time, perhaps decades, for policy outcomes to appear.
8. Foreign profits that establish future claims on the region's resources.
9. Domestic consumption needs that are not met because of growing income disparities.

Weisskoff had difficulty in accounting for the long-standing acceptance of labor surplus theory among economic development practitioners, except for its simplicity and timing, in view of its failure in Puerto Rico. The apparent successes that central governments in Europe and North America developed in coping with their problems of unemployment convinced many that this was also the route to corresponding successes for the less developed countries. In Puerto Rico, however, the growth of a modern export-producing industrial base and the concomitant demise of traditional agriculture brought only more poverty and greater dependence on U.S. transfer payments. Puerto Rico, with its extreme dependence on imported goods and services, is very much like the residential neighborhood of a modern city with a largely commuter population, where place of residence and place of work or source of income are separated.

Hirschman (1958/1978) proposed an alternative strategy for economic development in his unbalanced growth theory. The unbalanced growth theory focused on an industry rather than an entire sector or two of a developing nation's economy. Polenske (1993) noted in her review of the Hirschman proposal that he disagreed with both the supply-siders, who advocated a comprehensively balanced approach to economic growth among all sectors, and the demand-siders, who argued for the dualistic approach of labor surplus theory with its modern industrial sector absorbing the traditional one. Hirschman questioned the efficacy of the Keynesian solution of large injections of government spending to counter problems of lagging growth and high unemployment—a solution that seemed to work for the advanced countries at an earlier period in their history. This solution emphasized the role of central governments but minimized the role of the entrepreneur. In contrast, Hirschman envisioned, according to Polenske, efficient sequences of investment; creative, cooperative entrepreneurs; and strong linkages in production, consumption, and financing. Hirschman not only highlighted, but emphasized in significant ways, the role and importance of the private sector in economic development.

Polenske suggested that the occurrence of production linkages—backward to input supplies and forward to product markets—raises important questions about trade-offs among industry investments. She questioned, for example, the trade-off between investing in industries with high backward linkages (large purchase of intermediate inputs) versus investment in industries that have high direct job creation through large payments to labor (p. 13). Still missing in this questioning of industry investments is its relevance for economic development strategy.

Also missing is an assessment and understanding of the particular decision environment of the individual entrepreneur or organization making the location-specific business investment decisions. These decisions obviously have many different consequences for the immediate locality, among them job creation and enhancement of the local economic base, but also, at times, adverse local environmental effects—consequences that differ greatly from industry to industry and place to place.

Michael Porter (1990), in his book *The Competitive Advantage of Nations,* focused on the role of the entrepreneur and the innovative firm in the economic development process of advanced economies. He identified from his studies of more than 100 industries in ten countries four common characteristics of local decision environments that affect the competitive advantage of individual firms, These were the factor conditions; the demand conditions; related and supporting industries; and firm strategy, structure, and rivalry. These four characteristics, and the ways in which they connect to one another in the individual firm, serve as the basis of his competitive advantage strategy. This interconnectedness of the firm's activities forms the "value chain" that includes the firm's inbound logistics, operations, outbound logistics, marketing, and follow-up services. It also includes the firm's support activities, namely, its infrastructure (of finance and related activities), personnel management, technology, and procurement.

Porter's idea of the clustering of industries relates closely to Hirschman's emphasis on backward and forward linkages among industries. The occurrence of these linkages in the clustering of industries provides a favorable local environment for economic development. Industry clustering, claimed Porter, strengthens the individual firm in seeking its competitive advantage in global markets by increasing information flow and adding to the diversity and dynamism of the local labor market. This, in turn, facilitates the entry of new firms into the industry cluster and upgrades the competitive environment for all firms. In short, a competitively active local economy provides the training ground for local firms to succeed later in global markets.

Polenske concluded her review of Hirschman's unbalanced growth theory and Porter's competitive advantage strategy by challenging both contributions as they relate to our thinking about the economic development process. She found a problem with Porter's use of the linkage concept and his portrayal of it as an extension of Hirschman's view of this concept rather than simply its reformulation. She also questioned whether Hirschman was correct in saying that the production linkages are invariant to social and political change.

The introduction of property rights and geographic space into the formulation of a new strategy of regional development, Polenske (1993, p. 18) believed, is essential for policy makers to effectively deal with the related issues of power and control in the economic development process. She argued that the issues relating to property rights need systematic study. The distributive effects of changing property rights, the institutional changes they precipitate, and the consequences of these changes for the societal problems addressed by the economic development strategies are increasingly important to people and regions and, in Polenske's view, should be incorporated into these strategies. Geographic space and proximity to various producer services and markets also affect the exercise of the power and control emanating from certain property rights and should be included in the formulation of a new strategy of economic development.

Richard Weisskoff (1985), in his book *Factories and Food Stamps,* rejects the circularity of new exogenous industries providing jobs that replace the job losses of a declining traditional sector. This is based on his modeling of the Puerto Rico economy, in which production and consumption are disassociated. Local production accounts for only a small part of domestic consumption needs. This is especially the case where the domestic market is small and dispersed and both the rich and the poor depend on foreign imports as the economy develops. A similar paradigm would apply to the individual neighborhood that is largely a place of residence for a population of commuters, with only the essential resident-oriented services available in nearby shopping strips or malls.

The Weisskoff disassociated economy model differentiates one economic sector from another with detailed breakdowns of the individual industries and the corresponding commodity categories in each sector. It also uses a detailed breakdown of occupations in each industry. It thus shows the high dependence of the Puerto Rico economy on specific commodity imports. This allows the model to simulate industry-specific and commodity-specific effects of varying degrees of import dependency and show an increasingly "hollow" producing sector accompanying the foreign import dependency. The model thus represents the domestic economy more accurately and fully than the labor surplus traditional economy model, given different sets of assumptions about personal income distributions and public policy measures.

The Weisskoff model accurately predicted the state of the Puerto Rico economy in 1980 from its 1970 database. It did so by accounting for several of the factors omitted in the labor surplus model, including the increasing import dependency that accompanied the loss of small local businesses that had catered to local markets. This also contributed to growing income disparities and popular dissatisfaction with existing measures for creating jobs and improving the economic and social well-being of a majority of the Puerto Rican people. It thus realistically represented the domestic consumption and production—and even political—effects of a declining traditional sector and an expanding industrial sector.

Vernon Ruttan (1977; 1978) advanced a theory of "induced innovation" that he related to the agriculture sector of developing as well as developed economies. Technical change, in induced innovation theory, is endogenous to the development process. Induced institutional innovation accompanies the induced technical changes with the research support of agricultural experiment stations and research institutes. Ruttan suggested that "If research workers in the basic sciences are sensitive to the needs of applied researchers for new theory and new methodology they are in effect responding to the needs of society." He illustrated this point with the oil crisis of the 1970s, noting that the response of the scientific community in directing research to cope with the effects of the rise in fossil fuel prices "represents a dramatic example of the induced innovation process" (Ruttan, 1977, p. 210).

The sources of innovation for endogenous growth will vary with the type of enterprise. The manufacturer was the dominant source of innovation in engineering plastics in a survey reported by Eric von Hippel (1988) in his book *The Sources of Innovation* (p. 4). The supplier was the major source of innovation in wire termination equipment manufacturing, while the user was the major source in a large majority of scientific manufacturing enterprises. Since the innovative users, suppliers, and manufacturers depend on ready access to essential information and know-how, they congregate in the larger metropolitan areas. In these environments, a high level of networking among many different types of enterprises becomes an essential part of the innovation process. The examination of the sources of innovation adds another dimension to the theory of induced innovation and enriches our understanding of its role in local and regional economic development.

Concerns about the efficacy of depending on external sources of innovation and growth in lagging regions parallel important shifts in the focus and application of economic development theory. James Miller, for example, in a review of development prospects of small and midsize enterprises (SMEs) for nonmetropolitan areas, observed that "expanding world markets, improved communications and production technologies, and changes in corporate organization are leading towards a new spatial organization of production" (1993, p. 99). However, recent studies generally support the thesis that the new technologies and high-order producer services concentrate increasingly in metropolitan core areas, while low-technology manufacturers dependent on low-cost labor will have more success in nonmetropolitan areas. The metropolitan areas offer greater opportunities for hiring people with the appropriate skills and establishing strong and effective linkages with input suppliers and product markets. Miller suggested that "nonmetropolitan SMEs must either develop contractual relationships with larger enterprises or be linked to a network of SMEs in which resources and services are jointly purchased and shared" (p. 101). Endogenous growth theory thus emerges from the research findings about business behavior in a changing business environment, local and global.

Niles Hansen, in a review of the European historical experience in endogenous economic development, cited the example of Mediterranean France, formerly a peripheral area that is now the fastest-growing area in France. Hansen attributed "a combination of endogenous factors, including high levels of education and the elaboration of a network of SMEs for this growth" (1993, p. 72). Hansen noted further that "even in Denmark, it has been observed that the advice and counsel of public authorities (and other experts who design programs for potential entrepreneurs) often fail to have an effect because those who deliver such services often have different values, motivations, and terminologies from the person served in peripheral rural areas" (p. 85). Hansen appeared to view the job-creating and income-generating potential of the SMEs more favorably than Miller from his study of production/service complexes in Denmark. He observed that the cooperation among SMEs and local authorities in peripheral Jutland, for example, yielded "productivity advantages corresponding to external economies of agglomeration." They benefited from their own homegrown propensities for independent entrepreneurship and cooperation among their export-producing local industries.

Tracking Economic Base Changes

The findings reported in the early 1990s on business volatility and regional economic growth and related matters (Maki and Reynolds, 1994; Reynolds and Maki, 1990a, 1991, 1994a,b) addressed issues still current in studies and deliberations of the European Union. The findings on U.S. regional growth, for example, compare closely with more recent findings on the sources of European Union growth (Mack and Jacobson, 1996). Included in these studies is a shift-share analysis, based on annual time series from the U.S. Department of Commerce Regional Economic Information System, that yields a "regional-share effect" for each of 77 LMAs located in the thirteen-state, east-west transportation corridor region extending from Michigan to Oregon and Washington. Historically, the study region has experienced much economic volatility due to the many natural resource-based local economies in the interior states and cyclically sensitive durable goods manufacturing elsewhere. The remaining twenty-three LMAs in the current study (of 100 U.S. LMAs) include both rapidly growing and generally declining base economies that vary in income volatility and overall growth from the lowest to among the highest. A totally different data series, Duns Marketing Indicators, provides the estimates of employment change—by size, type, and location of firms—associated with the estimates of total earnings change, total earnings, business volatility, and spatial structure.

Change Sources for Local Labor Earnings

Sources of income volatility—that is, period-to-period shifts in labor earnings—are illustrated by the shift-share analysis that includes the two long periods of economic recovery: 1970–80 and 1982–86. These periods were separated by two recessions occurring in the 1980–82 period. Data for the recession from 1973 to 1975 are not included in the analysis. Both the income volatility index and the income growth index are based on income change over the entire 1970–86 period.

The summary results of the shift-share analyses show vastly different growth patterns for the four regional groupings. Over the sixteen-year period, total labor earnings—the principal source of personal income—increased by more than $782 billion (in 1982 dollars), from $1,426 billion in 1970 to $2,208 billion in 1986. The overall increases ranged from $40.5 billion in Mid-continent West to $50 billion in Mid-continent East, $132 billion in comparison LMAs, and $560 billion in the remaining LMAs in the United States. The comparison LMAs increased in importance from 10.6 percent of total U.S. labor earnings in 1970 to 12.8 percent in 1986. Mid-continent East dropped from 9.5 percent of the total to 8.4 percent.

The principal reasons for the contrasting growth patterns rest with the base economies of the two regions. The local base economies of Mid-continent East not only are dominated by below-average growth industries but also are marked by a continuing decline in the competitive position of their principal exports. The base economies of the comparison region are distinguished by an overall above-average industry-mix effect and an overall above-average regional-share effect.

A distinguishing difference between the high- and low-income-volatility LMAs is the direction of relative change. It is strongly negative for high-volatility areas and strongly positive for low-volatility areas. For most high-income-volatility LMAs, a positive regional-share effect in the 1970s turned negative in the 1980s, thus contributing to the strongly negative relative change in the 1980s.

The ranking of total change in labor earnings in the 1970–86 period confirms the unique role of the local base economy in accounting for regional job and income growth. For the 30 fastest-growing LMAs, total labor earnings increased from $182.6 billion in 1970 to $345 billion in 1986—an increase of 89 percent. During the same period, total labor earnings increased by only 22 percent for the 30 slowest-growing LMAs—from $96.1 billion in 1970 to $116.9 billion in 1986.

High growth in local labor income is as frequently associated with high as with low labor-income volatility—nine in both cases. In comparison, the low-income LMAs include thirteen of the highest-volatility and five of the lowest-volatility LMAs. Thus, the mid-range LMAs in labor income growth include twelve high-income-volatility and twelve low-income-volatility LMAs. The findings show a lack of strong correlation between income growth and income volatility when further differentiation of local base economies is lacking.

The excess-earnings variables in the regression analysis (cited earlier and presented in the next section) are used, also, in estimating the industry mix in the base economies of the 100 LMAs. Excess earnings of each two-digit industry group in the county-level labor earnings series compiled and reported by the U.S. Department of Commerce were calculated for each county and aggregated by LMA. The base economies of the high-income-volatility LMAs are marked by high levels of industry specialization in farming, mining, or manufacturing. In these areas, the high income volatility is associated with a high degree of vulnerability to the vicissitudes of cyclically sensitive export markets. Moreover, the extreme specialization of industry in the base economies of the high-income-volatility LMAs persisted through the 1970s and much of the 1980s. Where high-income volatility was accompanied by slow income growth, the local base economies also faced shrinking export markets.

High-income-growth areas differ from high-income-volatility areas and low-income-growth areas in the diversity of their base economies. Even specialized base economies support high income growth when the export-producing sectors remain competitive in their export markets and maintain their market shares. Generally, however, the specialized fast-growing economies had lost their earlier momentum by the mid-1980s and faced, instead, greatly reduced income growth.

Each measure of regional growth analysis varies in relative values from one period to the next. For some areas, the volatility in rates of regional growth is due to the cyclical sensitivity of the local economy. For others, the period-to-period changes in jobs and earnings are related to long-term changes in industry product cycles. Changes in industry mix reveal both short-term and long-term changes in the importance of individual industries in the U.S. economy. Changes in regional share reveal changes in the competitive position, or economic performance and importance, of a given industry relative to the corresponding industry in the United States. A distinguishing characteristic of declining and growing areas is the rapidity and direction of change in jobs and labor earnings. Once the volatility in jobs and income is removed, the residual "regional-share effect" becomes a measure of regional growth and decline.

Accounting for Local Employment Change

Regression models accounting for local employment change for each year and for their area orientations, based on the Duns Marketing Indicators series, start with the 1978–80 period. The findings for each model are reviewed in the context of the preceding discussion of the role and importance of the base economy in regional economic growth. Each measure of regional growth—excess-earnings change, sector size, business volatility, and spatial structure—varies in relative value from one period to the next. For some areas, the volatility in rates of regional growth is due to the cyclical sensitivity of the local economy. For other areas, the period-to-period changes in jobs and earnings relate to long-term changes in industry product cycles.

Excess Earnings Change

The excess-earnings-change variable is estimated from the two-digit, county-level wage and salary earnings series in the private sector. The statistical series was prepared by the U.S. Department of Commerce for 1970, 1975, 1980, 1982, 1985, and 1986. By straight-line data interpolation, intervening year estimates were obtained to complete the two-year, even-year change series.

Statistically significant (at a five-percent confidence level) estimates of the "All LMAs" regression model parameters (standardized beta-weights) are presented for the five two-year periods, as follows:

Industry Group	7880	8082	8284	8486	8688	E
Agriculture (1–9)	.396					R
Mining (10–14)	.345	.225			.272	R
Construction (15–17)	.164	–.076	.284	.119		U
Manufacturing, Nondurables (20–23, 26–31)	.165			.120		M
Manufacturing, Durables (24–25, 32–39)	.201	–.071	.094	.178		M
TCPU (40–47)						M
Wholesale (50–51)	.112			.090		R
Retail (52–59)	.216			.101		R
Business Serv. (60–67, 73, 81, 86)	.119					R
Consumer (70, 72, 75–76, 78–79, 84, 88)	.226	.065		.063	.144	R
Other Private Serv. (80, 82–83)	–.053					U

Most excess earnings variables are positively correlated with employment change. Exceptions occur in the 1980–82 period and in construction, durable goods manufacturing, and other services (health care, education, and social services) that relate to their role in the 1980–82 recessions. In the preceding two-year period, many LMAs peaked in total employment because of high levels of durable goods manufacturing in their local base economies. Peak employment levels in the 1978–80 period were followed, however, by large employment losses in the 1980–82 period.

The largest percentage change in total employment was associated with a given percentage change in agriculture sector labor earnings. On the other hand, the mining earnings-to-employment multiplier is large because of high earnings per worker in mining.

The series of eleven regression coefficients varied among LMAs because of primary economic emphasis (E)—urban metropolitan (U), rural (R), or manufacturing (M). Employment effects of changes in construction and other private services, like health care, education, and social services, were largest in the urban metropolitan areas. Employment effects of changes in manufacturing and the transportation, communications, and public utilities (TCPU) sector were large in LMAs with a manufacturing orientation. For the remaining industry groups, the employment effects were largest in LMAs with a rural emphasis.

Total Excess Earnings

The current-year values of excess earnings were included in the model to account for the differential effect of sector size, as well as rates of change, on total employment. Again, this measure of the base economy proved statistically significant in explaining model variance, as shown by the fitted regression coefficient values below:

Industry Group	7880	8082	8284	8486	8688	E
Agriculture (1–9)	.106			.066		R
Mining 10–14)			–.555	–.139		M
Construction (15–17)	.119	.288		–.211		R
Nondurables (20–23, 26–31)					.192	M
Durables (24–25, 32–39)		–.076	–.056			M
TCPU (40–47)			–.220		–.203	M
Wholesale (50–51)						R
Retail (52–59)	.211	.070		.225	.323	U
Business Serv. (60–67, 73, 81, 86)	–.105		–.113			U
Consumer (70, 72, 75–76, 78–79, 84, 88)						M
Other Private Serv. (80, 82–83)	.090					R

Sector size is related positively to employment change in agriculture, construction, nondurables manufacturing, retail trade, and other services. It is negatively related to employment change in mining, durable goods manufacturing, the TCPU sector, and business services.

Employment effects of sector size vary with economic emphasis. They are the largest in (1) the urban metropolitan emphasis for retail trade and business services; (2) the rural emphasis for agriculture, construction, wholesale trade, and other services; and (3) the manufacturing emphasis for mining and manufacturing; the transportation, communications, and public utilities sector; and consumer services.

Business Volatility

Business volatility is represented by changes in the number of establishments and related jobs due to the births and deaths, expansions, and contractions of the establishments. Firm volatility is represented by four variables—autonomous births and deaths and branch births and deaths. Job volatility is represented by eight variables—the factorial combination of autonomous and branch births and deaths, and autonomous and branch expansions and contractions. The employment effects of each of the twelve explanatory variables are summarized as follows:

Business Volatility Variable	7880	8082	8284	8486	8688	E
Branch Births		.099	–.135	.093		R
Branch Deaths			.100		–.103	R
Autonomous Births	.175	.312	.296		.264	U
Autonomous Deaths			.149			R
Job Growth, Auton. Births	.083					R
Job Loss, Auton. Contractions				.098		M
Job Loss, Auton. Deaths						R
Job Growth, Auton. Expansions					.240	M
Job Growth, Branch Births			–.094			R
Job Loss, Branch Contractions						
Job Loss, Branch Deaths					.097	U
Job Growth, Branch Expansions						R

Business volatility variables are positively associated with employment change, except for branch births and branch deaths in the 1982–84 period and job growth associated with branch births in the 1986–88 period. Autonomous firm births have the largest effect on total area employment.

Labor market areas with a rural emphasis are more strongly affected by the business volatility variables than LMAs with an urban metropolitan emphasis, particularly with autonomous births and autonomous expansions. While LMAs with a rural emphasis may experience more income volatility than LMAs with an urban metropolitan orientation, they also are more susceptible to the positive influences of increased business activity. One result of having a concurrence in firm births and job expansions as well as firm deaths and job contractions is an economic dynamism that shifts local resources into more productive enterprises.

Spatial Structure

Market access differences in the spatial structure of rural and metropolitan areas are represented by three dummy variables. The values of 1 or 0 depend upon the status of the LMA relative to the specified access attribute, population density, and location in or out of the Sunbelt (Texas, Oklahoma, and Florida). Distances from the principal urban center of the LMA to the nearest and the next nearest airline nodes are the measures of market access represented by the dummy variables. The importance of the three dummy variables, population density, Sunbelt location (also a dummy variable), and personal income in accounting for local employment is summarized as follows:

Market Access Variable	7880	8082	8284	8486	8688	E
Node 1 if less than 60 miles		.154	.070			R
Node 2 if less than 100 miles	.107	.099				R
Option if difference is less than 50	.188		.053			U
If LMA is in Sunbelt	.167	.404	–.183	–.173		U
Population Density					–.117	M
Personal Income Change	–.201			.107	.086	M

Market access as represented by proximity to one or two of the 29 U.S. airline nodes is a statistically significant locational attribute for differentiating among LMAs with reference to employment change. It helps articulate the role and dimensions of location in regional economic growth and change.

Each of the three economic orientations cited earlier has a different response to the market-access variables. Proximity to a primary and secondary airline node is positively correlated with employment change, especially for LMAs with a rural orientation. Proximity to two airline nodes is most important to LMAs with a metropolitan orientation.

A Sunbelt location was a positive factor in employment growth in the 1978–80 and 1980–82 period, but a negative factor in the 1984–86 and 1986–88 periods. Labor market areas with a metropolitan orientation were slightly more influenced by these factors in three of the four periods than were the combined LMAs. Population density was a negative factor for LMAs with a manufacturing orientation during the 1980–82 period but a positive factor in the 1984–86 period. It was a positive factor for LMAs with a rural orientation in the 1978–80 period. Finally, total personal income change was positively associated with employment change in the 1984–86 and 1986–88 periods and negatively in the 1978–80 period. Its largest effect was in LMAs with a manufacturing orientation.

New Firm Start-ups

We find several different explanations for new firm start-ups, that is, the appearance of new businesses. One explanation for firm foundings is the social and economic context that fosters business births; the other is the set of characteristics of the individuals who start the new firms. Firm births, deaths, and volatility are obvious indicators of economic growth and change. Above-average rates of firm births is a distinguishing characteristic of growing regions. While firm deaths as a percentage of total number of firms is about the same from one region to the next, an above-average percentage of new firm births accounts for the superior performance of the growing region. A dynamic regional economy has, of course, a large number of firm deaths. Hence, both firm deaths and firm volatility enter directly into any search to account for regional differences in firm births. The question thus remains: What features of an economic system lead to higher rates of firm births? We present the various study findings on the social and economic context under the headings of start-up processes, that is, the various causes of new firm births, the impact of the start-up processes on new firm births and deaths, and business volatility.

Start-up Processes

Reynolds, Miller, and Maki (1994a,b) list a number of postulated causes of new firm births, such as production input costs and access, and public infrastructure costs. However, testing the postulated causes, using multicounty LMA data for the United States, shows only a fraction of the total number of postulated causes with a major or strong impact on firm births, deaths, and volatility. Access to production inputs, customers and clients, and efficiency of public infrastructure fail to account for statistically significant regional differences in firm births, deaths, and volatility. Start-up processes that have the most impact include economic diversity, population growth, personal wealth, and employment flexibility, as well as low unemployment. For one of the causes—namely, unemployment desperation—the findings show a strong impact, but in the opposite direction to the one postulated. An explanation of this apparent disagreement with the current literature on the importance of public infrastructure spending lies in the interpretation of the findings. We start with the hypotheses rejected by the Reynolds, Miller, and Maki (RMM) findings.

We start with the *unemployment desperation* hypothesis, namely, that individuals unable to find careers in established organizations start new ones. Complementing this hypothesis is the notion that individuals, especially midcareer adults, start new firms to create *career opportunities* and thus achieve personal goals that could not be achieved otherwise. Some industries, like construction, are more volatile than others (as shown by higher turnover rates), which affects the industry mix and hence new firm births. That is, the larger the share of *volatile industries* in an area, the higher the rate of new firm births will be. Another hypothesis is that the lower the *factors of production costs,* the higher will be the rate of new business formation. A slightly different argument is that *factors of production access,* that is, having easy access to established sources of production factors, is important in starting new firms. Another argument is that the greater the tax payment for *public infrastructure costs,* the greater is the likelihood of starting a new firm. Also, the easier the *access to customers and clients,* the greater will be opportunities for starting new firms to successfully enter these new markets. *Access to research and development, information, and innovation* is an added incentive for new firm formation. *Personal wealth* allows individuals to seek greater diversity in their personal consumption, which, in turn, creates more opportunities for new specialized firms. *Social status diversity* also adds to the diversity of opportunities for new start-up firms. *Population growth* correlates with increased demand for goods and services, which, in turn, generates still more opportunities for start-up firms. *Size of an economic system* correlates with greater demand and hence more new firms. *Economic diversity,* not only in products produced, but workforce skills, correlates with new firm start-ups. *National transportation access* is important to new export-producing firms oriented to national markets, especially firms producing high-valued products for international as well as national markets. *Employment policy flexibility,* which allows businesses to respond quickly to product and market changes without the restraining influence of rigid work rules, favors new firm start-ups. The fifteen variables form a smaller set of clusters, according to the factor analysis performed on these variables by the Reynolds research team.

Table 4-3 lists the individual statistically significant items forming each of the start-up process variables as well as their factor clusters. The largest of the seven clusters, which includes four of the fifteen process variables, is labeled "agglomeration/access" to denote the agglomeration of a number of market access variables, like access to customers, clients, and information. Each process variable, in turn, is a composite of individual "items," that is, specific variables representing the "process." Population and establishments per square mile are measures of population and business densities, the argument being that the higher the density, the greater the market access. The "wealth/costs" cluster has the finance-related decision variables, while the "growth/turbulence" cluster has measures of business volatility.

All data series were prepared for individual counties, covering six two-year periods, starting in 1976, and then grouped into the LMAs. The U.S. Small Business Administration provided the data sources for the process variables. The firm births and deaths came from Dun and Bradstreet, the Dun's Market Identifier files.

Impact of Start-up Processes on Business Foundings

Table 4-4 summarizes the impact of the start-up processes on firm births, deaths, and volatility. The seven clusters are rearranged according to the level of impact on the dependent variables. Major impacts were confined largely to four factors: economic diversity, growth/turbulence, wealth/costs, and employment rigidities. The dependent variables are the total firm births and deaths, not differentiated by Standard Industrial Classification (SIC) Code, and an index of firm volatility. Births and deaths are computed as the number per 10,000 population per year for each LMA. The index of volatility is an average (standardized Z-scores) of

TABLE 4-3 Multi-item indices of variables affecting new firm formation in labor market areas, United States, 1976–1988

Factor and Process	Items	Factor and Process	Items
Agglomeration/Access		*Growth/Turbulence*	
Size of economic base	Total population	Career opportunity	Percent population age 35–44
	Total labor force		Percent some college
	Total establishments		Percent college degree
Availability of production factors	Per capita demand deposits		Percent managers
	Per capita savings deposits		Percent professionals
	Percent with high school diploma only		Percent technical occupations
	Percent adults age 15–64	Volatile industries	Percent workforce: construction
	Sales workers per sq. mile		Percent workforce: retail
	Clerical workers per sq. mile		Percent workforce: consumer services
	Service workers per sq. mile		Percent workforce: services
	Skilled craftsmen per sq. mile		Percent establishments: construction
	Machine operators per sq. mile		Percent establishments: consumer serv.
	Transport operators per sq. mile	Population growth	Ten-year population change
	Laborers per sq. mile		Percent living in same country five years earlier
Access to customers, clients	Population per sq. mile		
	Establishments per sq. mile		Percent in migration
Knowledge, R&D base	Post-college per 1,000 sq. miles	*Employment Rigidities*	
		Flexible employment policies	Labor force, percent unionized
	Professionals and technical employees per 1,000 sq. miles	Unemployment	Percent employees without state right-to-work laws
	Patents granted per 1,000 sq. miles		Annual unemployment rate
	Doctorates granted per 1,000 sq. miles		Transfer payments as percentage of total personal income

TABLE 4-3 Continued

Factor and Process	Items	Factor and Process	Items
Wealth/Costs			
Personal wealth	Personal income per capita; Income per household; Dividend, interest, + rent per capita	*Social Diversity* Social status diversity	Educational diversity index; Household income diversity index
Costs of factors of production	Business tax/worker; Local government revenue per capita; Local government debt per capita; Earned income per worker	*Economic Diversity* Economic diversity	Establishment/employee; Occupational diversity index
Efficient public infrastructure	Per capita govt. expense: education; Per capita govt. expense: highways; Per capita govt. expense: welfare; Per capita govt. expense: police	*National Transportation Difficulties* National transportation access	Distance to closest hub airport; Difference in distance to closest and second closest hub airports

Source: Reynolds, Miller, and Maki, 1994a, Table 6.

the standard deviations for four measures—birth and death rates for firms and branches. A firm may be a single establishment, a headquarters, or a branch with a headquarters in the same county. Branches, on the other hand (which were not included in either of the two tables), are establishments owned by firms outside the county.

The summary impact findings show a major importance for the start-up processes depicted by *economic diversity, career opportunity, volatile industries (industry mix),* and *greater personal wealth* on each of the three target variables. *Employment policy flexibility* is also of major importance in firm births and of strong importance in long-term volatility. *Population growth* also has a strong impact on each of the three target variables. *Unemployment desperation* has only a minor impact, and in the case of volatility, the short-term direction is the reverse of the one hypothesized. *Economic system size* has a minor impact on firm births and volatility, but none on firm deaths, while *social status diversity* has a minor impact on firm births and deaths, but none on volatility. The remaining process measures—*factors of production costs; public infrastructure costs; factors of production access; access to customers and clients; access to*

TABLE 4-4 Impact of start-up processes on new firm births and deaths, United States, 1976–88

Factor and Process	Differences Among Regions	Summary of Impact	
Economic Diversity			
Economic diversity	Higher firm birth rates in more diverse economic systems	Major	Firm births
		Major	Firm deaths
		Major	Volatility
Growth/Turbulence			
Career opportunity	More firm births where more midcareer, experienced adults reside	Major	Firm births
		Major	Firm deaths, long-term
		Major	Volatility, long-term
Volatile industries	More firm births where volatile industries are more prevalent	Major	Firm births
		Major	Firm deaths, long-term
		Major	Volatility, long-term
Population growth	More firm births in regions with more population growth	Strong	Firm births, short-term
		Strong	Firm deaths, short-term
		Strong	Volatility, short-term
Wealth/Costs			
Greater personal wealth	More firm births where greater personal wealth is present	Major	Firm births
		Major	Firm deaths
		Major	Volatility, long-term
Factors of production costs	More firm births where input costs are lower	None	
Efficient public infrastructure	More firm births where infrastructure is better	None	
Employment Rigidities			
Employment policy flexibility	More firm births in contexts with greater employment flexibility	Major	Firm births
		None	Firm deaths
		Strong	Volatility, long-term
Unemployment desperation	More firm births where more unemployment is present	Minor	Firm births
		Minor	Firm deaths
		Minor	Volatility, short-term, direction reversed
Agglomeration/Access			
Economic system size	Higher firm birth rates in larger economic systems	Minor	Firm births
		None	Firm deaths
		Minor	Volatility, long-term
Factors of production access	More firm births where production factors are accessible	None	
Access to customers, clients	More firm births where access to customers is convenient	None	
Information, R&D base	More firm births where access to R&D is better	None	
Social Diversity			
Social status diversity	More firm births where there is greater social status diversity	Minor	Firm births
		Minor	Firm deaths
		None	Volatility
National Transportation Difficulties			
National transportation access	Higher firm birth rates where access to national transportation is convenient	None	

Source: Reynolds, Miller, and Maki, 1994a, Table 6.

research and development (R&D), information, and innovation; and *national transportation access*—have no statistically significant effects on any of the three target variables. Lack of statistically significant results is attributed, in part, to the use of total firms rather than some sort of SIC code breakdown of these firms. We review, finally, each of the process variable findings with additional explanatory discussion.

Factors of production costs. Lower input costs provide a competitive advantage to firms when everything else is equal, according to this hypothesis. Lower costs of labor, land, capital, equipment, taxes, and other inputs should encourage business start-ups (Bartik, 1989; Hudson, 1989; Moyes and Westhead, 1990). The RMM findings, using business taxes per worker, local government revenue per capita, local government debt per capita, and earned income per worker as explanatory variables, show no impact on new firm births, deaths, and volatility.

Public infrastructure costs. As more of the infrastructure elements—such as transportation, communications, education, health care, police and fire protection, sanitation, and utilities—are provided in a form appropriate for businesses, they enhance the capacity to start a new firm by reducing the costs or complications of start-up, that is, more firm births where public infrastructure is better, according to this hypothesis (Bartik, 1989; Brusco, 1982). The RMM findings, using per capita government expenditures for education, highways, welfare, and police as explanatory variables, show no impact on new firm births, deaths, and volatility.

Factors of production access. New firms are more likely to be initiated where factors of production already exist, that is, more firm births where input access is easier, and hence information and assembly costs are lower (Bartik, 1989; Hudson, 1989; Moyes and Westhead, 1990). The RMM findings, using per capita demand deposits, per capita savings deposits, percent of individuals with high school diploma only, and workers per square mile (including sales, clerical, service, skilled, machine, transport, and laborers) as explanatory variables, show no impact on new firm births, deaths, and volatility.

Access to customers and clients. More firm births occur where access to customers and clients is more convenient (Bartik, 1989; Brusco, 1982; Mason, 1991; Moyes and Westhead, 1990). This is the best way to be informed about competitors as well as about customer and client needs and preferences. The RMM findings, using population per square mile and establishments per square mile as explanatory variables, show no impact on new firm births, deaths, and volatility.

Access to R&D, information, and innovation. More firm births occur where there is better access to R&D, information is readily available, and innovation flourishes (Oakey, 1984; Hugoson and Karlsson, 1996). The RMM findings, using measures of education and technical knowledge per square mile, that is, post-college enrollments, professional and technical employees, patents granted, and doctorates granted, as explanatory variables, show no impact on new firm births, deaths, and volatility.

National transportation access. Those areas with easy access to major airline hubs have an advantage over rural areas and smaller urbanized areas; that is, higher firm birth rates occur where access to a national transportation system is convenient (Hugoson and Karlsson, 1996; Irwin and Kasarda, 1991). The RMM findings, using airport distance measures (distance to closest hub airport and difference in distance to closest and second closest hub airports) as explanatory variables, show no impact on new firm births, deaths, and volatility.

Economic system size. Higher firm birth rates occur in larger economic systems, according to the economic system size hypothesis (Bartik, 1989). However, some findings show higher firm birth rates in rural areas (Johnson, 1986). This measure is distinct from related measures such as density, diversity, and complexity. The RMM findings, using total population, total labor force, and total establishments as explanatory variables, show minor impact on firm births and volatility, long-term; none on firm deaths.

Social status diversity. More firm births occur where there is greater ethnic-cultural diversity because of the wider range of demand it creates for various goods and services (Morky, 1988). The RMM findings, using an educational diversity index and a household diversity index as explanatory variables, show minor impact on firm births and firm deaths, and none on volatility.

Population growth. Most of those starting new firms have been residents of their county for a substantial period of time—70 percent for more than five years and 50 percent for more than fifteen years. That is, more firm births occur in regions with more population growth (Reynolds, 1994). This reflects a combination of processes, certainly an increase in demand, but also a concentration of talented, motivated people in some regions (Morky, 1988). The RMM findings, using ten-year population change, percent living in same county five years earlier, and percent in-migration as explanatory variables, show strong short-term impact on firm births, firm deaths, and volatility.

Unemployment desperation. Earlier studies in the United Kingdom provided some empirical support for the notion of "recession push," with more firm births being associated with more unemployment. That is, laid-off individuals become desperate enough to start new firms (Hamilton, 1989; Storey, 1982). Later studies in the United Kingdom contradict this (Keeble, Walker, and Robson, 1993). The RMM findings, using the annual unemployment rate and transfer payments as a percentage of total personal income as explanatory variables, show strong impact on firm births, firm deaths, and volatility, with direction reverse; that is, higher levels of unemployment depress firm births.

Employment policy flexibility. More firm births occur in contexts with greater employment flexibility; that is, managers and employees are able to respond well to changes in workforce requirements to compete successfully (Brusco, 1982; Plaut and Pluta, 1983). In regions where employment relationships are more formal and emphasize union involvement, firm births are reduced. The RMM findings, using labor force, percent unionized, and percent of employees without state right-to-work laws as explanatory variables, show major impact on firm births, none on firm deaths, and strong impact on volatility, short-term.

Industry mix (volatile industries). This hypothesis states that the higher the proportion of volatile industries (consumer services, construction, and retail) in the economic base, the higher the new firm birth rates will be. That is, more firm births occur where volatile industries are prevalent (Beesley and Hamilton, 1984; Moyes and Westhead, 1990; Storey, 1982). Some problems occur in separating turbulence that is a feature of an industry sector from turbulence representing growth and change. The RMM findings, using the percent of workforce in selected industries—namely, construction, retail, consumer services, and other services—and the percent of establishments in construction and in consumer services as explanatory variables, show major impact on firm births; firm deaths, long-term; and volatility, long-term.

Career opportunities. Regions with well-educated individuals in midcareer (age 25–45 years) with positions that provide experience valuable for successful entrepreneurial activity (managerial and administrative) have higher rates of new firm births. That is, more firm births occur where more midcareer, experienced adults reside, according to this hypothesis (Bartik, 1989; Johnson, 1986; Moyes and Westhead, 1990; Storey, 1982). The RMM findings, using percent population of individuals age 35 to 44 years, percent with some college, percent with college degree, percent managers, percent professionals, and percent in technical occupations as explanatory variables, show major impact on firm births; firm deaths, long-term; and volatility, long-term.

Personal wealth. More firm births occur where greater personal wealth is present. The basis for impact is probably a combination of the increased demand, which should increase birth rates in construction, retail, and consumer services; a wider range of demand; and the availability of capital for investment in new firm start-ups in all industry sectors (Brusco, 1982; Hudson, 1989; Storey, 1982, 1988, 1993). The RMM findings, using personal income per capita; income per household; and dividends, interest, and rent per capita as explanatory variables, show major impact on firms births; firm deaths; and volatility, long-term.

Economic diversity. Higher firm birth rates occur in more diverse economic regions that provide opportunities to develop new markets and identify suitable suppliers and workers (Chinitz, 1961; Morky, 1988). Also, a higher proportion of smaller firms is assumed to have a positive effect on new firm births. The RMM findings, using employees per establishment and an occupational diversity index, show major impact on firms births; firm deaths; and volatility, long-term. Even in smaller regions, economic diversity contributes to a greater number of firm births.

The summary findings also show substantial variations in new firm start-ups from place to place depending on the particular local social and economic environment. They suggest higher rates of business foundings in the larger metropolitan areas than in smaller ones, but much depends on the social and economic diversity of these areas. This includes the characteristics of the local LMA and the adaptability of the local economy to changing market conditions.

The findings on firm births, deaths, and volatility are consistent with the literature on the role and importance of the major metropolitan core areas as sources of new business formation and expansion. Business retention, a long-standing concern of declining rural areas and central-city neighborhoods, is more aptly viewed as business replacement in the major metropolitan areas. Also, an efficient public infrastructure, while not a sufficient condition for new business growth and development, remains a constraint on future business viability. Important also are the economic and business linkages of rural areas and smaller metropolitan areas with their core metropolitan areas. Measuring the importance of these linkages, for example, with input suppliers, local labor markets, and local and export product markets, requires a more detailed industry breakdown than is available in the studies cited earlier. The analytical frameworks also need reevaluation with reference to the driving forces and the constraining structures affecting business performance in the different local and regional economies.

Exogenous forces generate much of the physical and social infrastructure of an LMA. Revenues collected from the total economy of the governmental jurisdictions in the LMA and from transfer payments finance the infrastructure development. Public transportation and communications systems are part of the physical infrastructure. They are passive but constraining agents of local economic development. They are a necessary but not sufficient condition for economic development. They respond to, rather than lead, business and residential growth. Publicly supported health care and education are part of the social infrastructure of an LMA. Effective health care adds to the quality of life of local residents. It also adds to the productivity of the local workforce insofar as access to a superior health care system reduces lost time due to illness. An effective educational system, on the other hand, is also an active agent of economic development. Studies dealing with investment in the "human capital" document its contribution to the lifetime earnings of individuals.

Notes

1. We owe much of this discussion to Dr. Knut Ingar Westeren (1992).

2. *Understanding Your Community's Economic Base,* by Curtis Baschler, Department of Agricultural Economics, University of Missouri-Columbia; John A. Croll and Byran Phifer, Department of Community Development, University of Missouri-Columbia; and John A. Kuehn, President, Economic Futures, Hartsburg, Missouri. University of Missouri Extension, Community Decision Making publication DM3005. Reviewed October 1, 1993.

3. We validate the indirect measures of gross exports and gross imports with a variety of auxiliary measures of local economic activity and structure. First, product specialization and localization measures show the relative importance of the product in the region and the nation, as well as its geographic distribution within a region. Second, central places and their local labor market measures link the geographic localization of production within a region or an area to its industry structures. Third, regional advantage measures show the propensities to trade, both export and import, of individual, geographically differentiated regional industries. Study of the economic bases of rural and metropolitan areas thus reveals strong production and trade linkages between the two types of areas with the economic vitality of the one being in large measure dependent on the economic vitality of the other.

4. Applying the share of total employment engaged in producing an industry's exports as a link to the industry's total employment yields a measure of the industry's contribution to the local economic base. Using this measure, manufacturing accounts for 50 percent or more of the economic base of all but the North Central and North West LMAs of Minnesota and North Dakota. Agriculture is dominant in the North West LMAs, accounting for 46 percent of the subregion's economic base. Exports thus become the means to acquire an inflow of dollars into the area for purchasing (in large measure from its own metropolitan core area) the many imported goods and services sought by local producers and consumers.

Infrastructure and Transfer Systems

The infrastructure is the basic facilities and services of a community or society. This includes the transportation and communication systems, power plants, waterworks, waste disposal facilities, police and fire protection systems, schools, prisons, and post offices. Some experts extend this definition of infrastructure to include public housing, education, industrial parks, information technology, and various forms of human capital as represented by the total spending of state, local, and federal governments on education, science, and health research. About one-third of the nation's infrastructure is public, two-thirds private. Alternatively, about one-sixth of the total infrastructure is human capital, five-sixths nonhuman.

Using the popular economic jargon, we say the infrastructure of an area is a necessary but not sufficient condition for economic growth and viability. The infrastructure of rural America, for example, serves rural people by facilitating access to opportunity and choice, whether markets for local businesses or jobs, education, or personal consumption for local households. It links the rural community to the metropolitan core area of its region. Its critically important role demands a host of supporting institutions—economic, social, and governmental, among which state and local governments play an increasingly important role. Similarly, the infrastructure of urban America provides access for the suburban resident to jobs in the central city and its downtown district, as well as to the more distant destinations of its international airlines. Whether these facilities and information actually improve the productivity of work in households, businesses, and government depends on the ability and willingness of people and organizations to effectively use these resources and whether these resources are accessible to those who can benefit from their use.

Transfer systems—communication in its many forms as well as the use of air, ground, and water transportation—provide the means to benefit from the available infrastructure of an area or region. They also account for the flows of information, people, and freight originating from one locality and destined to another within a region and between regions, thus linking the economic activity at all origins and all destinations within a global system of interconnected communication and transportation centers. We address these issues by focusing on the demand for and supply of infrastructure investment, the generation of interregional business travel, and the role of the various transfer systems in the diffusion of information and knowledge within economic communities. The findings covered in this chapter, in turn, address the formulation of a framework and model for assessing transfer system performance.

Public and Private Infrastructure

We continue with the central place paradigm of farm-to-market roads, followed by the many forms of rapid travel and communication that support the growth of even larger central places. These are the very largest places, not only in size and areas served, but also in the variety and value of the services offered to their customers. For the most part, the commercial products of the largest places are services that involve face-to-face contact, rather than commodities readily traded by means of long-distance communications. Moving people around from service center to customers and clients and vice versa, and in the process moving information, are tremendous tasks, overshadowing in many ways the challenges and difficulties encountered in road and rail commodity transportation. These changes in economic activity call for corresponding changes in building a region's replacement and new infrastructure.

Building and Maintaining the Nation's Infrastructure

Much of the building of the nation's physical infrastructure was piecemeal. Only the influence of national defense immediately after World War II brought together the necessary support for a massive unified effort in transportation infrastructure investment—the building of the interstate highway system. This was a long-term program that affected the economic well-being of population and industry in both rural and urban areas. Establishment of the interstate highway system was followed by a second unifying influence in the transportation infrastructure—the work of the Federal Aeronautics Administration (FAA) in planning for and implementing an air transportation system, centered on metropolitan core areas remarkably similar in their economic structures. These two unifying accomplishments in transportation systems planning eventually solidified the central role of the two dozen or so high-order distribution and producer services centers in the U.S. economy. These centers now serve as the cores of their respective economic regions.

However, countervailing policies (the 90 percent federal financing of approved development of the local urban infrastructure, including highway and sewer and water systems, and the expensing of interest payments on home mortgage loans) subverted the density-increasing potential of concentrated urban settlement. This skewed residential development in the metropolitan region to the low-density periphery of its daily commuting area, thus greatly increasing the overall costs of metropolitan governance.

The 2.4 percent growth in residential capital stock in the 1980s was below the overall national growth rate of 2.7 percent, as shown in Table 5-1. It dropped to 1.2 percent in the 1990s. The overall growth rate in capital stock fell by a fourth of its 1980s rate, that is, from 2.7 to 2.0 percent per annum. Stocks of durable goods owned by consumers, on the other hand, grew at twice the rate of residential stock. During the 1990s, these differences became smaller, as they did generally among all categories except government. The federal rate dropped from 1.5 to 0.2 percent, while state and local capital stock grew from 2.1 to 2.6 percent. Meanwhile, residential stock and durable goods owned by consumers increased from 42.9 percent of the total in 1980 to 43.9 percent in 1990 and 44.4 percent in 1995.

According to some accounts, we face a "crumbling" and "decaying" infrastructure, which has become our "third deficit"—the first two being the budget deficit and the trade deficit (Aschauer, 1991). The trade deficit, according to

TABLE 5-1 Total capital stock, U.S., 1980–1995 (in 1992 dollars)

	Real Net Stock			Annual Growth Rate	
Type of Capital Stock	1980 (tril. $)	1990 (bil. $)	1995 (bil. $)	1980–90 (pct.)	1990–95 (pct.)
Total	14,269	18,586	20,556	2.7	2.0
Fixed private capital, total	9,950	12,890	14,171	2.6	1.9
Nonresidential, total	5,033	6,650	7,280	2.8	1.8
Equipment	1,855	2,507	2,886	3.1	2.8
Structures	3,177	4,142	4,399	2.7	1.2
Residential	4,921	6,240	6,891	2.4	2.0
Fixed government capital, total	3,124	3,778	4,147	1.9	1.9
Federal	969	1,126	1,137	1.5	0.2
State and local	2,156	2,652	3,014	2.1	2.6
Durable goods owned by consumers	1,198	1,919	2,241	4.8	3.1

Source: *Survey of Current Business* 77(5): 92, Table 15, May 1997.

Aschauer, is a manifestation of "a continuing slump in the growth rate of economic production." Other reasons for the trade deficit are "a low profit rate" during the 1970s and the 1980s and "a low rate of net private investment."

Aschauer viewed the third deficit as central to our long-term problem in declining productivity. His findings—based on a standard neoclassical production function of the form $y = f(k, k^*)$, where y denotes private sector output per worker, k denotes private capital per worker, and k^* denotes public capital per worker—show a close relationship between total productivity and public capital stock. He found that a 1 percent increase in core infrastructure capital (of highways, mass transit, airports, electrical and gas facilities, and water and sewer systems) will raise productivity by 0.24 percent. In 1987, gross domestic product was estimated at $4.5 trillion, while the core infrastructure was $1.2 billion, or 18 percent of total capital stock in 1987. Other public nonmilitary investment was estimated at $691 million, or 10 percent of the total. Aschauer constructed a minimal mathematical model for simulating the effect of higher public investment on gross domestic product. Of course, the capital stock included in the Aschauer model excluded residential and military fixed capital, and consumer durable goods.

The simulation data showed an apparent correlation among increases in three aggregate measures of investment and productivity. The level of public nonmilitary investment was increased by 1 percent of the private capital stock during the 1970–88 period. This amount was 125 percent greater than the actual amount, comparable to the actual rate during the 1950s and 1960s. The simulated results were 1.7 percent higher than actual (of 7.9 percent) on return to private capital, 0.6 percent higher than actual (of 3.1 percent) on private investment capital stock, and 0.7 percent higher than actual (of 1.4 percent) on productivity growth per annum. Comparing this period (1970–88) to the 1953–69 period, when the actual rates were consistently higher, Aschauer concluded that "the rate of return to private capital could have been only 1.1 percentage points lower (instead of 2.8 percentage points); private investment could have been only 0.1 percentage points lower (rather than 0.7 percentage points lower); and annual productivity growth could have been 0.7 percentage points lower (instead of 1.4 percent lower)" (p. 23).

Overspending in Public Infrastructure

Much of the criticism of the findings of an infrastructure gap stems from the highly aggregate and simplistic nature of the analytical models. The Aschauer two-equation model (Aschauer, 1991, p. 31) shows (1) the growth rate of the net private capital stock as dependent upon its own lagged value, the rate of return to capital, and investment in public nonmilitary capital and (2) the rate of return to private capital as dependent upon time, the private capital-to-labor

TABLE 5-2 Selected economic indicators, U.S., 1980–1990 (in 1987 dollars)

	Total *		Per Worker		Annual Chg.	Marg. Return
Economic Indicator	1980 (tril. $)	1990 (tril. $)	1980 (thou. $)	1990 (thou. $)	1980–90 (pct.)	1980–90 (pct.)
Total capital	10.8	14.4	109.0	123.0	2.5	15.8
Private capital	7.4	9.7	74.6	82.4	2.0	22.1
Public capital	3.4	4.8	34.4	40.6	4.0	7.0
Nonhuman capital	9.0	11.9	90.3	101.6	2.2	15.7
Human capital	1.9	2.5	18.7	21.4	4.9	13.8
Real gross domestic product	3.8	4.9	38.0	41.8	1.2	21.2*
	(mil.)	(mil.)			(pct.)	
Employed persons	99.3	117.2	NA	NA	1.8	NA

Source: Petersen, 1994.
*Employed persons times output per worker values. Marginal product of labor in thousands (1987) dollars.

ratio, the public nonmilitary capital-to-private-labor ratio, and the capacity utilization rate. One criticism refers to the model coefficients that yield four times the productivity impact of public over private infrastructure investment. The rate of return to public capital in the range of 50 to 60 percent is much higher than expected from related cost-benefit analyses. Another criticism is the assumed independence of changes in employment and capacity utilization. Public spending is likely to increase the general employment level through its demand-side multiplier effects that result in higher rates of capacity utilization.

Criticisms of the third-deficit hypothesis come from two sources—one that follows the aggregate approach in data and method, and one that uses detailed regional breakdowns of the nation's aggregate infrastructure. The aggregate approach followed in Table 5-2 starts with a breakdown of total capital stock into private and public, nonhuman and human, for the ten-year period from 1980 to 1990. The stock of private capital is the sum of real gross private domestic investment over the preceding fifteen years. Public investment by state, local, and federal governments consists largely of buildings, infrastructure, education, and science, all of which, according to Peterson (1994, p. 66), should be more durable than machines. Data for both private and public investment come from the Economic Report of the President (1982 and preceding years). The public capital stock constructed by state and local governments is the result of expenditures on education and highways, plus half of all other spending (namely, libraries, hospitals, health, employment security administration, veterans' services, air transportation, water transport and terminals, parking facilities, transit subsidies, police and fire protection, correctional institutions, protective inspection and regulation, sewerage, natural resources, parks and recreation, housing and community development, solid waste management, financial administration, judicial and legal services, general public buildings, other government administration, interest on debt, and other miscellaneous expenditures).

While the main difficulty of capital measurement is the assessment of depreciation, comparisons between studies also are difficult because of differences in the definitions of capital stock. Growth rates differ, in part, because of these differences in the composition of capital stocks and the measurements of individual components. The unique contribution of the Peterson data[1] and model[2] is the comparison of marginal returns from the two categories of capital stocks that make up the total, whether these are private and public or nonhuman and human. The marginal returns are highest for private capital and lowest for public capital. They are slightly higher for nonhuman than for human capital. Yet the share of total capital for two lower-ranking capital stocks was higher in 1990 than in 1980.

Studying the above findings, Peterson concluded (p. 71) that "The relative decline in the estimated marginal rates of return on human and public capital to levels below the return to nonhuman and private capital is cause for concern, particularly if one takes seriously the call for increase in spending on education and infrastructure." Peterson noted an imbalance in the mix of capital. "These results suggest," according to Peterson, "that policies that encourage the relative growth of private and nonhuman capital would do the most to stimulate economic growth and reduce unemployment. Such policies might include greater reliance on expenditure taxes and less on income taxes, and/or a reinstatement of the investment credit provision in the income tax code." The implications of these results, of course, run counter to the earlier studies that suggested an infrastructure deficit.

Infrastructure Spending Differences

The second source of criticism in the infrastructure deficit debate comes from an urban regional perspective based on a more disaggregative approach, which takes into account place-to-place differences in the levels and types of infrastructure investment. Sanders (1993, p. 4), for example, took a contrary view by asking "What infrastructure crisis?" With this approach, the salient arguments presented earlier are listed as four myths, starting with the myth of declining spending. He noted, first, that spending on core infrastructure is almost entirely an activity of state and local governments. Highways and streets make up by far the largest spending category. Yet they are only part of the large transportation infrastructure that, in turn, belongs to the even larger complex of transportation industries.

The commercially operated transportation industries—railroad transportation, local and interurban passenger transportation, trucking and warehousing, water transportation, air transportation, pipeline transportation, and other transportation services—produced $250 billion worth of services in 1990, valued in 1982 dollars. Moving supplies to producers accounted for $163 billion of the total $280 billion in transportation purchases from both private and common carriers, while moving people and products for final use and export accounted for the remainder. Household spending for personally owned transportation added another $253 billion to the 1990 figures. Finally, state and local governments spent $46 billion on transportation infrastructure and related services in 1990. Thus, the total transportation bill for the U.S. economy was at least $580 billion (in 1982 dollars); this estimate does not take into account much additional private transportation that escapes statistical tabulation. Even then, the U.S. transportation bill was equivalent to fourteen percent of the gross national product in 1990.

Because transportation is a necessary but not sufficient condition for economic growth, it lacks the urgency of concern expressed about other expenditures that directly enhance the success of business enterprise in competitive global markets or the quality of life for the residents of a regional community. Much of commercial transportation operates at levels that return revenues less than costs in highly competitive local and regional markets. Noncommercial transportation provided by households and government, including transportation infrastructure, accounts for 65 percent of total transportation outlays in the U.S. economy. Thus, nearly two-thirds of the transportation spending is determined by values outside of competitive business enterprise.

However, excessive spending on transportation adds to the costs of doing business and competes with other forms of spending by households and governments. An overriding concern affecting transportation spending, therefore, is its opportunity costs—the benefits lost from spending the nation's gross domestic product on excessive transportation facilities and services.

Institutional factors account for many of the current difficulties in optimizing public and private investment in transportation infrastructure. The institutional factors include (1) state and local subsidy of urban fringe infrastructure and federal tax expenditures (that is, deduction of interest payments on home mortgages) for residential housing, (2) exclusionary use of subdivision and zoning regulations, (3) public subsidy of long-haul trucking, (4) restrictive rules and practices of railroads, and (5) long-standing politicization of federal public works programs and spending. These intrusions into land use decision processes at the local level often result in inequitable and inefficient land uses that add measurably to transportation costs and, in turn, to the private and public costs of urban infrastructure and related services. The "crisis of the cities" is, in large part, the result of these policies.

Total government spending includes intergovernmental transfers as well as direct outlays for operations and physical capital. In the period from 1986 to 1991, for example, federal capital spending increased from $195 to $227 billion, but its share of all government spending, which increased from $1.8 to $2.4 trillion, dropped from 11 to 10 percent. Federal spending dropped from 63 to 62 percent of the total. Its capital spending dropped even more—from 50 to 42 percent of the total capital spending. Thus, state and local governments were left with an increased share of and responsibility for governmental capital spending.

Table 5-3 focuses only on the capital outlays—in the infrastructure of rural and urban America—and their growth and change over the four-year period from 1986–87 to 1990–91 (as defined by the listed spending categories). Sharp reductions in capital outlays for national defense and the postal service accounted for the absolute decline in federal government capital outlays during this period. State and local government capital outlays increased from $98.4 to $131.6 billion—a 34 percent increase. Federal government capital spending for

TABLE 5-3 Total capital outlays for specified activities, by type of government, 1987–88 to 1990–91

	Total		Federal		State		Local	
Item	1986–87 (bil. $)	1990–91 (bil. $)	1986–87 (bil. $)	1990–91 (bil. $)	1986–87 (bil. $)	1990–91 (bil. $)	1986–87 (bil. $)	1990–91 (bil. $)
Total	195.2	227.2	96.9	95.6	37.3	47.9	61.1	83.7
Selected federal programs	83.1	79.2	83.1	79.2	0	0	0	0
Education services	18.2	28.1	0	0.1	6	6.9	12.2	21
Social services and income maintenance	6.4	7.8	2.9	3	1.6	2.1	1.9	2.7
Transportation	33	43	1	1.4	21.5	27.4	10.5	14.2
Public safety	5	7.4	0.4	0.5	1.9	3.2	2.7	3.8
Environment and housing	24.7	31.5	7.9	9.4	2	3	14.7	19.2
Government administration	3.5	5.2	0.4	0.7	1	1.3	2.2	3.2
General expenditures not elsewhere classified	5.8	7.9	1.1	1.3	1.2	1.5	3.5	5.2
Other expenditure	15.4	17	0	0	2.1	2.5	13.3	14.5

Sources: U.S. Bureau of the Census. Government Finances in 1986–87, Series GF-87/5, and Government Finances in 1990–91, Series GF-91/5; U.S. Government Printing Office, Washington, D.C., 1988 and 1993.

the same items increased from $112.1 to $146 billion—a 30 percent increase—slightly less than the overall increases in direct spending. Intergovernmental outlays of the federal government, on the other hand, increased by 44 percent—from $111.4 to $160.1 billion—during the same period.

Sanders (1993) argued that the spending trends for the core infrastructure contradict arguments about public inattention and government failure. These spending trends are clearly positive for streets and highways, as well as sewer and water systems. Airport spending also is up. The clear exception is mass transit, which averaged about $1.5 billion in the 1980s, but even here the spending in 1991 was above that of previous years at $1.7 billion.

Table 5-4 focuses on the shares of total direct spending among governmental activities accounted for by capital outlays at each governmental level in the two periods (1986–87 and 1990–91). While total capital outlays as a percentage of all spending declined for both federal and state governments, it increased for local governments. However, share increases, representing budgetary reallocations, occurred for selected activities at all government levels. At the federal level, the environment and housing share increased from 8.4 to 14.9 percent. It

TABLE 5-4 Share of total spending for specified capital outlays, by type of government, 1987–88 to 1990–91

	Total		Federal		State		Local	
Item	1986–87 (pct.)	1990–91 (pct.)	1986–87 (pct.)	1990–91 (pct.)	1986–87 (pct.)	1990–91 (pct.)	1986–87 (pct.)	1990–91 (pct.)
Selected federal programs	23.1	18.7	23.1	18.7	0.0	0.0	0.0	0.0
Education services	7.5	8.4	0.0	0.5	9.7	8.5	7.3	9.0
Social services and income maintenance	3.2	2.6	4.6	3.5	1.8	1.5	3.9	3.9
Transportation	50.0	51.2	16.7	15.6	65.2	68.5	38.9	40.6
Public safety	8.1	8.4	8.0	6.3	11.2	12.3	6.9	7.2
Environment and housing	16.7	22.8	8.4	14.9	18.2	20.0	34.2	32.0
Government administration	8.0	8.1	4.4	4.4	7.7	6.8	10.0	10.7
General expenditures not elsewhere classified	9.5	6.2	4.6	1.6	7.1	7.1	17.5	19.3
Other (utility, liquor store)	22.3	21.0	0.0	0.0	26.3	25.0	22.2	20.1
Total	15.6	13.9	16.9	13.6	14.8	13.5	14.2	14.5

Sources: U.S. Bureau of the Census. Government Finances in 1986–87, Series GF-87/5, and Government Finances in 1990–91, Series GF-91/5; U.S. Government Printing Office, Washington, D.C., 1988 and 1993.

increased also for state government but declined for local government. At the state level, the transportation share increased from 65.2 to 68.5 percent. It also increased for local government but declined for the federal government.

The changing sources of funding fail to show the corresponding changes in the criteria for identifying and selecting high-priority infrastructure investment projects. The restructuring of the nation's economy and the accompanying changes in industry location and population distribution, especially in rural regions, leaves many roads, bridges, and other facilities largely unused. However, these are nonetheless maintained in top-notch condition. State and local infrastructure planning efforts remain piecemeal and uncoordinated. There still is no clear consensus about the implications of these changes for the varying local demands for the many different kinds of infrastructure investments that most clearly affect the well-being of rural residents and their local economic base. An understanding of the role of infrastructure in rural America is incomplete, therefore, without some sort of corresponding appreciation of the economic geography of rural America. Such an understanding can serve as a guide when sorting the infrastructure requirements of different rural areas located various distances from metropolitan areas.

Many, but not all, rural areas now face continuing disinvestment in rural infrastructure as farm numbers fall. In some areas, however, falling farm numbers foreshadow a new wave of local infrastructure investments. These investments link local manufacturing establishments with product markets and input supply sources in the metropolitan core area. They also connect with the individual households in the local labor market area (LMA). They include producer services, like banking, finance, management, and accounting, as well as new forms of transportation and communication that further facilitate the linkage functions of local infrastructure. The new technologies, global in scope yet local in consequences, radically change the mix of infrastructure investments being sought for rural areas.

Finally, the new infrastructure brings metropolitan area residents as seasonal visitors to rural areas with important recreational amenities and attractions. Some of these visitors build or buy second homes and even establish permanent residency. They thus contribute to an expanding local business economy. They also may engage in a common effort with other metropolitan area residents to conserve the nearby agricultural areas that are vulnerable to soil and wind erosion. They may even confront local residents with new efforts to preserve natural habitats for wildlife populations. Meanwhile, the overspill of population and industry from the metropolitan core area to its urbanized periphery continues, adding to the mounting costs of new infrastructure demands by increasingly powerful suburban constituencies.

Public infrastructure expenditures include federal, state, and local government spending on transportation facilities and services; education and training; water, sewer, and solid waste disposal systems; and essential municipal facilities and services. State and local infrastructure expenditures generally include other public facilities besides transportation. Public education also is viewed as an integral part of public infrastructure. Direct spending of $341.8 billion on state and local infrastructure over the eight-year period from 1992 to 2000 amounts to 5.5 percent of total state and local government expenditures or 3.5 percent of total federal government expenditures in the 1990s. It is also equivalent to 34 percent of total grants in aid from federal to state and local governments projected in the "moderate" Bureau of Labor Statistics scenario.

In 1990, education accounted for $187 billion (41 percent) of the $453 billion total state and local government purchases of industry output in 1990 (Table 5-5). Transportation and community infrastructure together accounted for only $71 billion (15 percent) of the total. Education sector spending is disbursed more widely than transportation and other infrastructure expenditures. The distinguishing characteristic of transportation and other infrastructure expenditures is their concentration in the construction industry. The construction

TABLE 5-5 State and local infrastructure purchases of specified industry output, U.S., 1991

Sector	Total (pct.)	Education (pct.)	Trans-portation (pct.)	Comm. Facilities (pct.)	Other (pct.)
Agriculture, forestry, fisheries	0.4	0.4	0.0	1.6	0.3
Mining	0.1	0.1	0.2	0.0	0.1
Construction	18.8	9.0	78.9	46.9	10.7
Manufacturing	11.5	12.3	5.3	10.2	12.4
Durable	4.4	4.1	4.2	5.9	4.6
Nondurable	7.1	8.2	1.1	4.7	7.7
Transportation, communications, utilities	4.9	6.3	2.2	5.5	4.2
Trade	0.8	–1.4	0.7	2.3	2.6
Finance, insurance, real estate	3.4	0.7	0.0	1.2	7.1
Private services	7.0	3.4	25.8	5.1	6.3
Government	53.2	69.3	17.3	45.3	47.0
Total	100.0	100.0	100.0	100.0	100.0
Sector distribution	100.0	41.3	9.9	5.7	43.1

industry and related input suppliers are the principal industry beneficiaries of increased state and local infrastructure spending.

The education infrastructure funding is largely in the elementary and secondary education category. Wage and salary payments account for the largest share of total outlays. These payments are recycled as personal income disbursements—personal consumption expenditures, personal taxes, and personal savings. Thus, each of the three education expenditure categories disburses its expenditures among (1) institutional and student and (2) faculty and other staff purchases of goods and services.

Highway construction and maintenance dominate transportation infrastructure spending. Purchases of construction industry products and services account for more than 92 percent of total expenditures.

Other infrastructure expenditures are for electric, gas, and sanitation utilities; water supply; urban facilities, and natural resource-related functions. Construction expenditures again dominate the total infrastructure spending. Natural resource-related infrastructure expenditures are more widely disbursed than utility, water, and urban facility expenditures because of the larger outlays for manufactured products and transportation, communication, and public utility services.

Infrastructure development also should face its own "means" tests in measurable contributions to productivity improvements—first, in its own backyard, and second, in the private sector. In such decision environments, short-term political choices would be compared with their long-term economic costs.

A second infrastructure myth, according to Sanders, is that the infrastructure deficit requires a national solution. The bridge problem is cited as an example of geographic concentration. About 13 percent of the 276,000 bridges in the federal aid system were described by the Federal Highway Administration as "structurally deficient" as of mid-1990. At least 40 percent of these bridges are located in only six states. New York leads the list with more than 55 percent of its 9,000 bridges being classified as structurally deficient. The same sort of uneven distribution exists for other highway needs, like pavement conditions, with only 11 percent of the nation's highways being of "poor" quality. For Idaho, Rhode Island, and Mississippi, however, the percentages are 27.7, 26.8, and 22.9, respectively.

A third infrastructure myth is that fixing the "crumbling" infrastructure will be very expensive. Needs estimates from groups such as the Associated Contractors of America are often in the trillions of dollars (Sanders, 1993, p. 8). An important problem is the difficulty of defining "need" when it changes from year to year depending on current preferences and priorities. The Federal Highway Commission estimates of highway investment needs completed in 1991, for example, added up to a total cost of $400 billion for eliminating the backlog of highway deficiencies. According to Sanders, annual spending to meet this need would exceed $60 billion per year—more than three times current levels (p. 10). However, 72 percent of the amount relates to congestion and traffic-carrying capacities—largely a function of the mismatch between place of work and place of residence. Only 28 percent of the spending would be directed to improving or maintaining existing pavement quality.

A final infrastructure myth is that state and local governments are too strapped for money to finance their infrastructure needs. As the earlier figures show, state and local capital spending is steadily increasing, even through recession, the result partly of the methods of financing public works (Sanders, 1993, p. 11). Highway building is financed largely by motor fuel and vehicle taxes that are tied to car ownership and mileage. Water and sewer system financing is tied to dedicated user fees. Bond issues that generally support infrastructure spending are tied to revenue sources that are independent of national economic conditions. State and local governments sometimes assert choices, like building new sports stadiums and business convention centers, given that long-term debt financing requires public approval. Approval for new infrastructure development may not be forthcoming because of a lack of community leadership or an overriding preference for the alternative spending choices. Sanders suggests that all levels of government "should invest seriously in reasonable renewal and rehabilitation efforts rather than in new development" (Sanders, 1993, p. 12).

Technology Transfers

Technology transfers occur largely within the market economy. Nonetheless, the esteemed research university retains its legendary value as a source of knowledge essential to community and national well-being. Research universities experienced tremendous growth and change in the years following World War II. People everywhere shared a common belief in a college education as a way to get ahead and enjoy the benefits of the new technologies that evolved, in part, from the land-grant research laboratories. The GI Bill supported the college education of many returning veterans from the armed forces, and national agencies were formed to support arts and humanities programs as well as science and health research. Rising personal incomes supported even higher rates of growth in state tax revenues, particularly the share going to the state and land-grant universities.

This common belief in a college education for all who can benefit from it goes even further within the university community, given the premise that a state must have a first-rate land-grant research university (1) to enrich our quality of life and allow us to compete effectively in the twenty-first century and (2) to serve as a unique meeting ground for scholars, students, and the general public, where learning involves the creation as well as the transmission of knowledge. Yet, an apparent lack of effective technology outreach programs for bridging the "lab-to-market" gap that is of great concern to many critics of state research universities. With peer review and frequent citation of publications in the prestigious journals of a particular discipline being the primary criteria for academic advancement, the possible intrusion of commercial market criteria into this process understandably alarms many currently successful academic researchers. Technology outreach efforts thus become of secondary importance to maintaining existing academic advancement criteria that have very little, if any, relevance to the underlying issues in technology transfer.

Assessing industry demand for technology research within the state or region served by the university and squaring the findings with the university's actual research products is, therefore, a first (although potentially controversial) step in building a technology outreach program. Once the technology outreach identifies an industry demand for a university research product, the bridging process is ready for the next difficult steps: communicating the findings of the university's technology research and organizing industry support systems for these efforts. Although difficult, the additional efforts appear consistent with a land-grant university's mission to improve the economic and social well-being of the people and institutions that support it, and to gain their confidence in the aims and efforts of the new land-grant university.

Forming university-business partnerships within a state or region is one way of gaining support for a land-grant university's research and technology outreach programs. These partnerships may occur around many researchable topics. The primary purpose of the partnerships is to provide the initial understanding of the critical issues confronting state or regional voters and residents, such as identifying ways to nurture export-producing businesses—particularly small and start-up firms. Success in achieving this primary purpose, more likely than not, becomes one of many steps before the combined efforts can lead to successful programs of technology research and outreach.

Transfer Systems in Urban Regional Growth and Change

Transfer systems serve directly and decisively in furthering urban-regional growth and change. For example, the construction of important local highway linkages between rural communities and their metropolitan core areas follows the urbanization of metropolitan regions. This process is accompanied by the gradual expansion of the urban periphery, the conversion of rural into urban land uses, and the disassociation of place of work and place of residence for a large proportion of the metropolitan area population. The devising of roadway development options at this stage of a metropolitan area's outward expansion is a process of best serving the interests of both local residents and the larger economic community, yet acknowledging the concerns of property owners adversely affected by one or more of these options. Concerns about urban sprawl should be addressed at an earlier stage of the metropolitan area's outward expansion. At this later stage, the people are in place, have jobs, generate above-average levels of disposable income, and support a growing number of local businesses, thus reducing their dependency and their neighbors' dependency on more distant shopping centers and places of work. The reconciling of development options with other local concerns thus starts with an awareness of what is already in place that may be affected by the proposed roadway development.

Competition and Development in Metropolitan Fringe Areas

This section addresses the growing concerns of local governments and residents about the high costs of urban spillover into adjoining rural areas. From a rural agricultural perspective, we focus on the loss of productive agricultural land and open space. From a metropolitan core city perspective, the focus shifts to the erosion of the city tax base and its fiscal capacity to pay for the associated high costs of neighborhood decline. From a personal and private perspective, however, urban growth creates new opportunities for residential and commercial development. Many sectors of the local economy share in these opportunities, including young families seeking their first single-family residence at a price they can afford. Given the multiplicity of concerns, the proposed solutions for managing urban metropolitan growth are many, but the outcomes are essentially the same: the converting of rural agricultural areas into suburban and open-country settlement continues unabated.

These issues bring along a host of concerns about changing rural land values and the methods of their appraisal and sale. Local regulatory measures like zoning and building codes, as well as distance from a metropolitan core area, affect land values in the urbanizing periphery even more than their agricultural productivity does. Because of the extreme difficulty of accurately anticipating the political and economic changes and weather conditions affecting local land values, property managers reduce their risks and uncertainties by diversifying their rural land portfolios. The individual farmer, of course, lacks this option.

While a large part of rural and metropolitan area infrastructure is local, the federal government has influenced its development in several ways, both nationally and locally. At the same time the interstate highway system was being built, several large automobile and tire manufacturers entered into a successful campaign to destroy the municipal trolley systems and replace them with private, individually owned automobiles. Thus, the combination of the federal government subsidizing interstate trucking and the loosely organized consortium of auto-related industries ended the city-building influence of fixed-rail transportation. The deregulation of the railroad industry came too late to allow the industry to acquire attitudes and competencies for coping with the new competitive forces affecting both short- and long-distance rail transportation.

Much of the current transportation infrastructure is based on farm-to-market roads and farm commodity shipments by truck, rail, and barge. Yet these shipments account for only a small part of the total value of all commodity traffic, and even a smaller part of all transportation expenditures. Simple market tests of value added by each transportation system investment clearly deserve consideration in setting system priorities. The alternative is a rapid rise in total transportation costs—direct and indirect, and especially those hidden in the escalation of personal travel expenditures of households and businesses that add to the total costs of exports produced in the metropolitan core area and of doing business in the area in competition with businesses elsewhere. For example, transportation investments in the periphery of the urbanized metropolitan area invariably subsidize urban sprawl at the expense of central cities and the oldest suburbs. This increases congestion and environmental pollution faster than it is reduced by energy-efficient, pollution-free automobiles, and amounts to further subsidy of fast-track access routes from exurbia to the city center, which, in turn, subverts the critical role of core areas as engines of economic growth by burdening them with high governmental costs supported by a shrinking tax base.

In the mid-1970s, as mentioned earlier in Chapter 2, the Twin Cities Metropolitan Council put into effect its Metropolitan Development and Investment Framework to manage fringe-area growth within its seven-county jurisdiction. Its plan constituted a contiguous growth development strategy focusing on managing fringe-area growth by confining metropolitan sewer, highway, and transit service to the Metropolitan Urban Service Area (MUSA). Its chief benefits were cost-efficient investment in public infrastructure and long-term preservation of agricultural areas. Overall, the plan appeared to work with 93 percent of the area's development in the 1980s occurring within the area intended for it. The area's growth in single-family households surged in the late 1980s and 1990s, however, particularly on the rural-urban interface. The council reevaluated its investment framework in the mid-1990s, replacing it with its Regional Blueprint. This plan established five goals for the area, including a "strengthened sense of community." Other goals were economic growth and job creation, reinvestment in distressed areas, preservation of the environment, and sound regional infrastructure investment. The council is now working on the setting of priorities among the goals and a physical development map for the long-term development and redevelopment of its area.

The current MUSA line is under growing pressure from area builders, with the strong likelihood that its extension would be supported by the next state legislature as well as courts of law. The council lacks access to effective market incentives that would help maintain the current MUSA line, given the increasing demand for residential space, coupled with rising land costs and other "push factors" operating from the two central cities and the first-ring suburbs. The likelihood of adopting various nonmarket incentives for containing urban sprawl, either directly (for example, by reducing residential mortgage interest payments and local infrastructure subsidies) or indirectly (for example, by improving schools, reducing crime, making streets safe, and rebuilding central city neighborhoods) is nil, at least in the next several years, when the pressures for extending the MUSA line will reach their crisis stage.

The current plans to widen and extend access roads to the downtown district from rural areas beyond the MUSA line, when mapped, however, seem to arbitrarily end at the boundary of the seven-county metropolitan area. We must remember, however, that the state transportation department has its own statewide roadway development plans for the seven-county area that extend beyond the seven counties. The council's roadway development plans thus account for only part of the roadway development plans covering the entire metropolitan commuting area. This is another manifestation of the council's difficulties in managing urban growth, given its present structure and responsibilities.

Alternative approaches to encouraging residential development within the MUSA line include various measures focusing on the environment and the adverse effects of commuter traffic and urban sprawl on air and water quality in the fringe localities. Such measures depend, however, on the availability of and access to rapid public transit, largely in lieu of the private automobile, for work-related and entertainment-related trips. Shopping by computer, coupled with low-cost delivery of groceries and other frequently purchased items, could reduce the current high dependency on the private automobile. However, measurement of these changes and the related costs and benefits and the assessment of their incidence among households and localities remains among the more challenging tasks of our times (Bromley, 1989, pp. 37–80).

The transfer of property rights from agriculture to other land uses is rampant among present and prospective land subdividers in the rural-urban periphery, as the business survey findings cited earlier show. Karen Polenske, regional economist and urban planner at Massachusetts Institute of Technology, believes that the introduction of property rights and geographic space into the formulation of regional development strategy is essential for policy makers to effectively deal with the related issues of power and control in the economic development process (1993, p. 18). She argues that the issues relating to property rights need systematic study. The distributional effects of changing property rights, the institutional changes they precipitate, and the consequences of these changes for the societal problems addressed by the area and regional development strategies are increasingly important to people and regions and, in Polenske's view, should be incorporated into these strategies. Geographic space and proximity to various producer services and markets also affect the exercise of the power and control emanating from certain property rights and should be included in the formulation of any new development strategy.

When we identified the key economic issues facing transitional rural agricultural areas, we found that many current owners of agricultural land no longer view themselves as commodity producers, but as specialty producers or even low-profile land speculators and developers. The real issue thus becomes one of managing the urban-rural interface in the "common interest." Similarly, the real issue in the rural-urban competition for labor is one of integrating the transitional rural agricultural area into the new economy that extends well beyond the metropolitan core area. The third real issue is agriculture's long-term viability and its successful integration into the daily life of the expanding economic community of which it is already an integral part.

We know too well the consequences of the changing economic conditions: rural depopulation, central-city implosion, and residential relocation to the expanding urban periphery, where property taxes are low, streets are safe (or at least perceived so), schools are good, and access is virtually free to many rural amenities and high quality-of-life attractions. Exacerbating all this is the push factor of land values being 100 times or more greater in the central city than at its periphery on one hand, and on the other, the mounting costs, for those left behind, of paying for a growing inventory of vacant land that is inaccessible or unavailable for new residential use.

The reported findings support the introductory statement that, given the multiplicity of concerns, the proposed solutions for managing urban metropolitan growth are many, but the outcomes are essentially the same: the conversion of rural agricultural areas into suburban and open-country settlement is likely to continue. New forms and strategies of regional governance that can more efficiently and effectively restrain individual opportunism and resolve problems of collective action are emerging in the United States and Italy, but they face formidable barriers in their nurture in each of the cited policy domains, whether metropolitan governance, infrastructure development and service delivery, environmental protection and property rights, or organizational structure. New market-driven organizational structures may emerge that combine the location diversity of their land holdings with the consolidation of small and abandoned farms into larger land units, thus reducing the risks and uncertainties of holding strictly agricultural lands in their portfolios. What remains from the sort of analysis and assessment that we have been presenting is a realistic understanding of the controlling market forces affecting the directions in which the agri-industrial sector is likely to evolve in the various economic regions of the United States. We can only hope that this understanding will provide opportunities for ameliorating the adverse effects of urban fringe development.

Transportation and Communication Systems

A region's location in the national and global settlement and trading systems imposes severe constraints on transportation and communication options. A rural LMA located well beyond the outer commuting limits of any metropolitan LMA, for example, has diminished prospects for long-term economic viability beyond the lifetimes of its principal product cycles. These findings come from comparisons of the contrasting labor earnings and employment experience of selected core versus selected peripheral LMAs in the United States. A series of statements contrasting the two types of areas—core and periphery—summarize these findings as follows:

1. Peripheral rural LMAs are overwhelmingly dependent on the utilization of local natural resources. Efficiency in the conversion of primary resources into finished products reduces the demand for primary products and places many peripheral areas at risk. Often cited also, but less evident, is the decoupling of advanced manufacturing from primary production. In any event, advanced manufacturing clearly is skill-dependent, which favors industry location in core metropolitan areas and adjoining rural areas and in new industrial spaces in formerly peripheral areas now anchored to cities that serve as small-scale metropolitan core areas.
2. Transitional LMAs are exceptions to the overall pattern of continuing industry specialization. They are close enough to the metropolitan core area to gain new industry, particularly new businesses of industries branching from the metropolitan core area to low-cost sites in nearby rural areas. Also, a new, diverse base economy is emerging in the transitional LMAs because of metropolitan core area businesses subcontracting with transitional area businesses. Thus, transitional rural areas experience high income growth, high income volatility, and high business volatility.
3. Metropolitan LMAs, with the exception of areas marked by negative industry mix and regional share values in a highly specialized base economy, generally are the fastest growing in labor earnings. At the same time,

income volatility may range from the lowest-order to among the highest-order LMAs. A high degree of dependency on a specialized base economy would still sustain high income growth, as shown by strongly positive industry mix and regional share effects. Business volatility is generally high in metropolitan areas.

Peripheral LMAs dominate the standardized and readily tradable products cluster. The metropolitan LMAs dominate the nonstandardized, less readily tradable products cluster. Successful strategies for maintaining and improving existing business locations, products, and technologies thus differ for the two types of industry clusters. Thus, the peripheral LMAs are most vulnerable to cyclically induced income volatility, while metropolitan core areas benefit most from business volatility. Transitional rural areas experience high income and business volatility as well as high income growth. Overall, reduced dependence on agricultural specialization balances increased dependence on manufacturing specialization. For most LMAs with a rural or manufacturing orientation, replacement of extreme dependence on industry specialization with a more diverse base economy seems unlikely, given the factual evidence presented earlier.

Improving access to decision information by the residents of a region is of overriding importance in building local infrastructure or supporting the base economy. However, available local resources limit access to information by local community leaders and resident small-business managers. The decision centers of the large corporations with branch plants and offices in the local community have the information access advantage. Access to information requires, of course, much more than an investment in communication system hardware and software. It requires certain capabilities for at least interpreting information received from various sources at some cost and then using the intellectually processed information in production, marketing, and other important business functions. Because of the strong likelihood that we may encounter several different ways of viewing this topic, we start with the idea of information transfers between different firms and different activities within a firm.

Information transfers characterized by a high level of uncertainty and competence involve a high frequency of business meetings. Conversely, those with a low level of uncertainty and competence require a low frequency of business meetings. The high-order firms engaged in these activities locate in the downtown districts of large metropolitan areas, but their face-to-face contacts may extend over a much larger geographic area. The distant contacts generate the long-distance business travel and the demand for cost-effective air access to the distant contacts and markets.

Small export-producing businesses in peripheral areas may gain access to markets through various contractual arrangements with core area businesses. These include outsourcing by core area producers during peak production periods, and promotion of training sessions sponsored by core area producers for input-supplying businesses. New public-private partnerships address the advantages of cooperation between rural and metropolitan area businesses and institutions in strengthening local and regional infrastructure and support industries for interregional and global competition.

Key sectors for improving local access to information include state and local educational institutions and related community functions, such as city and neighborhood libraries and social centers. Moreover, various information partnerships that involve local businesses and community leaders, as well as state and local governments, can become active participants in improving access to decision information. For example, local and regional postsecondary educational institutions with curricula and programs that address periphery-to-core-area linkages and information access may contribute directly to improving the quality of life and the economic well-being of local and regional residents. For many of these institutions, however, a radical change in the attitudes and values of its members and providers may be necessary to thus redefine the mission of the institutions of higher education.

The demand for interregional business communication and travel is a derived demand. It is derived from four categories of business activities: renewal, routine business service, market, and production (Hugoson and Karlsson, 1996). Renewal and market activities seem to be the most important factors for the generation of interregional business travel. The face-to-face contacts are primarily made within contact networks—as innovation, and as knowledge, customer, and supplier networks. The knowledge network is associated with the need to invest in new capacity and new products. Advanced business services are an important factor in these kinds of activities. Customer and supplier networks are associated with the need to sell a product or buy an input. This can, to large extent, be handled with electronic mail, but the part of this category that depends on nonstandardized information demands face-to-face contact.

Observed travel patterns show that business travel is primarily connected with business meetings between white-collar workers at high levels in the organization. This indicates that the uncertainty concerning the transmission of nonstandardized information demands high competence and that face-to-face contacts are primarily used for this kind of information exchange. This argument can also be used to explain why innovation and knowledge networks generate a high number of interregional business trips, together with the connected price negotiations and distribution of new products on the market associated with the customer and supplier networks.

An analytical framework for assessing the role and importance of air transportation to a locality or region must account for more than the passengers and freight originating from and arriving at a given airport. It must also include the air traffic to and from competing airports. Ultimately, such a framework must account for air transportation-related activity at all origins and all destinations within the national system of interconnected major airports. We address these issues from a theoretical perspective by focusing on the generation of interregional air travel and traffic. The findings presented in this chapter address the formulation of an economic model for assessing major airport facility requirements.

A *gravity model* is the most commonly used analytical framework for estimating the demand for air transportation in the context of the air traffic represented by enplanements, that is, aircraft takeoffs and landings. It takes the form (following Ghobrial and Kanafani, 1995),

$$T_{ij} = T(DM_i, DM_j, SU_{ij}) \quad (5\text{-}1)$$

where T_{ij} is the demand for air transportation between cities i and j, DM is a city-specific vector representing the socioeconomic characteristics of the passengers and their cities, and SU_{ij} is a vector of transport supply variables between cities i and j.

The socioeconomic characteristics usually include only total population and per capita income. Transport supply, on the other hand, includes a large number of variables (for example, the number of direct daily flights between city pairs during peak periods, the number of direct daily off-peak flights between city pairs, weighted average aircraft size during the peak periods between city pairs, weighted average aircraft during off-peak periods between city pairs, and average travel time between cities i and j).

The gravity model framework reappears in different studies, with important variations from the basic model in predicting air travel. The Jönköping International Business School studies (Hugoson and Karlsson, 1996), for example, incorporate industry-specific and area-specific explanatory variables in their accounting of the determinants of interregional business travel. The conceptual model is partitioned into four parts based on two sets of differentiating criteria. These are (1) within-firm and out-of-firm contacts and (2) lower and higher levels of uncertainty and competence. Unlike most other gravity models, this one captures the uniqueness of each industry and area with which each traveler, also differentiated by occupation, is associated. The Jönköping studies thus deal directly with the task of estimating the demand for transportation by all long-distance travelers, with business travelers being one segment of the long-distance traveler population.

Having well-targeted infrastructure investment calls for a special understanding of the linkage between a region's economic base and its economic performance. The value of product disbursements, for example, varies by industry and location, from one part of a region to another. They have high added value in the metropolitan core area, low in the rural periphery. They correlate closely with levels of industry investment per worker. Where expected profits are high, with little risk of failure, investment and production follow, along with jobs that generate above-average earnings. The location of the production activity in an economic region, whether in the core area, the periphery, or somewhere in between, has much to do with the earnings of its labor and capital resources and the region's long-term viability.

By extending an expanded economic base model to include transfer system linkages with the local and regional economic base, we introduce the feedback process using the backward linkages from producers to their respective suppliers. We focus on the performance of individual segments of one of the most critical transfer activities in an expanding regional economy, namely, air transportation and the access a region's air transportation system provides for its export-producing businesses. We review alternative approaches to understanding transfer system performance—not only the transfer system facilities but also the service providers using these facilities as they serve their particular customers and clients. We compare the performance of the air transportation system, for example, with alternative means of people and freight transfer between given origins and destinations.

Taking Another Look at the Infrastructure Debate

The inclusion of highway congestion as a measure of infrastructure deficiency turns the infrastructure debate on its head. The difficulty of separating cause from consequence in metropolitan governance stems, in part, from massive public subsidization of urban sprawl, thanks to the generous federal aid programs for urban highway construction and the deduction of interest on home mortgages when figuring federal income tax returns. Meanwhile, a chronic inability of local and metropolitan governments to correct the mismatch between place of work and place of residence nurtures a growing polarization of the extended metropolitan area. This is painfully demonstrated by a recent study of the concentrated crime and poverty in the central cities and first-ring suburbs in the Twin Cities Metropolitan Area and the concentrated wealth and resources of the rapidly growing newest suburbs (Orfield, 1997). Punitive regulatory restrictions on business expansion, along with failing public schools and homes, ensure continuing, if not worsening, prospects for many metropolitan governments.

Urban Sprawl and Spatial Mismatch

Urban sprawl results from many causes, including inner-city residents leaving to escape rampant crime, inadequate housing, dysfunctional schools, noise, congestion, and neighborhood deterioration. Others leave to gain more space, better schools, and a higher quality of housing in suburban areas. The lower crime rates, coupled with higher personal incomes that make suburban living affordable, attract city residents. The consequences of urban sprawl also are many—again, some are negative, like the further deterioration of the schools and neighborhoods left behind; others are positive, like less crime, better schools, and improved housing for those moving into the suburbs. In the long run, however, separation of place of residence and place of work, as well as place of shopping, contributes to increasing traffic congestion on local streets and roads, along with the continuing deterioration and poverty of the already declining neighborhoods left behind.

Some studies attribute the worsening crises of the poverty-stricken declining areas to a spatial mismatch between available jobs and unemployed job seekers (Ihlanfeldt and Sjöquist, 1991). Because the suburbs that dominate regional growth do not offer affordable housing, unemployed job seekers from the poverty-stricken areas lack access to the available jobs (Orfield, 1997, pp. 55–65). Other studies suggest the contrary, that spatial mismatch is not an important factor in concentrated urban poverty based on youth unemployment, as in Chicago. Still another study shows that spatial mismatch explains less than one-third of the gap between White employment rates and Black and Hispanic employment rates. Access to jobs apparently becomes a more important factor in explaining differential employment the larger the city or metropolitan area (Ihlanfeldt and Sjöquist, 1991).

Urban sprawl and spatial mismatch have common origins in the urban growth and development processes of present-day metropolitan area economies. Much of the development infrastructure in metropolitan areas is put into restrictively zoned communities, which results in land-use patterns that are "low density, economically inefficient, and environmentally dangerous" (Orfield, 1997, p. 10). This source included the earlier suggestion that "closing of vast numbers of schools in the core of the region, while scores are built in the periphery, symbolizes the magnitude of the flight to the edge and massive waste involved." These consequences of partially market-driven and partially policy-driven infrastructure investments in the urban periphery currently remain beyond the control of the central cities. The central cities of Minneapolis and St. Paul, for example, contribute to a system of sewer financing that was put in place in the mid-1980s through which the core of the region subsidizes the construction and operation of sewer capacity in the fringe, in the most exclusive suburban areas. By 1992, the central cities were paying more than $6 million a year to help move their middle-class households and businesses to the edge of the region (p. 8).

The central problem, as viewed by Orfield, is the concentrated poverty in the central city that does not receive the attention it deserves from either local school administrators or education committees of state legislatures. Studying the Minneapolis-St. Paul metropolitan area school systems, Orfield found that the central-city school districts "threaten the state legislature with a metropolitan desegregation lawsuit. In response, the education committees, with strong support from suburban members, approve increased funding for the city schools. Talk of the lawsuit cools for a time. This ugly bargain is repeated in region after region throughout the United States" (p. 45).

Education, Housing, and Environment

Local schools are the first victims as well as the most powerful perpetrators of metropolitan polarization. They are the first to suffer from the concentration of poverty in central-city neighborhoods, for poverty among a neighborhood's schoolchildren predicts the future of its adults, according to Orfield (1997, p. 3). As poverty concentrates in neighborhoods, so does crime. In the poorest neighborhoods in Minneapolis, for example, violent crime rates are ten times the average for the metropolitan area and thirty times the average for the suburbs (pp. 3–4). When neighborhood crime rates surpass a certain level, violent crime increases exponentially. This, in turn, has a devastating effect on property values, which in some neighborhoods of Minneapolis dropped 15 to 25 percent over a five-year period, while housing values soared in the exclusive suburbs.

Correctly identifying and measuring the causal factors accounting for low student attainment in basic skills tests and then implementing effective remedial measures remain extremely difficult tasks that leave inner-city parents with more questions than answers. All this, of course, feeds the flight from inner city to suburb. The 1980 Minneapolis Homeowners Survey, for example, reported that 14 percent of households with children planned to move out of the city within five years (Orfield, 1997, p. 44). By 1986, 23 percent and by 1993, 45 percent of the households with children planned to leave within the next five years. For the most part, these are White, middle-income, middle-class households.

Changing skill requirements of a technology-driven economy impose a heavy burden on employers and on the educational systems engaged in competency preparation. This calls for close collaboration between employers of prospective employees and educators. Most educational programs build on the interests and skills of available teachers and the funding employers provide for special programs. Preparing students for the skills sought in tomorrow's labor market depends on accurate forecasting of prospective job openings. These are defined in terms of skills and occupational titles rather than industry. However, most job forecasts still refer to industry rather than occupation. Few, if any, also provide forecasts of prospective earnings by occupation or skill level. Thus, an important dialogue must still take place between the forecast user and the forecast provider. A similar dialogue on skill-focused education, which is facilitated by client-driven curriculum planning, is starting to take place in our postsecondary educational institutions.

Implementation of demand-driven educational programs for central-city residents presumably would increase the number of available workers in an increasingly tight metropolitan labor market. Most of the new jobs exist in the surrounding suburbs; however, these suburbs have an even greater lack of affordable housing for potential workers than the central cities. The central cities in the Twin Cities region have almost twice their fair share of households below 30 percent of median income, but slightly more than 1.25 times their fair share of housing units affordable at this level (Orfield, 1997, p. 56). The net result is affordable housing for only two-thirds of the central cities' poorest households. The lack of affordable housing is much greater in the fast-growing suburbs, even though they have much less than their fair share of the poorest households. Concentration of the poorest households in the central cities is what proponents of "in place" strategies for dealing with housing for the poor prefer. Their supporters, according to Orfield (1997, p. 57), include suburban residents who want to keep affordable housing out of their backyards and city residents who, for "political or programmatic" reasons, insist that poor residents do not want to live in the suburbs.

Barriers to affordable housing in the suburbs occur in the form of zoning provisions, development agreements, and development practices. Zoning codes use lot sizes, minimum room sizes, fees, and development timetables to limit the expansion of affordable housing. Development agreements create additional impediments to affordable housing above and beyond those in the public codes. Local development practices block whatever openings for affordable housing might still remain in zoning codes and development agreements. In addition, some communities assess builders with construction, zoning, and planning fees that can add significantly to their building costs.

Business expansion that creates new jobs in the central cities remains limited or unlikely because of lack of space and high site costs. Much of available open space is vacant for a very good reason: It is polluted or believed to have toxic wastes on its premises, and few, if any, businesses are willing to risk incurring major clean-up costs, now or at some future date. Other site costs are higher than in the new suburban areas because of high property taxes and high levels of absenteeism and crime. Given these obstacles to business expansion near the downtown districts of the central cities, public subsidy of transporting central-city workers to suburban work sites—sometimes called reverse commuting—is offered as the only remaining means of correcting the mismatch between available jobs and available job seekers.

Metropolitan Governance

Strengthening metropolitan governance is one prescription for rationalizing urban infrastructure development. The prescriptions vary, with some calling for appointed metropolitan councils, others for their direct election. For most of the governing bodies, the focus is on the coordinated control of infrastructure development and maintenance. For the Twin Cities Metropolitan Council, the infrastructure under its jurisdiction includes transit and transportation, sewers, solid waste management, land use, airports, and housing. Neither education nor environmental quality management are in any way part of its responsibilities, although both are critically important to the council's success. Naftalin and Brandl (1980, p. 35), in their study of regional governance, cited the importance to the governing body of a regional strategy, which they defined as capacity building. Implementation of a capacity-building strategy involves creating a council and various metropolitan commissions, and the sharing of property tax bases and state revenues by local governments.[3]

Economic Revitalization

Currently, the Metropolitan Council's regional strategies call for "a holistic approach to stabilize the core and ensure economic vitality" (Metropolitan Council, 1994, p. 1). The Council recognizes that the "entire region plays a part in shaping solutions" that include measures for "economic revitalization, matching workforce to jobs, maintaining quality communities, and renewing neighborhoods." The council also recognizes that a "combination of incentives, rather than a reliance on strict regulation or penalties, is necessary to shape workable solutions" (p. 3). The "fully developed area" within the seven-county Metropolitan Council region is the focus of the council's current efforts in regional governance. The *Regional Blueprint*—the new regional planning guide replacing the Regional Development Planning and Investment Framework—broadens the council's role in addressing economic and social issues of the council region, as well as its infrastructure needs, but with emphasis on the core area.[4]

The Twin Cities Metropolitan Council is one of many such entities established in the 1960s. Currently, it is a body of officials and citizens appointed by the governor to represent the local governments and populations in the seven-county Twin Cities Metropolitan Region. Growth management, particularly the use of sewer and water access to guide the location of population and economic activity, and the reduction of "regional disparities" were the council's key issues in the 1960s and 1970s (Kolderie, 1972). Policy issues now facing this regional body pertain to the viability of its economic base and infrastructure, including air transportation, technical education, and the acquisition of fiscal resources for maintaining the region's essential public services. These differ, however, from the immediate policy issues being addressed by the council's current membership and staff. These issues include "balanced" growth and development, orderly expansion of the metropolitan area, affordable housing, and poverty (Metropolitan Council, 1994).

In 1994, the Minnesota Legislature passed the Metropolitan Reorganization Act, which made the Metropolitan Council, with its new jurisdiction, the second most powerful government entity in the state. An appointed council now has under its direct operational authority the council planning staff, the operation of metropolitan transit, waste control, and a combined yearly budget of more than $600 million. These operations are financed by state grants, user fees, and annually levied property taxes. Council members still face the possibility of being periodically elected for the office. Critics of an appointed council claim that the presence of developers on the council has weakened its growth-management efforts. An elected council would face close scrutiny as to occupation and sources of contributions. On the other hand, rural counties, fearful of a loss of legislative power against a unified metropolitan region, are strongly opposed to an elected council. A coalition of rural and suburban legislators still retains control over enough votes to defeat any bill to establish an elected metropolitan council.

Coalition Building

Urban regional coalition building is one means of designing and implementing programs for regional revitalization, as proposed by Minnesota State Representative Byron Orfield (1997), in *Micropolitics: A Regional Agenda for Community and Stability,* published by the Brookings Institution and the Lincoln Land Institute. Such efforts have multiple goals. Among those cited are (1) uniting central cities with the middle-class and lower-middle-class voters in the declining and low property tax value suburbs; (2) demonstrating that tax-base sharing lowers taxes and improves local services, particularly schools; and (3) convincing voters that fair housing will limit their commitment to poor citizens to manageable regional standards and thereby stabilize residential change in their communities (p. 37). These goals are not readily attained without the resources and desire for gaining a more general and greater understanding of the determinants and consequences of attaining any or all of these goals.

From an urban regional economics perspective, the series of steps outlined by Orfield for reaching the goals of regional coalition building fit nicely into a program of education and outreach on the workings of a metropolitan regional economy. Orfield labeled the individual steps that are summarized from *Micropolitics* (pp. 167–171) as "lessons," starting with information and data on the region's population, labor force, and employment, plus a host of other regional indicators that come under the "demographics" umbrella:

- *Understand the region's demographics.* Develop the most accurate and comprehensive picture of the region possible.
- *Reach out and organize the issues on a personal level.* Invite broad input from individuals and groups affected by identified regional trends and then lay out the areas where regional progress is necessary, like affordable housing, tax base sharing, and land use planning.
- *Build a broad, inclusive coalition.* Stress two broad themes: It is in the long-term interest of the entire region to solve the problems of polarization, and it is in the immediate short-term interest of the vast majority of the region.
- *Join forces with the older suburbs.* Orfield contends that the inner and low-tax suburbs are the pivot point in American politics and the reformers' key political allies.
- *Reach into the central cities to be sure the message is understood.* Metropolitan reforms must be presented as complements to existing programs that would gradually reduce overwhelming central-city problems to manageable size.
- *Seek out the region's religious community.* Churches can provide legitimacy for its message in distrustful blue-collar suburbs, and understanding and a sense of responsibility and fair play to more affluent ones.
- *Seek out the philanthropic community, established reform groups, and business leaders.* Every day, philanthropic organizations face the consequences of regional polarization, and their mission statements are often in line with regional reform.
- *Draw in distinct but compatible issues and organizations.* Regionalism is a multifaceted gemstone that can show a bright face to many different constituencies to build broad support.
- *With the coalition, seek out the media.* Reporters who have covered the same political stories over and over will be interested in something new and potentially controversial.
- *Prepare for controversy.* Reform never happens effortlessly or overnight. It entails building coalitions, creating power, and engaging in strenuous political struggle.

- *Move simultaneously on several fronts and accept good compromises.* Get as many issues moving as can be effectively managed, but not so many that nothing happens. Keep opponents busy and on the defensive.

Orfield placed great emphasis on having the facts, along with an understanding of how these facts relate to the urban crisis of concentrated poverty and its many adverse consequences. He noted that the "crisis of city and inner-suburban schools often proceeds without any relevant facts. In older metropolitan areas post hoc analysis has revealed that test scores ratchet downward in synchronization with increasing poverty. By 1994 the central cities, with more than half of their students living in poverty, were experiencing a huge loss of middle-class students. During the 1980s, this growing instability coincided with enormous parental and public concern about the quality of the central-city schools. Perhaps because they were afraid of the potential reaction, school administrators did not forthrightly discuss changes in their school populations or their implications, and neither system employed testing systems that provided accurate information on student performance. Experts told the school board and parents that test scores were stable. In reality, the testing data were so incomplete and so heavily manipulated that no one really knows how these changes were affecting students" (pp. 45–46).

Orfield continued: "This dearth of relevant facts combined with growing and often explosive parental discontent. The ensuing reform discussion, centered on curriculum, bureaucracy, and teacher accountability, was in many ways isolated from the reality of what was occurring in the schools in the 1980s. A series of reforms, some very expensive, were instituted to correct the 'problems' of the central-city schools. None of these reforms directly addressed the schools' overwhelming transformation to majority poverty status as a result of polarization in the regional housing market. Whether the reform had any effect at all on slowing the loss of middle-class families was hard to know, so quickly were they abandoned and often without any evaluation. In any event they clearly did not turn the tide" (p. 46).

Tracking Transfer Systems Performance

Monitoring the performance of transfer systems is a widely dispersed responsibility, given the variety of private and public transfers and transfer system environments. For any economic region, the monitoring involves both metropolitan urban services and core-periphery linkages. The metropolitan urban services refer to the use of the metropolitan urban infrastructure for moving people, freight, and information within the metropolitan core area and between metropolitan core areas. A formally defined MUSA is one means of controlling the deployment of these services within the jurisdictions of a metropolitan area governing entity. The core-periphery linkages refer largely to business relationships, especially between businesses in the core area and the rest of a region.

Metropolitan Urban Services

Deliberate government efforts to increase population densities in urbanized metropolitan areas will likely change the infrastructure debate from one of increasing the rate of infrastructure spending to one of reducing it. There is, also, the repeated recommendation to redirect at least part of the "savings" to more highway maintenance and improved urban services delivery within the designated MUSA. Without corresponding restrictions on the extension of metropolitan urban services beyond the council's jurisdiction, the surrounding counties will likely become the new destinations of metropolitan area residents on the move. For the Twin Cities Metropolitan Council, the new strategy would require close working relationships with the central cities, the surrounding suburbs, and both the state legislature and the governor in supporting urban services containment policies over the entire commuting area.

Relaxing regulatory constraints on affordable housing and assisting central city businesses in reducing the risks and costs of expansion in the central cities would help directly in reducing current mismatches between place of work and place of residence. Data and analyses for assessing the costs and benefits of these sorts of efforts are not available currently. Nor are data and analyses generally available for assessing the spillover costs to metropolitan area residents of infrastructure subsidies that are comparable to the current data and analyses on the demand for and supply of affordable housing. These additions to a council's monitoring capabilities would require a shift from a largely demographic approach for regional forecasting to one that starts with the economic base of the region and its subregions, taking into account several different perspectives on the problems and potential facing the individual industries in the region's economic base.

Core-Periphery Linkages

The increasing importance of interregional trade and the building of a regional infrastructure for sustaining the region's export-producing industries suggests forms of interstate cooperation based on some sort of political legitimization of these efforts. Local economic conditions, of course, influence the changes that will result in new institutional arrangements. The changes take several forms. Bromley (1989) views these changes as institutional transactions that (1) increase productive efficiency, (2) redistribute income, (3) reallocate economic opportunity, and (4) redistribute economic advantage. For example, improving highway access to the downtown district that results in an increase in its daytime population increases the district's land-use efficiency. The increase in productive efficiency may result in less funding for public transit and reduced access by low-income workers to available jobs in the suburbs. This would lead to a redistribution of income from the poor to the rich. Changes in institutional arrangements thus account for the possibility of several kinds of changes in commuter and taxpayer behavior.

The widespread reshaping of regional infrastructure systems is both a cause and a consequence of increasing global competition. State and regional transportation is thus no longer an activity isolated from events beyond state and regional borders. Yet, much of transportation systems planning is held captive by the durability of its own legacy in maintaining costly but unneeded roads and bridges, and sustaining a world-class transportation network to almost every corner of its jurisdiction.

From an economic perspective, a regional strategy for capacity building must eventually focus on the region's economic base and the economic environment for successful export-producing enterprise. Skill-based education and employment counseling activities, for example, enhance the availability of various labor skills that, in turn, improve the competitive position of an area's export-producing businesses. Readily accessible and well-managed private and public delivery systems that provide adequate health care at competitive prices also contribute to a favorable local economic environment. In addition, quality-of-life concerns loom large for a core metropolitan area with a diverse resident population, including interest groups with above-average personal income levels who have moved from the central city to the newest commuter suburbs. A regional policy affecting quality of life would provide access to the education, health care, housing, and recreational and cultural activities sought by the region's residents. In central cities and rural areas with many dislocated workers and businesses that lack easy access to export markets, quality-of-life concerns remain secondary. Such areas are likely candidates for community-based economic development.[5]

Establishing the neighborhood as "a good place to live" strengthens that portion of its economic base represented by the "outside" income receipts of its resident population. The export-producing businesses in the neighborhood account for the remainder of the local economic base. By starting with the local economic base, community-based economic development addresses the sources of the neighborhood's economic well-being and viability. The determinants of neighborhood well-being are the quality and price of housing and services, ease of access to local parks and amenities, safety of person and property, quality of schools, and amount of annual property taxes. Dislocated workers and single-person households without income from steady employment and lacking other income sources thus become a heavy burden on any neighborhood, but especially those bordering the downtown district.

The core LMA provides, among its many other contributions, a favorable environment for increasing collaboration among regional organizations that depend upon one another for jobs and services. These include state and local governments and their special agencies for intraregional cooperation, even across state boundaries. The central city particularly seeks greater collaboration with surrounding suburban governments in providing community services and sharing the expenses incurred in supporting impoverished and dislocated households now concentrated in the central city. However, individual municipalities view their economic development efforts as competitive, particularly with neighboring communities (Goetz and Kayser, 1993, p. 76).

Notes

1. Based on: U.S. Department of Commerce. 1992. Fixed reproducible tangible wealth series. *Survey of Current Business* 72(8): 39–42.

2. Estimated by log X(t) = A + bT, where A is the intercept, X(t) is the item under consideration, b is the annual percent rate of growth, and T is time.

3. The Twin Cities Metropolitan Council was authorized by the Minnesota State Legislature to adopt a comprehensive development guide. This guide has twelve chapters, with each chapter adopted only after many open hearings and much agency and public review. The ten substantive chapters are health, airports, housing, protection of open space, recreational use of open space, transportation, solid and hazardous waste management, sewage disposal, water resources, and law and justice. The remaining two chapters—the development framework and the investment framework—are the implementing components. The development framework breaks the Twin Cities Metropolitan Region into five planning zones: the region's two metropolitan centers, the fully developed area of twenty older suburbs surrounding the metropolitan centers, the area of planned urbanization lying beyond the fully developed area, the thirteen freestanding growth centers, and the remaining agricultural areas. In spite of its many chapters, the development guide is not fully comprehensive. Its emphasis is on physical development and is concerned much less with social and economic planning. At no point is distress or a distressed area singled out for attention. As noted by Naftalin and Brandl, "The council's most effective fiscal power is its approval authority over the capital budgets of its metropolitan commissions for sewers and transportation" (1980, p. 44). The authors concluded that without this control, the council's policies for guiding regional development would be ineffective. A 1977 report of the Fully Developed Area Task Force emphasized a new urban philosophy of "(1) intensive reuse of the existing public and

private investment structure, and (2) region-wide sharing of the costs of such redevelopment." Moreover, the council strongly favors an open planning process at all governmental levels that would come to have "a commonly accepted set of policies reflecting neighborhood and private sector inputs." In their commentary, Naftalin and Brandl suggested that the proposed metropolitan reinvestment fund would be a major tool for the implementation of the redevelopment plans.

4. The 1991 Minnesota State Legislature directed the Twin Cities Metropolitan Council "to examine changes that encourage the economic and societal strengthening of the fully developed area." Legislation called for the analysis of economic and societal trends affecting the fully developed area, specifically development patterns, migration, household composition, demographics, economic and societal conditions, comparative costs of redevelopment and new development, and effects of light rail transit, if implemented. The study findings identify broad regional strategies "geared to growth and recovery" that include specific policy recommendations. These are strategies (under each of the four issues cited earlier) to (1) recycle contaminated sites back into productive commercial and industrial use, with cities playing a stronger role in site assembly and reuse of existing buildings and improving public funding tools for their redevelopment efforts; (2) increase people's skills to match job needs through various measures that reallocate available funds, improve transit and transportation to better link job seekers in the fully developed area with job opportunities in the developing areas, and provide better access to affordable housing throughout the council region; (3) make communities safer, maintain high quality housing stock, ensure the quality of city schools, "appropriately" involving public, private, and neighborhood groups to increase confidence in schools and neighborhoods; and (4) eliminate policies that restrict housing choices of low-income people, link economic development with community development in larger neighborhoods, and encourage cities and counties to work together in forming common strategies (Metropolitan Council, 1994).

5. In keeping with the "wave" terminology cited earlier, community economic development is viewed here as the "fourth wave" of economic development. Earlier discussions of state economic development cited "smokestack chasing" and "helping firms compete in global competition" as the first and second waves, with the second wave including state and local government incentives ranging from financing and export initiatives to technology transfers. The "third wave" focused on long-term investments in technical education and business infrastructure. This rationale justified subsidy of education that could result in a more productive workforce which, in turn, would lead to improved economic growth and individual well-being (Mattoon and Testa, 1993).

PART 2
Tools

Forecasts and Forecast Methods

While the location decision and its related investment decisions are central to the study of urban regional economies, underlying both are the alternative futures we envision both for ourselves and for our decision environments. Our understanding of the determinants of these alternative futures is tested in varying degree with every location or investment decision, but most of all in our efforts to forecast the likely future course of a particular line of business or industry. To forecast is ". . . to estimate or calculate something in advance; predict the future."[1] We begin our task with a review of alternative futures identified with particular regions and areas and the resources and time involved in preparing them and understanding their uses and limitations.

Forecasting Regional Economies

Forecasting regional economies without some sort of forecasting model is like traveling from Boston to New York without any means of transportation. Momentarily, at least, we stay in the realm of speculation by simply identifying the various forecasting models, saving their applications for later. Among these models are location quotients, shift-share coefficients, inter-industry tables, and statistical forecasting models. We compare the predictive qualities of each tool against a reference, for example, a naïve model of the form

$$v_{(t+1)} = v_{(t)} + e_{(t)} \qquad (6\text{-}1)$$

where $v_{(t+1)}$ is some value in period t + 1 that is equal to the same value in period (t) plus an error term. It would be like predicting weather by saying that tomorrow's temperature is like today's plus or minus some unspecified, usually small, but unpredictable, difference. Each tool, of course, depends on its own special database for actually forecasting regional economic activity.

Economic forecasting models come in many sizes and shapes for any region. The various models contain many different values, including estimates of population, employment, income, product outputs, inputs, and prices. They sort into both one-period static models and multiperiod dynamic models. Even the one-period models range in complexity from a rather straightforward use of the inter-industry transactions table, as in a simple economic impact model, to highly sophisticated representations of supply-and-demand relationships of product prices and quantities, as in a computable general equilibrium model.

Several empirical approaches exist for measuring the economic base of a region. Location quotients are among the easiest to use. Shift-share analysis is a bit more complex, bringing into account change from one period to the next and for a given area or region relative to a larger region. Input-output analysis is still another approach. Hybrid forecasting systems generated from various land use planning scenarios provide various measures of local economic growth, like small-area population projections for planning future extensions of a metropolitan urban services area. Finally, statistical analysis can be employed as an alternative to the strictly mathematical approaches of location quotients and input-output tables. Altogether, the different approaches fall into two categories of modeling systems—static and dynamic—static in the sense of dealing with only one period at a time, as in economic base forecasts, dynamic when dealing with several periods in continuity, as in economic growth forecasts.

Economic Base Forecasts

Economic base theory and models were cited in Chapter 4. Both location quotients and input-output approaches to measuring this base derive their numerical values from historical data, adjusted for whatever changes are appropriate when manipulating one or more variables in the model to estimate changes in related variables. These changes are the forecast effects, assuming all other variables remain the same. In reality, any or all variables may change. Nonetheless, the input-output model and the various location quotient approaches, including an approach we call "excess employment coefficients," provide useful estimates of an area's economic base that serve as a starting place in preparing economic base forecasts.

Location Quotients

The most simple and yet widely used forecasting framework is the location quotient in its various manifestations. This measure is nothing more than a ratio of ratios, one being the proportion of a region's employment in an identified industrial sector, say automobile production, the other being the proportion of a nation's employment in the same industry. If the resulting value is greater than one, the inferred conclusion is that auto production is more concentrated in that region than in the nation as a whole.

Location quotient models refer to individual industries in an area's economic base. Each so-called basic industry has its own unique multiplier components—direct, indirect, and induced. In practice, the percentage contribution of each industry to the economic base usually serves as a measure of its contribution to an area's overall economy. This is not necessarily the case, given the variance from one industry to the next in each multiplier component.

Location quotient models lend themselves to the practice of comparative statics, that is, comparing various measures of an area's economic base from one period to the next. This provides a visual, as well as a numerical, measure of year-to-year variability in the contribution of each industry to an area's economic base. Moreover, the year-to-year variability may show secular and cyclical patterns of change that can be attributed to overall national growth and change, or the general business cycle. Variability might also relate to each area's competitive position within a given industry's product cycle, thus leaving a final "unexplained" variation that is only a small fraction of the initial year-to-year variation.

This is the simplest approach to measuring a region's economic base, using a ratio of two ratios—one for the area or region, the other for the nation. When the location quotient's value is equal to one, the region is said to mirror the larger region for the given attribute. Implicit in this statement is some very restrictive assumption, particularly that the regional economy is very much like the larger region, except for scale.

While the assumption of homogeneity leads to fundamental weaknesses in location quotient methodology, these ratios can tell a researcher much about a region in a very short time and with minimum data requirements. This ease of use accounts for its continuing popularity in regional analyses. Numerically defined, the location quotient is equal to the percentage of a reference region's activity in a particular industry divided by the percentage of activity in that same industry for the larger region (usually the nation). Assuming the nation to be the larger region, the form of the quotient, Q, is

$$Q_i = (R_i/R)/(N_i/N) \quad (6\text{-}2)$$

where Q represents the location quotient, R_i represents regional activity in industry i, R represents total regional activity, N_i represents national activity in industry i, and N represents total national activity.

Given the model's assumptions, the following applies:

$Q_i = 1$ indicates that this industry produces at the same level as its national counterpart. If the nation is self-sufficient in the production of this commodity, then so is the region. Therefore, no exports or imports are associated with this regional industry.

$Q_i > 1$ indicates that this industry is producing output at a level that is greater than self-sufficiency requires. This excess in activity must be exported. For example, if the location quotient value is equal to two, then 50 percent of the regional activity is excess and is assumed to be exported.

$Q_i < 1$ indicates that this regional industry is producing at a level that is less than self-sufficiency requires. Any deficit in supply is assumed to be imported. Thus, if the location quotient value is equal to half (.5), the region produces half of its needs and imports the remainder.

Remember that the simple economic base multiplier is calculated as total divided by basic activity (T/B). If the assumptions of the location quotient hold, the B of this ratio can be estimated as the value of excess activity in the region.

The assumptions associated with the location quotient are restrictive. However, if nothing else, the location quotient provides an indication of the relative importance of a particular industry to a region compared to the larger region of which it is a part. The greater the location quotient value, the more important this industry is to the economic base of the region. What is more, changes in location quotient values over time represent changes in a defined industry's importance to an area. As such, location quotients are quite useful as rough approximations of the local economic base.

This is illustrated in Table 6-1, which is a continuation of the comparisons of metropolitan statistical areas (MSAs) and primary metropolitan statistical areas (PMSAs) presented in Chapter 1. Even as approximations, they quantify the relative importance of individual industries in the economic base of each metropolitan statistical area (MSA). They clearly identify the federal government—with a location quotient of 6.0, or six times the industry share of total earnings in the nation—and its ancillary private services as one of the dominant basic industries of the District of Columbia MSA (listed as the Washington, DC-MD-VA-WV PMSA in Tables 6-3 through 6-17). Similarly, they identify durable goods manufacturing, even as an aggregation of both basic and nonbasic industries, as an important part of the Detroit economic base.

The location quotient approach to the quantification of an area's economic base is only a step removed from an even more complete measure of a locality's basic industries—the excess activity approach. The added step is simply multiplying, not a ratio of ratios, as in the location quotient approach, but the difference between the two ratios.

TABLE 6-1 Location quotients for total earnings of employed workers in specified industry, 1969 and 1994

Industry	D.C. PMSA		MnStP MSA		Detroit PMSA	
	1969	1994	1969	1994	1969	1994
Earnings, total	1.0	1.0	1.0	1.0	1.0	1.0
Agriculture, agricultural services, forestry, fisheries, other	0.3	0.4	0.4	0.3	0.1	0.2
Mining	0.2	NA	0.1	0.1	0.1	0.1
Construction	0.9	0.9	1.2	0.9	0.9	0.8
Manufacturing, total	0.2	0.2	1.0	1.2	1.6	1.8
Transportation and public utilities	0.7	0.8	1.2	1.0	0.8	0.8
Wholesale trade	0.6	0.6	1.5	1.4	1.0	1.2
Retail trade	0.9	0.8	1.0	0.9	0.9	0.9
Finance, insurance, real estate	0.9	0.8	1.1	1.3	0.8	0.8
Services, total	1.4	1.3	0.9	0.9	0.8	0.9
Hotels and other lodging places	NA	1.5	0.9	0.5	NA	0.3
Personal services	1.0	0.9	1.1	1.1	1.0	0.9
Business services	2.1	1.7	0.9	1.2	0.9	1.1
Health services	0.8	0.7	1.0	0.8	0.9	1.0
Legal services	2.1	2.0	1.1	1.0	0.8	0.8
Educational services	1.8	1.3	0.6	0.8	0.4	0.4
Social services	NA	1.0	NA	1.3	NA	0.9
Membership organizations	2.0	2.9	1.0	1.0	0.9	0.8
Engineering and management services	NA	2.4	NA	0.8	NA	1.0
Government and govt. enterprises, total	2.7	1.9	0.8	0.8	0.6	0.6
Federal, civilian	7.2	6.0	0.6	0.6	0.4	0.5
Military	2.7	2.3	0.2	0.2	0.1	0.1
State and local	0.8	0.7	1.0	0.9	0.8	0.7

Source: U.S. Department of Commerce, Regional Economic Information System, 1996.

Excess Activity Values

Variations of the location quotient approach occur when the ratio of ratios is separated and joined with measures of the absolute levels of regional activity. Thus, the "more than one" finding for labor earnings, for example, would provide an absolute measure of regional specialization when expressed in absolute values rather than ratios. That is, the difference between a region's industry labor earnings ratio and the corresponding U.S. industry ratio, when multiplied by the region's total industry labor earnings, is a measure of the "excess" or "deficit" labor earnings for that industry in the region. The total excess labor earnings (which equals total deficit labor earnings) is an alternative, although not a totally accurate, measure of the region's economic base.

An excess activity value is determined from the form,

$$X_i = R_i \times [(R_i/R) - (N_i/N)] \quad (6\text{-}3)$$

where X_i represents the activity level in regional industry i; all other activity variables are the same as in Equation 6-2. The excess activity approach, as presented in Table 6-2, yields monetized values of labor earnings, rather than simply a number. These values can be sorted, aggregated, and compared with other values. All positive values, when added together, provide another measure of an area's economic base. For any one area, the positives and negatives balance, as shown for the total earnings entry in Table 6-2. Typically, a smaller number of industries show an excess value than show a deficit value because of regional specialization. In the case of labor earnings, the 1994 excess for the federal government in the District of Columbia PMSA was more than $16.8 billion. It was more than $14.3 billion for durable goods manufacturing in the Detroit PMSA.

We now shift to another database with much more detail on individual industry earnings, employment, and other activity measures. This is the University of Minnesota regional economic modeling system called IMPLAN (*Im*pact Analysis for *Plan*ning). This system is now available commercially from the Minnesota IMPLAN Group (MIG). It tracks the key economic measures of over 500 industries, mostly in the four-digit classification.[2]

Table 6-3 lists the calculated excess activity values—both positive and negative, as well as other measures—for the top twenty industries in the first of three different rankings, namely, excess employment. The other two rankings are for excess value added and gross commodity exports. In the District of Columbia PMSA, of the top twenty industries ranked according to excess employment, eight are in the private services sector (covering SIC codes 70–89), three in the transportation, communications, and public utilities sector (covering SIC codes 40–49), two in manufacturing (covering SIC codes 20–39), two in contract construction (covering SIC codes 15–17), one in trade (covering SIC codes 50–59), one in finance, insurance, and real estate (covering SIC codes 60–67), two in the federal government, and one in the household workers sector. Also listed are the top twenty in the value-added and export rankings, total commodity exports, net commodity exports, total commodity imports, and the import dependency values. The excess employment ranking includes sixteen of the top twenty in the value added ranking and twelve of the top twenty in the total exports ranking. The import dependency value represents the proportion that commodity imports are of total commodity inputs of the corresponding industry. The number of industries equals the number of commodities. A given industry may produce more than one commodity. It is represented in name by its dominant commodity.

TABLE 6-2 Excess earnings activity of employed workers in specified industry, by place of work, 1969 and 1994

Industry	D.C. PMSA 1969 (mil.$)	D.C. PMSA 1994 (mil.$)	MnStP MSA 1969 (mil.$)	MnStP MSA 1994 (mil.$)	Detroit PMSA 1969 (mil.$)	Detroit PMSA 1994 (mil.$)
Earnings, total	0	0	0	0	0	0
Agriculture, agricultural services, forestry, fisheries, other	–304	–1,260	–144	–749	–493	–1,258
Mining	–107	NA	–64	–444	–152	–672
Construction	–73	–576	84	–351	–82	–972
Manufacturing, total	–2,958	–14,811	–86	2,322	2,777	11,877
Transporation and public utilities	–235	–1,216	84	0	–256	–955
Wholesale trade	–305	–2,321	217	1,510	–15	864
Retail trade	–158	–2,328	39	–295	–152	–1,106
Finance, insurance, real estate	–96	–1,418	56	1,039	–191	–1,363
Services, total	711	9,820	–63	–1,068	–438	–1,836
Hotels and other lodging places	NA	480	–4	–241	NA	–533
Personal services	–1	–125	5	38	0	–91
Business services	279	3,402	–9	596	–27	597
Health services	–117	–2,841	9	–807	–95	–250
Legal services	132	2,281	4	–4	–31	–358
Educational services	114	409	–32	–152	–114	–507
Social services	NA	37	NA	161	NA	–86
Membership organizations	178	1,789	3	–20	–15	–115
Engineering and management services	NA	5,334	NA	–424	NA	55
Government and govt. enterprises, total	3,524	14,681	–296	–1,974	–1,038	–4,587
Federal, civilian	3,249	16,821	–135	–740	–401	–1,311
Military	512	1,561	–142	–520	–349	–817
State and local	–236	–3,701	–19	–713	–288	–2,459

Source: U.S. Department of Commerce, Regional Economic Information System, 1996.

Table 6-4 lists the same economic activity measures as in Table 6-3, but ranked according to excess value added. The excess value added ranking includes sixteen of the top twenty in the excess employment ranking and twelve in the total exports ranking. The excess value added for the top twenty in this ranking is slightly larger than the excess employment ranking, but excess employment is slightly less. The three commodity measures also are slightly larger in this ranking. The excess value added correlates roughly with gross commodity exports, with several important exceptions, namely, federal government, civilian and military, state and local government, and federal utilities. While the three government sectors depend on intermediate inputs imported from outside the MSA or PMSA, the special procedure adjusts the exports entry to the required level of imports since the three activities are outside the interacting sectors in the inter-industry transactions table. The lack of an entry in the

TABLE 6-3 Excess employment activity in specified industry, Washington, DC-MD-VA-WV PMSA, 1990

Rank	Database Sector Name	SIC Code	Ranking		Excess Activity		Total Exports (mil.$)	Net Exports (mil.$)	Tot.Com. Imports (mil.$)	Import Depend. (pct.)
			VA	Exp.	Employ. (no.)	Val. Add. (mil.$)				
1	Federal Government–Nonmilitary		1	10	293,141	14,377	848.8	0.0	848.8	NA
2	Computer and Data Processing Services	7370	4	2	65,706	3,608	4,137.5	3,962.0	175.4	25.3
3	Management and Consulting Services	8740	6	4	45,261	1,852	2,735.2	2,662.1	73.2	25.3
4	Research, Development, and Testing	8730	8	7	40,200	1,402	2,034.6	1,976.9	57.7	25.3
5	Federal Government—Military		5	33	40,048	2,611	225.2	0.0	225.2	0.0
6	Business Associations	8610 8620	9	5	32,510	1,259	2,282.9	2,247.5	35.4	39.3
7	New Residential Structures	Part 15, 16, 17	27	83	32,024	90	51.8	51.8	0.0	52.2
8	Engineering, Architectural Services	8710	10	16	28,626	957	479.9	338.7	141.2	7.6
9	Real Estate	6500	2	1	28,394	7,125	8,301.4	7,906.1	395.3	10.1
10	Legal Services	8110	7	3	23,904	1,789	2,796.7	2,680.2	116.4	13.7
11	U.S. Postal Service	4311	11	11	21,674	938	797.0	743.1	53.9	28.4
12	Services to Buildings	7340	19	45	19,829	220	159.9	127.4	32.5	27.8
13	Communications, except Radio and TV	4810 4820 4840 4890	3	6	19,676	3,818	2,089.1	1,760.8	328.3	20.7
14	New Industrial and Commercial	Part 15, 16, 17	31	26	11,624	64	304.4	304.4	0.0	53.4
15	Household Industry—Low Income	–	33	58	8,661	57	99.9	90.7	9.3	NA
16	Federal Electric Utilities	Part of 4910	16	50	6,475	285	135.0	0	0	46.6
17	Radio and TV Communication Equipment	3663	12	21	6,227	787	382.7	202.9	179.8	45.4
18	Colleges, Universities, Schools	8220	17	13	4,853	285	577.7	523.5	54.2	15.8
19	Apparel and Accessory Stores	5600	38	62	4,740	44	86.1	37.0	49.1	17.0
20	Periodicals	2720	20	14	3,894	217	486.7	307.5	179.2	43.1
	Top 20, total				737,468	41,786	29,012	25,923	2,955	–

Source: University of Minnesota IMPLAN Model System.

TABLE 6-4 Excess value-added activity in specified industry, Washington DC-MD-VA-WV PMSA, 1990

Rank	Database Sector Name	SIC Code	Ranking		Excess Activity		Total Exports (mil.$)	Net Exports (mil.$)	Tot.Com. Imports (mil.$)	Import Depend. (pct.)
			Emp.	Exp.	Employ. (no.)	Val. Add. (mil.$)				
1	Federal Government—Nonmilitary		1	10	293,141	14,377	848.8	0.0	848.8	NA
2	Real Estate	6500	9	1	28,394	7,125	8,301.4	7,906.1	395.3	10.1
3	Communications, Except Radio and TV	4810 4820 4840 4890	13	6	19,676	3,818	2,089.1	1,760.8	328.3	20.7
4	Computer and Data Processing Services	7370	2	2	65,706	3,608	4,137.5	3,962.0	175.4	25.3
5	Federal Government—Military		5	33	40,048	2,611	225.2	0.0	225.2	0.0
6	Management and Consulting Services	8740	3	4	45,261	1,852	2,735.2	2,662.1	73.2	25.3
7	Legal Services	8110	10	3	23,904	1,789	2,796.7	2,680.2	116.4	13.7
8	Research, Development, and Testing	8730	4	7	40,200	1,402	2,034.6	1,976.9	57.7	25.3
9	Business Associations	8610 8620	6	5	32,510	1,259	2,282.9	2,247.5	35.4	39.3
10	Engineering, Architectural Services	8710	8	16	28,626	957	479.9	338.7	141.2	7.6
11	U.S. Postal Service	4311	11	11	21,674	938	797.0	743.1	53.9	28.4
12	Radio and TV Communication Equipment	3663	17	21	6,227	787	382.7	202.9	179.8	45.4
13	State and Local Government—Noneducation	–	374	29	–3,262	767	272.3	0.0	272.3	0.0
14	Accounting, Auditing, and Bookkeeping	8720 8990	354	38	–2,225	387	213.0	143.1	70.0	12.8
15	Labor and Civic Organizations	8630 8640	31	23	2,365	307	345.6	320.5	25.1	22.3
16	Federal Electric Utilities	Part of 4910	16	50	6,475	285	135.0	–	–	46.6
17	Colleges, Universities, Schools	8220	18	13	4,853	285	577.7	523.5	54.2	15.8
18	Radio and TV Broadcasting	4830	38	106	1,172	260	30.5	28.7	1.8	26.8
19	Services to Buildings	7340	12	45	19,829	220	159.9	127.4	32.5	27.8
20	Periodicals	2720	20	14	3,894	217	486.7	307.5	179.2	43.1
	Top 20, total				678,470	43,251	29,332	25,931	3,266	–

Source: University of Minnesota IMPLAN Model System.

imports column denotes a lack of these imports for MSA industries and institutions. The four industries are important parts of the local economic base despite a lack of commodity exports.

Table 6-5 introduces a new type of measure for the local economic base, namely, gross commodity exports. In this case, only twelve of the top twenty industries ranked by excess employment are among the top twenty in the exports ranking. The federal government sectors, which are important parts of the local economic base, receive lower rankings based on exports than does either excess activity measure. The totals for both exports and imports, however, are larger for this ranking than for the others. Advertising and wholesale trade account for the largest differences between the export ranking and the two excess activity rankings. In this case, the ranking differences relate to alternative measures of the economic base. They do not, however, address the critical role of exports in bringing new dollars into a local economy for the purchase of the many imported goods and services.

Table 6-6 summarizes the "bottom line" for each of the three sets of activity rankings. The top two rows of totals represent reference values for the various activity comparisons. The IMPLAN totals refer to the 1990 IMPLAN database and model results for the District of Columbia PMSA. These rows show totals for the employment and value added series used in calculating excess activity values as well as gross commodity exports and imports for the metropolitan economy. Only the IMPLAN totals show a negative difference between gross exports and gross imports.

The excess activity totals refer to the 57 industries with positive excess employment and value added activity values. The commodity exports and imports totals refer only to the values associated with the excess employment ranking. The corresponding 57 commodity exports represent a large portion of the aggregate of commodity exports from the IMPLAN totals, while the corresponding 57 commodity imports represent only a small portion of the aggregate of commodity imports. An IMPLAN activity total divided by the corresponding excess activity total is the excess activity economic base multiplier value. These totals are 2.145 (that is, 2,905.7/789.2) and 2.863 (that is, 130,440/45,561) for employment and value added, respectively. The top 20 and top 40 excess activity totals relate to the IMPLAN totals in the last four columns of activity shares.

The occurrence of negative net commodity imports is commonplace for local economies with much capital investment and other forms of deficit spending. For the District of Columbia, the inflow of federal income taxes generates the new dollars to pay for the excess imports.

TABLE 6-5 Gross exports of specified commodity groups, Washington, DC-MD-VA-WV PMSA, 1990

Rank	Database Sector Name	SIC Code	Ranking Emp.	Ranking TV	Excess Activity Employ. (no.)	Excess Activity Val. Add. (mil.$)	Total Exports (mil.$)	Net Exports (mil.$)	Tot.Com. Imports (mil.$)	Import Depend. (pct.)
1	Real Estate	6500	9	2	28,394	7,125	8,301.4	7,906.1	395.3	10.1
2	Computer and Data Processing Services	7370	2	4	65,706	3,608	4,137.5	3,962.0	175.4	25.3
3	Legal Services	8110	10	7	23,904	1,789	2,796.7	2,680.2	116.4	13.7
4	Management and Consulting Services	8740	3	6	45,261	1,852	2,735.2	2,662.1	73.2	25.3
5	Business Associations	8610 8620	6	9	32,510	1,259	2,282.9	2,247.5	35.4	39.3
6	Communications, except Radio and TV	4810 4820 4840 4890	13	3	19,676	3,818	2,089.1	1,760.8	328.3	20.7
7	Research, Development, and Testing	8730	4	8	40,200	1,402	2,034.6	1,976.9	57.7	25.3
8	Advertising	7310	359	369	–2,405	–208	1,587.8	1,505.7	82.1	20.8
9	Wholesale Trade	5000 5100	412	412	–48,148	–2,007	1,143.4	891.2	252.2	20.1
10	Federal Government—Nonmilitary		1	1	293,141	14,377	848.8	0.0	848.8	NA
11	U.S. Postal Service	4311	11	11	21,674	938	797.0	743.1	53.9	28.4
12	Credit Agencies	6100 6710 6720 6733 6790	34	374	1,545	–254	783.0	751.8	31.1	22.7
13	Colleges, Universities, Schools	8220	18	17	4,853	285	577.7	523.5	54.2	15.8
14	Periodicals	2720	20	20	3,894	217	486.7	307.5	179.2	43.1
15	Commercial Printing	2750	63	25	–19	118	483.8	–371.9	855.7	59.1
16	Engineering, Architectural Services	8710	8	10	28,626	957	479.9	338.7	141.2	7.6
17	Maintenance and Repair Other Facilities	Part 15, 16, 17	21	394	3,752	–502	477.4	477.4	0.0	63.1
18	Hotels and Lodging Places	7000	28	21	2,794	155	466.8	392.3	74.5	17.4
19	Other Business Services	7320 7331 7338 7383 7389	30	26	2,384	116	418.8	355.2	63.6	27.8
20	Book Publishing	2731	46	41	472	23	391.8	141.2	250.6	41.1
	Top 20, total				568,216	35,068	33,320	29,251	4,069	–

Source: University of Minnesota IMPLAN Model System.

TABLE 6-6 Summary comparisons of economic models, Washington, DC-MD-VA-WV PMSA, 1990

Model	Excess Activity		Commodity Exports		Commod.	Relative to IMPLAN Totals			
	Employ. (thou.)	Val. Add. (mil.$)	Total (mil.$)	Net (mil.$)	Imports (mil.$)	Ex. Emp. (pct.)	Val. Add. (pct.)	Tot. Exp. (pct.)	Tot. Imp. (pct.)
IMPLAN 528 Sector, total	2,905.7	130,440	45,742	–7,959	53,566	100.0	100.0	100.0	100.0
Excess Activity, total	789.2	45,561	34,379	28,313	5,931	27.2	34.9	75.2	11.1
Top 20: Excess Employment	737.5	41,786	29,012	35,921	2,955	25.4	32.0	63.4	5.5
Top 20: Excess Value Added	678.5	43,251	29,332	25,931	3,266	23.3	33.2	64.1	6.1
Top 20: IMPLAN Exports	568.2	35,068	33,320	29,251	4,069	19.6	26.9	72.8	7.6
Top 40: Excess Employment	783.0	41,234	33,219	27,814	5,271	26.9	31.6	72.6	9.8
Top 40: Excess Value Added	757.2	44,842	32,827	27,314	5,378	26.1	34.4	71.8	10.0
Top 40: IMPLAN Exports	578.2	36,476	38,700	30,166	8,533	19.9	28.0	84.6	15.9

Source: University of Minnesota IMPLAN System.

Econometric Analysis

Econometric methods can also be used to estimate a region's economic base. The simplest form of a multiple regression equation is

$$Y = \alpha + \beta_1 X_1 + \beta_2 X_2 + \varepsilon \tag{6-4}$$

where α, β_1, and β_2 represent the parameters of the model; ε represents the error term; Y represents the dependent variable, and X_1 and X_2 represent the independent variables.

In the economic base model, Y represents a change in one of the measures of economic performance in the region (such as employment, income, and value added) and X_1 and X_2 represent export activity in regional industries 1 and 2. There could be more than two industries in the equation, of course.

This method is particularly helpful when one is trying to determine the impact of a particular industry on the region. In such a case, X_1 might be representative of export activity in the industry being analyzed and X_2 might be export activity by all other regional industries. The βs become a type of export base multiplier that can be applied to the industry.

More complicated econometric methodologies are often used in export base studies. Most of them are based on either the Keynesian notion of the multiplier—including estimates of the marginal propensities to consume, import, and tax—or on neoclassical notions of regional growth.

An alternative view is suggested by another national model on economic development, the so-called Harrod-Domar model of economic growth. Harrod-Domar results, simplified and applied to a regional economy, seem to imply that imports rather than exports provide the development potential for a region. The following equations summarize this view:

$$I = b(dY/dt) \tag{6-5}$$

$$S = sY \tag{6-6}$$

$$M = mY \tag{6-7}$$

where I represents sustainable investment, b represents the capital-to-output ratio for the region, dY/dt represents the time growth path of regional income, Y represents regional income, S represents savings (regional), s represents the marginal propensity to save (1 – b), M represents regional imports, X represents regional exports, and m represents the marginal propensity to import.

$$S + M = I + X \tag{6-8}$$

$$(s + m)Y = b(dy/dt) + X \tag{6-9}$$

$$(s + m)Y - X = b(dy/dt) \quad (6\text{-}10)$$

$$(s + m) - X/Y = b(dy/dt)(1/Y) \quad (6\text{-}11)$$

$$(s + m) - X/Y = dy/dt(1/y) = g \quad (6\text{-}12)$$

Equation 6-8 suggests the time path growth of regional income is some function of the internal rate of savings in the region coupled with supplemental sources of saving through import financing. The m – (X/Y) factor highlights this import factor for the region. This model implies there is an equilibrium rate of growth for the region that will just absorb the region's inputs of finance capital. One might term this as a supply-side or import base model that is used somewhat in contrast to the notion of export base.

Regional Growth Forecasts

The preparation of economic base forecasts is only a step or two removed from the preparation of regional growth forecasts. The major missing element is time. The economic base forecasts pertain to a single time period, with only the results for different periods being compared, one with another. The results are not linked together recursively with the ending calculations for one period being the starting values for the next period. We present two approaches to regional growth forecasts: shift-share analysis and regression analysis.

Shift-Share Analysis

We view the shift-share coefficient—a measure of changes in sources of individual industry and regional growth from one year to the next—as a further step in computational and conceptual complexity. Usually the activity changes are partitioned into three parts: a national-growth effect (NGE), an industry-mix effect (IME), and a regional-share effect (RSE). The shift-share model used in calculating the three change sources of the forecast (t + 1) value of the ith regional activity, R_i, is given by the form

$$R_{i(t+1)} = NGE_{i(t+1)} + IME_{i(t+1)} + RSE_{i(t+1)} \quad (6\text{-}13)$$

$$NGE_{i(t+1)} = R_{it} \times [(N_t + 1/N_t) - 1] \quad (6\text{-}14)$$

$$IME_{i(t+1)} = R_{it} \times [(N_{it+1}/N_{it}) - (N_{t+1}/N_t)] \quad (6\text{-}15)$$

$$RSE_{i(t+1)} = R_{it} \times [(R_{it} + 1/R_{it}) - (N_{it+1}/N_{it})] \quad (6\text{-}16)$$

While the delineation of the three "effects" is somewhat arbitrary, they still serve the purpose of identifying some likely sources of a region's changing economic performance. The national-growth effect is the proportional change in measure of aggregate national economic activity, say, total industry employment, multiplied by the region's measure of specific industry employment, say, the federal government, in the base year t. The industry-mix effect is the difference in the proportional change in the nation's employment and total industry employment, multiplied by the region's specific industry employment in the base year t. The regional-share effect is the difference in the proportional change in the area's employment share of the region's total, multiplied by the region's specific industry employment in the base year t. Thus, the shift-share coefficients show the relative differential change from the overall national average for each industry, that is, the industry-mix effect, and each region, that is, the regional-share effect.

Once the coefficients are given for future periods, they provide a means of forecasting future employment in our example (the auto industry) and in other industries. Given a set of national projections for deriving the national growth and industry mix coefficients, only the regional share coefficients remain unknown. These values for future years are forecast from a simple regression model that calculates future values—subject, of course, to a forecast error that would differ from one industry forecast to the next.

An application of the shift-share approach to regional growth forecasts starts with the historical data series, like the labor earnings presented earlier (Table 6-1). For the present application, we show the reported employment of the roughly one-digit classification code industries in the United States and the District of Columbia MSA cited earlier. We present these calculations only for selected years representing the peaks and troughs of three business cycles, starting with the peak year, 1974, and ending with 1994 (Table 6-7). The object of this application is to extend the historical employment series to the year 2010 using the shift-share approach. For this reason we present both the U.S. and the District of Columbia forecast series. The U.S. forecasts provide us with the given values for estimating the national-growth and industry-mix effects.

The first step in a shift-share analysis is calculating the national growth coefficients for all industry and the individual industry-mix and regional-share coefficients for each industry. The results (Table 6-8) show tremendous variability in the individual coefficient values, with wide differences between the industry-mix and regional-share coefficients from one period to the next. For the District of Columbia MSA, the variability was the least during the period of greatest growth from the recession year 1982 to the peak year 1989. Most of this growth can be attributed to the growth in private services, especially the more specialized, high-order business-related and government-related services.

Table 6-9 provides estimates of the relative importance of the three change sources. National growth is, by far, the largest source of employment change in the District of Columbia MSA, particularly in the period from a recession trough to the peak year of the business cycle, such as 1975–79, 1982–89, and

TABLE 6-7 Total employment in specified industry, U.S. and Washington, DC-MD-VA-WV PMSA, 1974–94

Industry	1974 (thou.)	1975 (thou.)	1979 (thou.)	1982 (thou.)	1989 (thou.)	1992 (thou.)	1994 (thou.)
United States							
Total (full-time and part-time)	99,993	98,674	112,963	114,152	136,414	138,473	144,391
Farm employment	3,926	3,905	3,764	3,657	3,196	3,044	3,001
Agricultural services, forestry, fisheries, other	624	656	864	957	1,341	1,459	1,694
Mining	817	875	1,152	1,503	1,040	928	912
Construction	5,031	4,651	5,890	5,353	7,235	6,763	7,287
Manufacturing	20,397	18,651	21,493	19,266	20,030	18,711	19,025
Transportation and public utilities	5,135	4,971	5,617	5,641	6,375	6,499	6,923
Wholesale trade	4,655	4,870	5,671	5,721	6,709	6,686	6,774
Retail trade	15,264	15,137	17,750	18,172	22,746	23,030	24,276
Finance, insurance, real estate	7,312	7,304	8,506	8,943	10,720	10,324	10,635
Private services	19,407	19,852	23,720	26,398	36,265	39,577	42,239
Federal, civilian	2,902	2,912	2,951	2,940	3,199	3,165	3,080
Military	2,734	2,656	2,425	2,613	2,763	2,638	2,371
State and local	11,790	12,236	13,160	12,987	14,796	15,650	16,174
Washington, DC-MD-VA-WV PMSA							
Total (full-time and part-time)	1,844.0	1,858.9	2,084.8	2,161.1	2,941.5	2,932.8	3,021.3
Farm employment	18.9	17.6	17.3	17.8	14.0	13.9	12.8
Agricultural services, forestry, fisheries, other	8.8	9.3	12.4	15.1	24.0	48.4	52.7
Mining	2.2	2.0	2.3	3.6	4.2	3.0	3.1
Construction	114.8	100.7	115.6	97.2	196.1	140.7	152.3
Manufacturing	73.1	71.1	80.0	83.4	109.0	97.8	101.3
Transportation and public utilities	76.2	73.4	82.0	88.8	129.1	125.7	128.3
Wholesale trade	50.7	51.7	62.6	68.4	91.7	86.5	89.1
Retail trade	251.3	253.7	290.7	303.4	426.9	413.5	428.8
Finance, insurance, real estate	161.2	160.7	178.5	176.7	242.4	222.9	232.6
Private services	428.3	449.5	560.6	643.4	983.3	1,039.6	1,094.0
Federal, civilian	362.7	367.8	389.1	370.9	387.9	406.3	393.1
Military	97.7	100.3	79.3	89.5	95.4	98.6	93.1
State and local	198.0	201.0	214.2	202.5	237.0	249.9	252.9

Source: U.S. Department of Commerce, Regional Economic Information System.

TABLE 6-8 Employment change coefficients, Washington, DC-MD-VA-WV PMSA, 1974–94

Industry	1974–75 (pct.)	1975–79 (pct.)	1979–82 (pct.)	1982–89 (pct.)	1989–92 (pct.)	1992–94 (pct.)
National-Growth Coefficient: U.S.						
Industry-Mix Coefficient: U.S.	-1.3	14.5	1.1	19.5	1.5	4.3
Farm employment	0.8	-18.1	-3.9	-32.1	-6.3	-5.7
Agricultural services, forestry, fisheries, other	6.4	17.3	9.7	20.6	7.3	11.8
Mining	8.3	17.2	29.4	-50.3	-12.2	-6.0
Construction	-6.2	12.2	-10.2	15.6	-8.0	3.5
Manufacturing	-7.2	0.8	-11.4	-15.5	-8.1	-2.6
Transportation and public utilities	-1.9	-1.5	-0.6	-6.5	0.4	2.3
Wholesale trade	5.9	2.0	-0.2	-2.2	-1.9	-3.0
Retail trade	0.5	2.8	1.3	5.7	-0.3	1.1
Finance, insurance, real estate	1.2	2.0	4.1	0.4	-5.2	-1.3
Private services	3.6	5.0	10.2	17.9	7.6	2.5
Federal, civilian	1.7	-13.1	-1.4	-10.7	-2.6	-7.0
Military	-1.5	-23.2	6.7	-13.8	-6.0	-14.4
State and local	5.1	-6.9	-2.4	-5.6	4.3	-0.9
Regional Share Coefficient: PMSA						
Farm employment	-6.1	2.0	5.6	-8.7	3.7	-6.2
Agricultural services, forestry, fisheries, other	0.0	1.5	11.4	18.7	93.2	-7.2
Mining	-16.2	-17.7	26.7	46.5	-17.9	5.5
Construction	-4.7	-11.9	-6.8	66.5	-21.7	0.5
Manufacturing	5.8	-2.7	14.6	26.8	-3.7	2.0
Transportation and public utilities	-0.5	-1.3	7.9	32.3	-4.6	-4.4
Wholesale trade	-2.6	4.7	8.4	16.8	-5.3	1.7
Retail trade	1.8	-2.7	2.0	15.5	-4.4	-1.7
Finance, insurance, real estate	-0.1	-5.4	-6.2	17.3	-4.4	1.3
Private services	2.7	5.2	3.5	15.5	-3.4	-1.5
Federal, civilian	1.1	4.5	-4.3	-4.2	5.8	-0.6
Military	5.5	-12.2	5.1	0.9	7.9	4.5
State and local	-2.3	-1.0	-4.1	3.1	-0.3	-2.1

Source: U.S. Department of Commerce, Regional Economic Information System.

1992–94. During the downturn of the business cycle, the industry-mix effect and/or the regional-share effect will exceed the national-growth effect, often by a very large margin. Private services again are the largest beneficiary of both the national effects (that is, national growth and industry mix) and the regional effects.

Our next step is to extend the historical employment series for the District of Columbia to the year 2010, using the U.S. forecast employment series from the U.S. Department of Commerce. Our first challenge is to extend the 1994 industry employment estimates to 1995. We do this by simply extending the 1993–94

TABLE 6-9 Employment change effects in specified industry, Washington, DC-MD-VA-WV PMSA, 1974–94

Industry	1974–75 (thou.)	1975–79 (thou.)	1979–82 (thou.)	1982–89 (thou.)	1989–92 (thou.)	1992–94 (thou.)	1974–94 (thou.)
National-Growth Effect	–24.3	269.2	21.9	421.4	44.4	125.3	858.0
Industry-Mix Effect							
Farm employment	0.1	–3.2	–0.7	–5.7	–0.9	–0.8	–11.1
Agricultural services, forestry, fisheries, other	0.6	1.6	1.2	3.1	1.8	5.7	13.9
Mining	0.2	0.4	0.7	–1.8	–0.5	–0.2	–1.3
Construction	–7.2	12.3	–11.8	15.2	–15.7	4.9	–2.3
Manufacturing	–5.3	0.5	–9.1	–13.0	–8.8	–2.5	–38.2
Transportation and public utilities	–1.4	–1.1	–0.5	–5.8	0.6	2.8	–5.4
Wholesale trade	3.0	1.0	–0.1	–1.5	–1.7	–2.6	–1.9
Retail trade	1.2	7.1	3.8	17.2	–1.1	4.7	32.9
Finance, insurance, real estate	1.9	3.2	7.3	0.6	–12.6	–2.8	–2.4
Private services	15.5	22.5	57.4	115.0	74.9	25.5	310.8
Federal, civilian	6.0	–48.3	–5.5	–39.7	–10.0	–28.3	–125.7
Military	–1.5	–23.2	5.3	–12.3	–5.8	–14.2	–51.7
State and local	10.1	–13.9	–5.1	–11.3	10.1	–2.3	–12.4
Industry-Mix effect, total	23.3	–41.3	43.0	60.1	30.2	–10.0	105.3
Regional-Share Effect							
Farm employment	–1.2	0.4	1.0	–1.5	0.5	–0.9	–1.7
Agricultural services, forestry, fisheries, other	0.0	0.1	1.4	2.8	22.3	–3.5	23.2
Mining	–0.4	–0.4	0.6	1.7	–0.8	0.2	1.0
Construction	–5.4	–12.0	–7.8	64.7	–42.6	0.7	–2.4
Manufacturing	4.2	–1.9	11.7	22.3	–4.1	1.9	34.2
Transportation and public utilities	–0.4	–0.9	6.5	28.7	–5.9	–5.6	22.4
Wholesale trade	–1.3	2.4	5.2	11.5	–4.9	1.5	14.4
Retail trade	4.5	–6.8	5.8	47.1	–18.7	–7.1	24.8
Finance, insurance, real estate	–0.2	–8.7	–11.0	30.6	–10.6	3.0	3.1
Private services	11.4	23.5	19.5	99.4	–33.5	–15.5	104.8
Federal, civilian	3.8	16.4	–16.8	–15.6	22.5	–2.3	8.0
Military	5.3	–12.3	4.0	0.8	7.5	4.5	9.9
State and local	–4.5	–2.1	–8.8	6.2	–0.7	–5.4	–15.2
Regional Share Effect, total	15.9	–2.1	11.2	298.8	–69.0	–28.5	226.4
Total change	14.9	225.8	76.2	780.3	5.6	86.9	1,189.7

Source: U.S. Department of Commerce, Regional Economic Information System.

TABLE 6-10 Total employment in specified industry, U.S. and Washington, DC-MD-VA-WV PMSA, 1990–2010

Industry	Reported 1990 (thou.)	Estimated 1995 (thou.)	Projected 2000 (thou.)	Projected 2005 (thou.)	Projected 2010 (thou.)
United States					
Total (full-time and part-time)	138,981	148,098	157,656	167,817	176,164
Farm employment	3,168	2,957	3,018	2,962	2,886
Agricultural services, forestry, fisheries, other	1,420	1,802	1,875	2,117	2,318
Mining	1,038	899	829	796	768
Construction	7,208	7.682	7,861	8,373	8,803
Manufacturing	19,756	19,314	18,890	18,887	18,850
Transportation and public utilities	6,595	7,143	7,352	7,766	8,098
Wholesale trade	6,716	6,986	7,406	7,822	8,128
Retail trade	23,020	25,124	26,402	28,020	29,450
Finance, insurance, real estate	10,670	10,740	11,485	12,179	12,737
Private services	38,188	43,696	49,474	54,883	59,379
Federal, civilian	3,289	3,045	3,052	3,007	2,962
Military	2,673	2,259	2,390	2,390	2,390
State and local	15,241	16,451	17,623	18,616	19,398
Washington, DC-MD-VA-WV PMSA					
Total (full-time and part-time)	2,985	3,069	3,242	3,405	3,548
Farm employment	14	12	12	12	11
Agricultural services, forestry, fisheries, other	26	43	44	50	55
Mining	4	3	3	2	2
Construction	181	161	167	180	191
Manufacturing	108	104	104	106	108
Transportation and public utilities	129	130	128	130	130
Wholesale trade	91	93	101	109	115
Retail trade	427	442	461	485	506
Finance, insurance, real estate	237	238	262	285	307
Private services	1,022	1,119	1,236	1,335	1,405
Federal, civilian	397	380	359	334	329
Military	98	91	101	106	106
State and local	250	254	265	271	283

Source: U.S. Department of Commerce, Regional Economic Information System.

employment growth rate for each industry to 1995. We then use the 1990–95 growth rates to extend the 1995 District of Columbia employment forecasts to 2000 and beyond. We summarize the results of these calculations in Table 6-10.

For the purpose of forecast evaluation, we list the change sources for the District of Columbia MSA forecasts in Table 6-11. These findings show a declining positive national-growth and industry-mix effect and a persistently negative

TABLE 6-11 Change sources for employment in specified industry, Washington, DC-MD-VA-WV PMSA, 1990–2010

Industry	1990–95 (thou.)	1995–00 (thou.)	2000–05 (thou.)	2005–10 (thou.)	Total (thou.)
National Growth + Industry Mix					
Total (full-time and part-time)	192.8	227.0	217.8	176.8	814.4
Farm employment	–0.9	0.3	–0.2	–0.3	–1.2
Agricultural services, forestry, fisheries, other	7.0	1.8	5.7	4.8	19.3
Mining	–0.5	–0.2	–0.1	–0.1	–1.0
Construction	11.9	3.9	10.9	9.2	35.9
Manufacturing	–2.4	–2.2	0.0	–0.2	–4.8
Transportation and public utilities	10.8	3.9	7.2	5.5	27.5
Wholesale trade	3.6	5.7	5.6	4.2	19.2
Retail trade	39.1	23.0	28.3	24.8	115.1
Finance, insurance, real estate	1.6	16.7	15.8	13.1	47.2
Private services	147.4	149.1	135.1	109.3	541.0
Federal, civilian	–29.4	1.3	–5.4	–5.0	–38.5
Military	–15.1	5.3	0.0	0.0	–9.8
State and local	19.8	18.4	14.9	11.4	64.5
Regional Share					
Total (full-time and part-time)	–108.1	–54.0	–55.3	–33.9	–251.2
Farm employment	-0.7	–0.4	–0.4	–0.3	–1.7
Agricultural services, forestry, fisheries, other	9.3	0.1	0.0	0.0	9.5
Mining	–0.4	–0.2	–0.1	–0.1	–0.8
Construction	–32.2	1.7	1.8	1.9	–26.7
Manufacturing	–2.3	2.2	2.2	2.2	4.3
Transportation and public utilities	–9.9	–5.7	–5.6	–5.7	–26.8
Wholesale trade	–1.4	2.2	2.4	2.5	5.7
Retail trade	–24.3	–3.9	–4.1	–4.3	–36.7
Finance, insurance, real estate	–0.4	7.2	7.9	8.7	23.4
Private services	–50.5	–32.6	–36.0	–38.8	–157.9
Federal, civilian	11.9	–21.5	–20.4	0.0	–29.9
Military	8.1	4.6	5.1	0.0	17.9
State and local	–15.4	–7.9	–8.2	0.0	–31.4

Source: U.S. Department of Commerce, Regional Economic Information System.

regional-share effect that contributes to a declining of overall employment growth and an absolute decline in private services and government.

The forecast approaches presented to this point use simple mathematical models in the derivation of forecast relationships. This is coupled with use of external data sources on national-growth and industry-mix effects and a procedure for deriving the individual industry regional-share coefficients for each forecast period. Various statistical methods also provide forecasts of many economic activity variables and relationships, like those cited in Chapter 5 relating to an alleged infrastructure gap and its implications for national and regional economic growth. We summarize these additional approaches under the headings of regression analysis and hybrid forecast systems.

Multiperiod Econometric Analysis

Econometric methods, starting with simple least squares regression analysis, can be used to estimate period-to-period changes in a region's economic growth, as well as a region's economic base, and to identify the influence of various factors accounting for these changes. Possible factors explaining regional growth changes in regional economic base are described next.

Access to markets by firms and to jobs by household members varies greatly from area to area and region to region because of the high cost of access, coupled with lack of information and available resources to reduce such costs. When market and job access costs are reduced for individual firms and households, improvements in the competitive position of both eventually follow. Business and personal moving and other relocation costs, however, are important enough for many to deter quick and easy adjustment to changing local economic conditions. Interest rates, which adjust more quickly than wage rates to changing economic conditions, may remain high in some regions because of the uncertainty associated with almost any form of investment. Widely perceived difficulties of access to product markets and a productive local labor force reinforce the initial conditions and legacy of a lagging local economy. A regression analysis of a region's growth prospects that successfully incorporates various measures of risks and uncertainty facing local businesses is difficult to accomplish. Partly for this reason, we present the two models cited earlier that offer an introduction to this topic.

Peterson (1994), cited in Chapter 5, estimated the annual percentage rates of growth of capital per worker (in constant dollars) using a standard production function of the form

$$X_t = A + bT \tag{6-17}$$

X_t, the item under consideration, is the estimated real capital stocks per worker—a number derived from several data sources and related theoretical constructs (Peterson, 1994, pp. 66–67). T is time, and b is the annual rate of growth.

Aschauer (1991, p. 32) also used a simple regression model—a more complex extension of the regression equation (6-4) presented earlier—to estimate the growth rate of net private capital stock, but with more than one equation and several explanatory variables in each equation. This study fitted a highly aggregated U.S. data series on capital stocks and employment to the form

$$Y_{it} = a_i + b_{il}X_{it} + \ldots + b_{mn}X_{nt} + e_{il} \qquad (6\text{-}18)$$

where Y_{it} is the dependent variable (such as real capital stocks per worker or real capital stock annual growth rate), e_{il} is its error term, X_{it} is the first of several explanatory variables with b_{il} being the parameter representing the relationship between the dependent variable and the first explanatory variable, X_{nt} is the n-th explanatory variable, and $b_i \ldots$ are the corresponding regression coefficients.

Aschauer fits his model to three data series from essentially the same sources and related theoretical constructs as Peterson, using the forms

$$dk/k = 0.04 + 0.60 \times dk(-1)/k(-1) + 0.79 \times r - 0.99 \times dkg/k \qquad (6\text{-}19)$$

$$r = 2.52 + 0.006 \times time - 0.27 \times \log(k/n) + 0.09 \times \log(kg/n) + 0.19 \times cu \qquad (6\text{-}20)$$

$$dy/y - dn/n = 0.008 + 0.26 \times (dk/k - dn/n) + 0.39 \times (dkg/kg - dn/n) + 0.43 \times 9dcu/cu) \qquad (6\text{-}21)$$

where dk/k, the growth rate of the net private capital stock, is dependent on its own lagged value, that is, dk(–1)/k(–1), the rate of return to capital, r, and net investment in public nonmilitary capital, dkg/k; r, the net rate of return to private capital stock, is dependent on time, the private capital to labor force ratio, k/n, the net public nonmilitary capital to private labor ratio, kg/n, and the capital utilization rate, cu; and dy/y – dn/n, the difference in annual growth rates between private business sector output and labor force, is dependent on the lagged values of the difference in annual growth rates between private capital stock and labor force, the lagged values of the difference in annual growth rates between nonmilitary capital and labor force, and the annual growth rate in capital utilization.

Adapting capital stock models using national data series, as shown in the preceding examples, introduces additional difficulties in accounting for place-to-place differences in the discounting of interest rates for the perceived levels of risk and uncertainty associated with both private and public investment. For some peripheral areas and places, this means a complete cutoff from available capital resources in the region's financial centers. Even for core areas like the Detroit MSA cited earlier, which has a legacy of high wages and high dependency on a single industry, capital rationing may occur while not occurring in areas for somewhat comparable industry without such a legacy.

Hybrid forecast systems provide some new approaches in forecasting the alternative futures of locally unique economies. They also fit well into metropolitan planning processes of locally unique economies, but with an entrepreneurial legacy favored by the region's financial community.

Hybrid Forecast Systems

A variety of population and economic forecast systems have currency in state and regional planning and administration and in the marketing departments of the larger private corporations. These range from one- and two-year revenue forecasts of state finance departments to five- and ten-year and even more distant population forecasts of state transportation departments and metropolitan regional councils. They include five- to ten-year employment forecasts of state departments of labor, or their equivalents, and the multiyear market forecasts of the larger manufacturers. Public sector forecasts serve special agency purposes and hence lack overall coordination of methodology and results. The private sector forecasts achieve some degree of consistency among themselves with the use of common data sources and forecast providers. They achieve varying degrees of credibility depending upon the agency using the forecasts.

Given the tools and concepts for measuring a region's performance, the next step is one of really understanding how a region's performance changes over time and then being able to accurately anticipate these changes. Earlier we noted that the export-producing sectors of a region's economy are extremely important for its long-term viability. Loss of export markets means a loss of jobs, not only in the export-producing industries, but also in other industries dependent on the purchases of these industries. Indirectly, this includes businesses catering to the households and families of the workers in both the export-producing and supporting industries. We might envision still another scenario involving efforts to intentionally change a region's performance by introducing various incentives and penalties affecting households, businesses, and even government agencies. This scenario, while politically controversial and often difficult to achieve, can be represented by predictable changes in an urban regional economy.

Economic System Forecasts

Economic system forecasts in the U.S., like hybrid forecast systems, are of many types, ranging from the forecasts of rural local economies to large metropolitan regions. The common denominator in these forecasts is a common database, compiled on a county basis. They differ in being single-period or multiperiod, readymade or proprietary, partial equilibrium or general equilibrium. We start with the single-period, partial equilibrium model of inter-industry transactions.

Inter-industry Transactions and Output Multipliers

An input-output, or inter-industry transactions, model shows the economic linkages among industries within a given economy. Each industry (or groups of similar industries combined into industry sectors) not only produces goods or services, but is also a consumer itself, purchasing other goods and services for the production process. Input-output analysis permits the simultaneous determination of all of these product flows—both sales and purchases among industry sectors.

The first applications of inter-industry transactions tables of the United States were developed by Wassily Leontief in the 1940s for fewer than 50 industry sectors. He used a 1939 table to predict post–World War recovery capacity and demand. A larger and more detailed table for 1947 was developed with support from the U.S. Air Force. Since then, inter-industry transactions, or input-output, tables with more than 500 industry sectors have been made available. In theory, the model attempts to make operational the concept of general equilibrium.

Figure 6-1 illustrates the aggregate structure of inter-industry transactions. It shows all income payments, that is, purchases made by all economic activities and their distribution among all institutions. Its numbers represent individual sector purchases of inputs and disbursements in four quadrants. Quadrant I contains a common measure of inter-industry transactions, namely, the proportion of total industry purchases (columns) from each input-supplying sector (row), including both intermediate and primary. Quadrant II shows—as a proportion of the total—the final demand sector purchases (columns), namely, households, government, business investment, and exports, from each producing sector (the same as the input-supplying industry in Quadrant I) disbursing its final product (row) to the final demand sectors. The final demand coefficients represent the proportion of total final product purchases from each producing sector. Quadrant III shows the proportional purchases of the primary inputs of labor and capital (rows) by the final demand sectors (columns). Quadrant IV shows the proportional purchases of primary inputs (rows) by each of the producing sectors, namely, industries (columns).

Purchases from the input-supplying industries are partitioned further into locally produced inputs and imported inputs. By introducing a change into the

FIGURE 6-1 Structure of inter-industry accounts, by originating and receiving sector

		Institutions				
Originating Sector	Industries	Private	Govt.	Capital	World	Total
Commodities	T_{11}	T_{12}	T_{13}	T_{14}	T_{15}	T_{16}
Value added	T_{21}	T_{22}	T_{23}	T_{24}	T_{25}	T_{26}
Rest of world	T_{31}	T_{32}	T_{33}	T_{34}	T_{35}	T_{36}
Total	T_{41}	T_{42}	T_{43}	T_{44}	T_{45}	T_{46}

final demand sectors (Quadrant III), the effects from this change on individual sectors and the total economy are readily forecast, again, subject to the forecast error. This type of forecast differs from those presented earlier in several ways, including the assumption that all other values remain the same, except for the postulated change in final demand and the calculated impact of this change on both the individual sectors and the total economy.

A description of each transaction submatrix in Figure 6-1 provides further clarification of the content and uses of the input-output model. The individual model elements, by row, are as follows:

T_{11}, the industry use matrix to show the commodity composition of industry use of intermediate inputs from local sources.

T_{12}, the commodities final demand account to show the purchases of locally produced commodities by resident firms and households.

T_{13}, the commodities final demand account to show purchases of locally produced commodities by resident government offices—federal, state, and local.

T_{14}, the commodities final demand account to show purchases of locally produced commodities in private capital formation and inventory additions.

T_{15}, the commodities export trade account to show purchases of locally produced commodities by nonresident firms, households, and governments, including visitors.

T_{16}, the total output of locally produced commodities.

T_{21}, the value-added account purchases of labor and capital inputs by local firms.

T_{22}, the value-added account purchases of labor inputs by local households.

T_{23}, the value-added account purchases of labor inputs by non-enterprise local government.

T_{24}, the value-added account purchases of capital inputs by resident firms.

T_{25}, the value-added account adjustments for purchases of capital inputs locally by nonresidents.

T_{26}, the total value added by resident labor, capital, and entrepreneurial resources.

T_{31}, the industries import account to show the purchases of imported goods and services by resident industries.

T_{32}, the private institutions import account to show transfers from rest of world to resident private institutions.

T_{33}, the government institutions import account to show transfers from rest of world to resident government institutions.

T_{34}, the private capital sector import account to show transfers from rest of world to resident private capital formation sector.

T_{35}, the rest of world import account to show transfers from rest of world import account to rest of world export account.

T_{36}, the total imports from rest of world.

T_{41}, the total industry output (and purchases).

T_{42}, the total private institutions output (and purchases).

T_{43}, the total public institutions output (and purchases).

T_{44}, the total capital account output (and purchases).

T_{45}, the total rest of world purchases from resident industry (as exports), value-added account, and institutions.

T_{46}, the total value of all accounts.

While input-output analysis consists of a number of different types of accounts, the transactions accounts are the most commonly used. These accounts show the flows of commodities from each of a number of producing sectors to all other consuming sectors, both intermediate and final. Intermediate transactions represent the flow of goods and services that are produced and consumed in the process of current production. Final demand represents the ultimate purchases by consumers or government inside or outside of the region for which the table has been constructed. Final inputs include payments to households (as wages and salaries), proprietor income, business taxes, and imports.

The use accounts are also of interest in this study in that they show the value of each commodity used by each sector. The term commodity is used here to represent the output of a given industry, since some industries may produce more than one commodity. Input-output models assume the existence of a *general equilibrium* for each regional economy. Probably the more realistic assumption is that of disequilibrium from one period to the next, with general equilibrium being the tendency, but not the reality, at any given moment.

General equilibrium recognizes that every economic unit in an economy relates, directly or indirectly, with every other unit. In partial equilibrium analysis, when the demand for corn increases, for example, the price of corn is expected to rise, while everything else is held constant. Everything is not constant for very long, however. The rising price of corn entices entrepreneurs to put additional resources into the production of that commodity. Farmers respond to higher corn prices by bidding resources out of non–corn-related production. This means that the price of resources in corn production will increase relative to other production options. It also means a shift of resources out of alternate production options in favor of corn-growing activities.

The demand for corn might have increased for any number of reasons. There might have been an increase in the price of wheat, causing consumers to switch to corn because of its relatively low cost. Or, there might have been an increase in tastes related to corn consumption. Or, new customers might have entered the market for corn, as was the case when the (former) Soviet Union began to purchase grains from the United States.

Whatever the reason, relative prices of corn and other commodities change, relative productions change (as the resources are taken out of other forms of production and put into corn), and relative prices of resources change, all as a result of the initial change in demand. In fact, if there are unemployed resources, or if the new resources are put into the market because of higher resource returns, there is even a change in the general level of resource use associated with this change in demand.

In short, virtually every sector of the economy is changed, at least somewhat, due to this initial change in demand for one commodity. These changes continue to occur until the total adjustment to the initial change has had time to play itself out and the economy settles down into a new "equilibrium" position. General equilibrium would simultaneously solve for all of these changes and would simultaneously estimate new variable values associated with the shifting demands and supplies for all goods, services, and resources.

Although the general equilibrium model is conceptually complete, it has the deficiency of being nonoperational for empirical analysis. Even if there were a computer large enough to solve for all prices and quantities in an economy, the data requirements of such a model would be out of reach. However, the alternative of looking at a change in one industry or resource market at a time, assuming all other markets to be constant and unaffected, creates an extremely unrealistic estimate from another direction.

Input-output attempts to move toward general equilibrium while staying within the data and computational capabilities of most research projects. Certain assumptions have to be made in order to make general equilibrium more operational:

1. Prices are assumed to be constant. This means that the model no longer solves for this variable, but, rather, makes price a parameter to the equation system, thereby eliminating the solution requirements for one-set variables.
2. Individual firms are assumed to be capable of being meaningfully aggregated into industrial sectors. This eliminates the requirement of solving for output levels of every firm in the economy and reduces the number of production equations to a manageable size.
3. The production function for the region is assumed to be linear and homogeneous. This simply means that the production-input pattern is constant

for all levels of regional output. A doubling of output requires that all inputs also be doubled, a tripling of output requires that all inputs be tripled, and so on.

4. Production coefficients, per dollar of output, are assumed to be constant. This means that if industrial sector A requires $0.10 of input from sector B, in order to produce $1 worth of sector A's output, it will require $0.10 per dollar no matter how many total dollars of A's output are produced.
5. Intermediate production requirements by local industries from other local industries are assumed constant. This means there is no substitution through changing trade patterns between regions.

These assumptions have the effect of greatly reducing the number of computations necessary for the implementation of a full-scale general equilibrium model. Now, instead of having a separate equation for every firm in the market, there are only as many equations as there are identified aggregate industrial sectors. Instead of a separate equation for every resource market, the resources are lumped into an exogenous final payments sector.

Table 6-12 provides an overview of the resident commodity-related economic accounts of a District of Columbia MSA input-output table. The total locally produced commodity output (T16) is $189.7 billion, of which $37.2 billion is the total of inter-industry transactions (T11). Total local final demand (T42 + T43 + T44) is $167 billion, while the export demand (T45) is $45.1 billion, of which only 11 percent is from foreign demand. Imports (T36) total to $59.8 billion with local final demand (T32 + T34 + T35) accounting for $30.9 billion, 57.3 percent of the total. The bottom row lists the total industry and institutional (that is, household and government, not listed under industry) employment.

Table 6-13 presents a one-digit industry breakdown of intermediate and primary input purchases (T11, T21, and T31 accounts). Local manufacturing is the largest market for imported goods and services, followed closely by private services. Government is the largest market for primary inputs of labor, followed closely by private services. Private services are the largest market for locally produced intermediate inputs, followed by finance, insurance, and real estate. The latter is also the third largest market, of the nine one-digit industries, for primary inputs, largely capital inputs. Government and related services thus account for the unique and specialized industry structure of the District of Columbia MSA.

Table 6-14 lists the goods and services purchased by the local final demand sectors—households, government, and private investment. It also lists the total employment attributed directly to these sectors rather than the individual industries listed in Table 6-13. Local final purchases exceed $138 billion, with households

TABLE 6-12 Intermediate and final demand sector purchases of specified industry output, Washington, DC-MD-VA-WV PMSA, 1990

				Final Demand				
Commodity	SIC Code	Total Purchases (mil.$)	Intermed. Demand (mil.$)	Total (mil.$)	Local (mil.$)	Export (mil.$)	Dom.Exp. (mil.$)	For.Exp. (mil.$)
Local and imports, total	01–89	374,475	161,090	213,385	166,972	46,413	40,664	5,749
Local	01–89	189,690	37,215	152,475	107,334	45,141	40,664	4,477
Agriculture, forestry, fisheries	01, 02, 07–09	1,782	438	1,344	88	1,256	1,183	73
Mining	10–14	379	24	355	8	347	340	8
Construction	15–17	19,301	2,917	16,384	15,458	926	922	4
Manufacturing	20–39	13,629	3,139	10,490	3,341	7,149	5,793	1,356
Transportation, communication, public utilities	40–49	16,614	5,755	10,860	7,631	3,229	2,599	63
Trade	50–59	18,173	2,365	15,808	13,759	2,049	1,229	819
Financial, insurance, real estate	60–67	33,339	7,812	25,527	15,760	9,767	8,980	787
Private services	70–89	53,684	13,222	40,462	22,702	17,760	16,985	775
Government enterprise	4311,pt41,pt491	32,790	1,544	31,245	28,587	2,659	2,634	25
Primary inputs, local	–	130,718	101,462	29,256	28,694	563	0	563
Employee compensation	–	89,834	60,942	28,892	28,892	0	0	0
Indirect business taxes	–	7,763	7,763	0	0	0	0	0
Proprietary income	–	5,920	5,920	0	0	0	0	0
Other property income	–	27,201	26,837	364	–198	563	0	563
Imports, total	–	53,851	22,197	31,654	30,944	710	0	710
Domestic imports	–	47,818	20,005	27,813	27,112	702	0	702
Foreign imports	–	6,033	2,192	3,841	3,833	8	0	8
Employment (thousand)	–	3,045.5	2,351.6	693.9	693.9	0.0	0.0	0.0

Source: University of Minnesota IMPLAN system.

TABLE 6-13 Industry payments for specified intermediate and primary inputs, Washington, DC-MD-VA-WV PMSA, 1990

Input Source	Total (mil.$)	Agric. (mil.$)	Mining (mil.$)	Constr. (mil.$)	Manuf. (mil.$)	TCPU (mil.$)	Trade (mil.$)	FIRE (mil.$)	Priv. Serv. (mil.$)	Govt. (mil.$)
Industry input purchases, total	161,090	1,782	379	19,301	13,635	16,614	18,176	33,339	53,684	4,180
Intermediate inputs, local	37,214	528	39	4,763	2,885	3,690	2,212	8,145	14,096	856
Agriculture, forestry, fisheries	438	97	0	29	37	1	13	215	41	4
Mining	24	0	1	10	6	4	0	0	1	1
Construction	2,917	31	2	25	56	724	53	1,012	685	328
Manufacturing	3,139	21	2	596	787	326	261	182	939	25
Transportation, communications, public utilities	5,755	75	14	678	627	1,047	493	534	2,006	279
Trade	2,365	59	2	1,257	302	107	82	281	448	27
Finance, insurance, real estate	7,812	91	9	233	174	403	385	3,911	2,563	43
Private services	13,222	138	9	1,878	697	932	813	1,913	6,752	92
Government	1,544	16	1	59	198	145	111	296	661	57
Primary inputs, local	101,462	495	260	7,827	6,502	10,815	14,974	23,688	33,980	2,920
Employee compensation	60,942	494	68	6,421	4,067	4,449	11,027	5,039	26,659	2,718
Indirect business taxes	7,763	17	39	75	113	695	2,333	3,925	565	1
Proprietary income	5,920	160	46	870	60	470	324	71	3,918	0
Other property income	26,837	–175	107	461	2,261	5,202	1,290	14,652	2,838	200
Intermediate inputs, nonlocal	22,197	756	80	6,659	4,229	2,094	989	1,444	5,541	405
Domestic imports	20,005	729	47	6,095	3,648	1,639	870	1,415	5,189	372
Foreign imports	2,192	26	33	563	581	456	120	29	352	32
Employment (thousand)	2,352	29.4	2.5	230.7	106.9	117.3	507.5	206.0	985.2	166.0

Source: University of Minnesota IMPLAN System.

TABLE 6-14 Local final purchases of specified industry output, Washington, DC-MD-VA-WV PMSA, 1990

Industry	Local Fin.Dem. (mil.$)	Pers. Cons. Expend.			Federal Govt.		State/Local Govt.		Private Invest. (mil.$)
		Low (mil.$)	Medium (mil.$)	High (mil.$)	Non-mil. (mil.$)	Military (mil.$)	Non-educ. (mil.$)	Education (mil.$)	
Local final purchases, total	138,278	5,832	24,077	46,576	19,222	14,157	8,014	2,989	17,410
Local, total	107,334	3,998	16,369	32,056	17,703	11,260	6,838	2,712	16,398
Agriculture, forestry, fisheries	88	7	23	41	0	1	11	3	2
Mining	8	1	2	4	0	0	1	0	0
Construction	15,458	0	0	0	80	786	620	387	13,586
Manufacturing	3,341	98	423	713	101	428	152	15	1,412
Transportation, communication, utilities	7,631	313	1,243	2,297	817	2,323	339	50	248
Trade	13,759	885	3,817	7,938	103	333	109	27	546
Finance, insurance, real estate	15,760	1,194	5,323	8,720	18	6	40	60	400
Services	22,702	1,393	5,104	11,590	771	3,175	378	95	195
Government	28,587	107	433	754	15,811	4,208	5,190	2,075	10
Primary inputs, local	28,694	278	0	0	4,420	16,659	2,174	5,345	–183
Employee compensation	28,892	293	0	0	4,420	16,659	2,174	5,345	0
Indirect business taxes	0	0	0	0	0	0	0	0	0
Proprietary income	0	0	0	0	0	0	0	0	0
Other property income	–198	–15	0	0	0	0	0	0	–183
Imports, total	30,944	1,834	7,708	14,521	1,519	2,897	1,176	278	1,102
Domestic	27,112	1,637	6,857	12,830	1,470	2,263	997	150	907
Foreign	3,833	196	850	1,691	49	634	179	128	105
Employment (thousand)	693.9	36.2	0.0	0.0	96.1	337.0	88.1	136.5	0.0

Source: University of Minnesota IMPLAN System.

accounting for the largest share of the total—$75.5 billion, or 54.6 percent of the total. Federal government accounts for $33.4 billion, or 24.1 percent of the total, while the state and local government accounts for $11 billion, or 8 percent. Private capital formation accounts for the remaining $17.4 billion, or 12.6 percent of the total. The federal government, as an important part of the local economic base, contributes even more to the local economy through its various income and employment multipliers as well as its inherent capacity to bring new dollars into the local economy.

A comparison of several of the accounts in Tables 6-12 and 6-14 reveals the importance of private services and government in the local economy. While the private services sector accounts for 28.3 percent of the total area output, its purchases locally are second to those of the government sector at 21.2 percent of total local purchases of all goods and services produced locally. The government sector—federal, state, and local—accounts for 26.6 percent of the local purchases of all locally produced goods and services.

The next series of tables derive from the inter-industry transactions listed in Table 6-13. The first of the series, Table 6-15, is a table of technical coefficients showing the value of the given input purchases per $1 of total input purchases by the given industry. The larger the input purchase per $1 of total purchases, the greater the indirect impact of this industry's input purchases on the local economy. The inter-industry input, or backward, linkages are quite small, with private services having the largest total. The low values are due, of course, to the large labor component in local services and the lack of a large and diverse manufacturing sector.

Table 6-16 lists the Leontief inverse derived from the technical coefficients. The total value for each column is the simple output multiplier consisting of its two parts—a direct and an indirect effect. The direct effect is represented by a diagonal value one in the $n \times n$ matrix (n being nine in this example). The actual diagonal value is the indirect effect derived from intra-industry transactions. The off-diagonal values are the indirect effects of transactions with other local input-supplying industries.

Table 6-17 list two sets of multiplier values for the local economy—the industry output multipliers and the combined industry and household multiplier values. The industry (output) multipliers are the column totals in Table 6-16. The combined multipliers incorporate the effects of employee and proprietor spending of their earned personal income, that is, the household indirect effects, according to the given personal consumption expenditure schedule.

The final two tables present the predicted effects of a $10 billion change in two industry outputs—government and private services—using the technical

TABLE 6-15 Technical coefficients of direct purchases per $1 of industry output, Washington, DC-MD-VA-WV PMSA, 1990

Industry	Agric.	Mining	Constr.	Manuf.	TCPU	Trade	FIRE	Priv.Serv.	Govt.
Agriculture, forestry, fisheries	0.055	0.000	0.001	0.003	0.000	0.001	0.006	0.001	0.000
Mining	0.000	0.002	0.001	0.000	0.000	0.000	0.000	0.000	0.000
Construction	0.018	0.005	0.001	0.004	0.044	0.003	0.030	0.013	0.010
Manufacturing	0.012	0.005	0.031	0.058	0.020	0.014	0.005	0.018	0.001
Transportation, communications, public utilities	0.042	0.036	0.035	0.046	0.063	0.027	0.016	0.037	0.009
Trade	0.033	0.005	0.065	0.022	0.006	0.004	0.002	0.008	0.001
Finance, insurance, real estate	0.051	0.024	0.012	0.013	0.024	0.021	0.117	0.048	0.001
Private services	0.077	0.022	0.097	0.051	0.056	0.045	0.057	0.126	0.003
Government	0.009	0.003	0.003	0.015	0.009	0.006	0.009	0.012	0.002
Total	0.297	0.102	0.247	0.212	0.222	0.122	0.244	0.263	0.026

Source: University of Minnesota IMPLAN System.

TABLE 6-16 Leontief inverse of technical coefficients matrix, Washington, DC-MD-VA-WV PMSA, 1990

Industry	Agric.	Mining	Constr.	Manuf.	TCPU	Trade	FIRE	Priv.Serv.	Govt.
Agriculture, forestry, fisheries	1.058	0.000	0.002	0.003	0.001	0.001	0.008	0.002	0.000
Mining	0.000	1.002	0.001	0.001	0.000	0.000	0.000	0.000	0.000
Construction	0.025	0.008	1.006	0.009	0.049	0.006	0.037	0.019	0.011
Manufacturing	0.018	0.007	0.037	1.065	0.026	0.017	0.010	0.024	0.002
Transportation, communications, public utilities	0.056	0.041	0.047	0.057	1.075	0.033	0.025	0.050	0.010
Trade	0.039	0.006	0.068	0.025	0.012	1.006	0.007	0.012	0.002
Finance, insurance, real estate	0.070	0.031	0.024	0.022	0.035	0.029	1.139	0.065	0.002
Private services	0.108	0.032	0.122	0.070	0.079	0.057	0.082	1.156	0.005
Government	0.012	0.004	0.006	0.017	0.011	0.008	0.012	0.016	1.002
Total	1.387	1.131	1.314	1.268	1.288	1.158	1.320	1.342	1.034

Source: University of Minnesota IMPLAN System.

TABLE 6-17 Output multipliers for specified industry, Washington, DC-MD-VA-WV PMSA, 1990

Industry	Direct Effect	Indirect Effect		Output Multiplier		Per Capita
		Industry	Household	Industry	Combined	PCE ($)
Agriculture, forestry, fisheries	1.000	0.387	0.776	1.387	2.163	17
Mining	1.000	0.131	0.277	1.131	1.409	2
Construction	1.000	0.314	0.602	1.314	1.916	0
Manufacturing	1.000	0.268	0.390	1.268	1.658	292
Transportation, communications, public utilities	1.000	0.288	0.363	1.288	1.650	912
Trade	1.000	0.158	1.053	1.158	2.211	2,993
Finance, insurance, real estate	1.000	0.320	0.335	1.320	1.656	3.608
Private services	1.000	0.342	0.812	1.342	2.154	4,283
Government	1.000	0.034	0.787	1.034	1.821	306
Total	1.000	0.335	0.800	1.335	2.136	12,413

Source: University of Minnesota IMPLAN System.

coefficients and multipliers cited in Table 6-17. Table 6-18 lists the combined industry and household spending impacts on each of the nine industry groups representing the District of Columbia economy on two industry variables—total industry output and total value added. The industry changes differ for the two sets of industry output changes—an $18.3 billion overall impact for the government output change and $21.7 billion overall impact for the private services change. For value added, the totals are $15.4 billion and $14.3 billion, respectively, a reversal of the earlier changes in magnitude because of the higher value added per worker for government than for private services. The percentage changes in output and value added are the same for a given industry. They differ, however, by source of change, whether government or private services. The relative importance of a given input for the industry introducing the change in output makes a difference, finally, in the overall change, as shown by the differences in the bottom percentages in Table 6-18.

Table 6-19 lists the individual industry changes in employment and employee compensation attributed to the two sets of changes in total industry output. The patterns of changes cited earlier occur again with the individual industry changes in employment and employee compensation because of the constancy of industry production relationships, along with the initial differences in these relationships from one industry to the next. The employment impact for government is less than for private services, for example, as a result of the relatively high employee compensation per government workers, coupled with a ratio of employee compensation to total output and total value added.

TABLE 6-18 Base year values and impacts of a $10 billion change in specified industry output, D.C. PMSA, 1990

	Industry Output*			Value Added			Chg. Ind. Out.		Chg.Val.Add.	
Industry	Base (bil.$)	Govt. (bil.$)	Pvt.Serv. (bil.$)	Base (bil.$)	Govt. (bil.$)	Pvt.Serv. (bil.$)	Govt. (pct.)	Pvt.Ser. (pct.)	Govt. (pct.)	Pvt.Ser. (pct.)
Agriculture, forestry, fisheries	1.8	0.0	0.0	0.5	0.0	0.0	1.7	2.5	1.7	2.5
Mining	0.4	0.0	0.0	0.3	0.0	0.0	0.4	0.5	0.4	0.5
Construction	19.3	0.3	0.3	7.8	0.1	0.1	1.3	1.8	1.3	1.8
Manufacturing	13.6	0.3	0.5	6.5	0.1	0.2	2.1	3.7	2.1	3.7
Transportation, communications, public utilities	16.6	0.8	1.2	10.8	0.5	0.8	4.9	7.4	4.9	7.4
Trade	18.2	1.6	1.7	15.0	1.3	1.4	8.7	9.5	8.7	9.5
Finance, insurance, real estate	33.3	2.3	3.0	23.7	1.6	2.1	6.9	9.0	6.9	9.0
Private services	53.7	2.8	14.4	34.0	1.8	9.1	5.3	26.9	5.3	26.9
Government	32.8	10.2	0.4	31.5	9.9	0.4	31.3	1.2	31.3	1.2
Total	189.7	18.3	21.7	130.1	15.4	14.3	9.7	11.4	11.8	11.0

Source: University of Minnesota IMPLAN System.
*Change in total industry output of $10 billion in government and private services, respectively.

TABLE 6-19 Base year values and impacts of a $10 billion change in specified industry output, DC PMSA, 1990*

Industry	Employment			Employee Compensation			Chg.Employ		Chg.Emp.Comp.	
	Base (thou.$)	Govt. (thou.$)	Pvt.Serv. (thou.$)	Base (bil.$)	Govt. (bil.$)	Pvt.Serv. (bil.$)	Govt. (pct.)	Pvt.Ser. (pct.)	Govt. (pct.)	Pvt.Ser. (pct.)
Agriculture, forestry, fisheries	29	0.5	0.7	0.5	0.0	0.0	1.7	2.5	1.7	2.5
Mining	3	0.0	0.0	0.1	0.0	0.0	0.4	0.5	0.4	0.5
Construction	231	3.0	4.1	6.4	0.1	0.1	1.3	1.8	1.3	1.8
Manufacturing	107	2.2	4.0	4.1	0.1	0.2	2.1	3.7	2.1	3.7
Transportation, communications, public utilities	117	5.7	8.7	4.4	0.2	0.3	4.9	7.4	4.9	7.4
Trade	508	44.2	48.4	11.0	1.0	1.1	8.7	9.5	8.7	9.5
Finance, insurance, real estate	206	14.2	18.5	5.0	0.3	0.5	6.9	9.0	6.9	9.0
Private services	985	52.0	264.7	26.7	1.4	7.2	5.3	26.9	5.3	26.9
Government	720	225.1	8.7	31.3	9.8	0.4	31.3	1.2	31.3	1.2
Total	2,906	346.9	357.8	89.5	12.9	9.7	11.9	12.3	14.4	10.8

Source: University of Minnesota IMPLAN System.
*Change in total industry output of $10 billion in government and private services, respectively.

Thus, the input-output model represents a detailed accounting of the economic base of a region. It can be used to delineate the export structure of the regional economy and the multipliers that emerge from that structure. It also identifies, in final demand, the relationship between local activity, investment, and export activity in relation to the identified industrial structure. As in most models, its weakness is in its assumptions. But at the least, the input-output system can be used for simulations and sensitivity analyses for a regional economy.

Computable Spatial Price and General Equilibrium Models

More than a quarter century ago, Judge and Takayama edited *Studies in Economic Planning over Space and Time* (1973). Most of the 35 chapters in this volume discussed various forms and uses of computable spatial price equilibrium models already in existence and implemented with available statistics.

More than a decade earlier, North-Holland published Volume 21 in the same series—*A Multi-Sectoral Study of Economic Growth* by Leif Johansen (1960). The Johansen study presents one of the first computable general equilibrium models—in this case, for Norway's economy. It was, like the later computable spatial price equilibrium models, constructed ". . . with an eye to the possibility of it being implemented by existing statistics" (p. 1). We reference these studies as reminders of the highly specialized and sophisticated modeling studies already in existence decades ago as contributions to economic analysis and planning.

Often cited in a critique of economic predictions based on complex modeling systems is the lack of standard statistical measures of variance and error, other than observing the difference between the predicted (or simulated) values and the corresponding actual values (Miller and Blair, 1985). More critical, however, to the "real world" practitioner is the adequacy and relevance of the forecasts for their intended clients and decision applications.

A growing body of literature addresses this question by comparing the results of several regional modeling systems (Brucker, Hastings, and Latham III, 1987; 1990). The results reviewed include measures of local industry output and employment, as well as population and income, responses to given changes in demand for the locally produced output (Crihfield and Campbell, 1991, 1992; Grimes, Fulton, and Bonardelli, 1992). They include also some comparisons of the critical model parameters, like import propensities (Rickman and Schwer, 1993). We cite these technical issues again with reference to the construction and use of readymade modeling systems.

Multiperiod Forecast Models

Multiperiod models go the additional step of taking the model outputs of one period and introducing them as inputs for the next period, along with inputs representing external conditions affecting a region's economic activities. This requires, first, the linking of one-period-at-a-time calculations into a multiperiod series of changes from the initial period. It also requires the building of some sort of a model for generating these changes, taking into account interdependencies among variables. These interdependencies are found both within the model, that is, the area economy being modeled, and outside the model.

These changes can be calculated simultaneously, that is, by one large equation system, or recursively, by a series of smaller equation systems linked together within a single time period and from one time period to the next. The differences between the two approaches become quite apparent in their application, as well as in their construction and calibration. The one approach requires a high level of aggregation that leaves very little room for real-world application, while the other allows the highest levels of disaggregation in the critical economic series, like industry employment, earnings, and value added.

The multiperiod models usually are calibrated—first, to the preceding historical period and second, to the given forecasts of external conditions. The calibration step is critical in designing systems such as regional forecasting tools. They, of course, lack the capability for making statistically valid estimates of forecast error and variance, which is a capability that is easily overlooked, given its stringent condition of all other things remaining the same.

The recursively solved, multiperiod forecast model provides an analyst with flexible, interactive techniques for generating a series of forecasts with different sets of assumptions. Examples include changes in federal forest-related policies or local land use density regulations that are linked to certain variables and parameters in the modeling system. Each new series of forecasts are simulations of alternative futures. They show how a particular area economy will react to supply-side and demand-side changes associated with the different policies and regulations (Treyz, 1993; Treyz, Rickman, and Shao, 1992).

One such modeling system, called IPASS (Interactive Policy Analysis Simulation System), for example, has eight basic elements that are recursively linked around an inter-industry transactions table—a Leontief matrix that is the only simultaneously solved element (Maki, Olson, and Schallau, 1985). Besides the inter-industry transactions table, which serves as the production module, there are seven other modules: investment, final demands, regional output, employment, labor force, population, and primary inputs.

The investment module calculates investment needed to expand capacity in order to produce more goods and services. It is connected to the final demand module, which incorporates the postulated policy changes and forecasts of external final demand, that is, exports.

The production module performs the conventional multiplier calculations of the individual industry impacts of changes in the demand for an area's industry output.

The regional output module responds to the production constraints emanating from the demand side via the final demand module and the supply side via the investment and labor-force modules.

The employment module updates model parameters that influence labor productivity. The labor-force module calculates the supply of labor by occupation classes. The population module uses migration and cohort survival rates, as well as age-specific birth rates, to forecast year-to-year changes in an area's population.

Components of value added, including components of personal income, are calculated by the primary input module. This module includes all value components—employee compensation, proprietary income, indirect taxes, depreciation allowance, and profits.

A Readymade Input-Output Model

The point of departure for this assessment is the suggestion that ". . . a truly flexible readymade model will enable the introduction of survey-based trade coefficients in some sectors while continuing to balance the rest of the sectors in a truly unbiased manner" (Brucker, Campbell, and Latham III, 1990, p. 136). System effectiveness requires not only a truly flexible model but one that invites "coefficient fix-up" with superior information, coupled with ". . . software and/or handbooks that guide the user (professional or lay) through the intricacies of final demand determination" (p. 137).

Forecasting Area Economic Impacts

Use of the IMPLAN regional modeling system as an impact prediction model starts with the existing database. The U.S. Department of Commerce Regional Economic Measurements Division Annual Regional Economic Information System (REIS) series covering industry employment, labor earnings, total population, and total personal income is a common starting place.[3] The historical (REIS) series include every county in the United States. They cover total employment and total labor earnings in a two-digit industry breakdown based on the 1987 Standard Industrial Classification Manual.

The U.S. IMPLAN database calibrates to the REIS series. The IMPLAN series also use the individual state ES-202 covered (by the cooperative federal-state unemployment insurance program) employment and payroll files, especially for the three- and four-digit industry groups that are not available in the REIS database. IMPLAN has a 528-sector industry breakdown for each of 3,120 counties in the United States.

The 1988 U.S. Department of Commerce Office of Business Economics Regional Series (OBERS) on industry employment, labor earnings, total population, and total personal income extend the corresponding 57-industry REIS series to 2040.[4] The 1988 OBERS series calibrate to the 1988 U.S. Bureau of Labor Statistics (BLS) moderate projection series. High and low projection series, which are derived for individual states, MSAs, and the Bureau of Economics economic areas in the auxiliary IMPLAN database, correspond to the U.S. BLS high and low projection series (Kutcher, 1991).

The IMPLAN database extends the OBERS series to equivalent measures of industry output and commodity production in a long-term forecast mode. It further allocates the commodity production to intermediate and final demand sectors in the United States and in each of the 50 states. The intermediate demand sectors include the two-digit industry groups in the OBERS sectors. Individual industries in the 528 sectors of the IMPLAN database aggregate to the three- and four-digit BLS sectors, the two-digit OBERS sectors, and many other combinations of two- and three-digit industry groups.

The final demand sectors in the IMPLAN database include (1) personal consumption expenditures, (2) gross private capital formation, (3) change in business inventory, (4) federal government purchases, (5) state and local government purchases, (6) exports, and (7) imports. Regional purchase coefficients (RPCs) that allocate imports to each local purchasing sector are calculated for each IMPLAN "model" (that is, a county or multicounty impact assessment). The uniquely estimated RPCs produce estimates of local exports and imports that are consistent with levels of industry output and commodity production in each IMPLAN impact assessment.

The IMPLAN-based regional forecast methodology presents a series of readily reproducible steps for converting BLS and OBERS projections to corresponding sets of county forecasts of industry employment, labor earnings, resident population, and personal income. The individual county series track their respective state projection series. Each state has a set of high, low, and moderate projections based on the 1988 OBERS projection series and the corresponding high, low, and moderate 1988 and 1990 BLS projections series for the United States. This method of approach to county-level forecasting thus extends the BLS and OBERS forecasting methods and results. It introduces the BLS county-level modeling capabilities and database for use in industry-specific assessments of local resource requirements and the effects of these requirements on local and state economies.

State, regional, and county projection series relate directly to corresponding data series from the IMPLAN models of one or more counties. Individual IMPLAN regional reports, for example, expand the number of variables that correlate with the two-digit employment and earnings projections, including commodity exports and commodity imports. They also provide a framework for assessing the differential rates of growth of individual counties and regions. Each IMPLAN model takes given changes in final demands and derives the effects of these changes on the local economy and its institutions. Included with each IMPLAN model is a social accounting matrix (SAM) for tracking changes in local income distributions in the local economy.

The IMPLAN input-output model has been constructed using 528 industry sectors, although the model can be run for any level of aggregation of these sectors. The underlying coefficients in the model are derived from the U.S. input-output accounts. Flows of goods and services in the Minnesota model are derived from commodities produced and consumed in Minnesota as well as those that are imported into the state and exported to areas outside of the state. The system is run for all regions together to ensure consistency with both U.S. and individual regional input-output accounts.

One very useful aspect of the IMPLAN model is the IMPACT module. It permits the user to evaluate the effects of changes or variations in economic activity. For example, the impact of the direct purchase of goods and services by the air transportation industry can be traced through the economy as a series of spending iterations among all sectors, including households. The long-term multiplier used in the model includes indirect effects (to which multipliers are normally limited) as well as induced effects related to employment and population change.

The U.S. Departments of Commerce, Labor, and Agriculture maintain the *reference data systems* for Micro-IMPLAN. The Department of Commerce houses the periodic censuses of population and employment, agricultural, manufacturing, wholesale and retail trade, and selected business services, as well as the annual statistical series on personal income and industry employment and earnings of the employed industry workforce. State- and county-level data sources most critical for early fix-up and updating of the current database are the individual state reports on county business patterns, ES-202 files on covered industry employment and payroll, and the agriculture censuses.

A common problem in using each of these data sources is the occurrence of nondisclosures. Use of supplementary information in the biproportional adjustment procedures for filling in the missing data, for example, allow for closer correspondence of the remaining calculated values with values reported by the U.S. Department of Commerce.

Delays in the reporting cycles for reference data systems result in two- to three-year lags in the availability of each new update of the county-level Micro-IMPLAN database. Reducing lags in data availability is probably a less feasible alternative, however, than forecasting new control totals for the biproportionally adjusted U.S., state, and county input-output tables.

A hybrid approach that combines local surveys of critically important industries with the forecast approach facilitates the likelihood of attaining both greater timeliness and greater accuracy in regional impact assessments. Such an approach incorporates various measures of linkage between core and peripheral labor market areas, like survey-based estimates of the physical volume and market value of commodity shipments between the core area and periphery.

Delineation of the LMAs within an economic region introduces a spatial structure into the organization of the Micro-IMPLAN database. This helps address the twofold problem focus—system bias and specification error. Each of the problem sources, whether industry production functions, RPCs, marketing margins, or industry output, varies between center and periphery. Investment per worker is lower in the periphery, and rate of return on investment also is lower when discounted for perceived investment risk. However, high levels of commodity trade occur between center and periphery. This emanates from the unique competitive advantage of each of the two types of export-producing systems, with the center specializing in high-order, high-profit services, and the periphery specializing in standardized commodity production.

The use of LMAs and the center-periphery structure of these areas applies especially well to the organization of transportation and local land use impact assessments. Commodity transportation originates from dispersed farms, mines, and factories. It concentrates in major shipping centers that also are the primary and secondary core LMAs of the U.S. trading regions. Air transportation concentrates even more than commodity transportation in the primary core areas. This concentration of high-order economic services near the globally connected air nodes of core metropolitan areas apparently accounts for the higher productivity of both labor and capital in the core areas.

Modeling System Formulation

The first step in model reformation is to calculate *total regional commodity demand*. We multiply the regional absorption matrix by the regional industry output to obtain the intermediate input purchases of each industry. We add our estimate of gross final commodity demand to the estimate of intermediate demand to obtain total commodity demand. The U.S. estimates of industry purchases include both domestic production and foreign imports. Thus, the input profile for each industry includes all commodity inputs of that industry. In addition, each industry may produce more than one commodity. The estimates of gross domestic exports relate to both the commodity production and the regional demand for this production.

The next step relates to the calculation of *total regional commodity supply*. Again, the estimate of total regional industry output enters into the calculation, but in multiplication with the industry byproduct ratios from the U.S. byproduct matrix. The result is the regional matrix that shows the commodity production (columns) by each industry (rows). These estimates, together with the estimates of institutional commodity output (commodity sales by government and from inventory depletion), yield the total commodity output for the regional economy.

Finally, to estimate trade flows, the *RPC* is the key parameter.[5] The RPC value times the corresponding value in the regional gross use matrix yields the regional industry use of the locally supplied commodity. Similarly, the *import propensity* for a given commodity times the corresponding value in the regional gross use matrix yields the regional *industry domestic imports* of each commodity. This procedure applies also in estimating regional institutional use and regional institutional imports, that is, the commodity purchases for local final demand.

The calculation of *domestic commodity exports* results from subtracting regional commodity demand from regional gross commodity supply. The individual commodity imbalances in the U.S. estimates of foreign exports and imports carry through to the individual county or multicounty Micro-IMPLAN models. Domestic exports and imports theoretically balance for the domestic economy as a whole, but not for individual counties or multicounty areas. However, the criteria for allocating the two sets of exports and imports differ greatly. Micro-IMPLAN allocates U.S. *foreign commodity exports* to regions according to their share of U.S. commodity production. It also allocates U.S. *foreign commodity imports* to regions according to the same rule. Estimates of a region's total imports and total exports thus derive from a variety of data sources and allocation criteria.

While local commodity production provides the basis for allocating foreign exports and imports, uniquely generated local RPCs provide the basis for estimating domestic exports and imports for each county or multicounty area. These estimates of gross domestic imports relate to both commercial production and the demand for this production in a given region. Model reformulation calls for similar criteria in allocating U.S. foreign imports to individual industries and regions.

Interregional trade is synonymous with commodity shipments. Most commodity shipments move from producing areas to export markets by truck, rail, and barge. However, an increasing volume of high-value manufactured products move by air transportation to and from the designated air transportation nodes. These shipments typically move by truck to the larger air transportation nodes, such as Chicago. Micro-IMPLAN currently fails to account for such multimodal shipments.

Technology transfer is an increasingly important form of interregional trade. It is also a singularly important factor in accounting for a region's competitive advantage in specialized production and its export to other regions. It is associated, in part, with the total value of technology-intensive manufactured products in a given region. Again, Micro-IMPLAN, when conjoined with an optimizing transportation network model, can simulate the local economic effects of technology transfer. This application may extend to the role of a state's research universities in the formation and strengthening of spatially separated, functionally integrated industry clusters. These clusters are viewed by at least one student of regional growth and change as the new industrial systems of the emerging information economy (Saxenian, 1994).

Refinements and Applications

Several types of refinements are available for the outcomes of the preceding steps (Alward et al., 1989). These include (1) changing regional supply, (2) modifying industry production function, (3) editing RPCs, and (4) controlling for induced effects once better information becomes available. Superior local knowledge warrants changing the readymade database values in each category. Superior local knowledge also warrants changing regional purchase coefficients, by institution, industry, or commodity. The RPC adjustments for an industry or institution result in the given change being applied to all commodities, by industry or institution. Overlooked, however, is the further regionalization of the final local sales accounts and the industry margins that convert industry output from producer prices to purchaser prices. This process requires detailed, regionally differentiated estimates of final product sales to households, governments, and businesses. Furthermore, input-output models generally are demand-driven with no supply constraints.

The lack of capacity limits for industry expansion and the assumption of full resource use or availability, including labor, result in overestimating industry production response to demand changes. Fixed-price multipliers add to this problem by overestimating multiplier effects and underestimating the substitution effects from exogenous changes (Koh, Schreiner, and Shin, 1993). Also, the current modeling system sidesteps the issue of commuting effects. These attributes of input-output models ultimately result in underestimating or overestimating factor income responses to market changes.

A Simple Input-Output Model

The model is triggered by changes in final demand; that is, demand for goods or services related to final uses. The components of final demand are exogenous to the model's structural characteristics in much the same way as final payments are, but the role of final demand as an initiator of impacts gives it a unique role in the input-output scheme.

The basic input-output model consists of a series of three separate tables. The first is called the transactions table. The transactions table lists all industrial sectors defined for the purposes of the analysis being conducted. It should be noted that these sectors have to be defined so as to account for every firm in the region. The individual sectors should be relatively homogeneous in terms of their input requirements and output distributions. They should generally be disaggregated enough to highlight the true structure of the region without being so disaggregated as to cause significant problems in data collection or in disclosure of the operations of any one firm in the region.

The transactions table also contains values for final demand, as discussed earlier, as well as the values for final payments. The grand totals of such a table contain the gross outputs for each industrial sector and the gross inputs required to produce those outputs.

Table 6-1A represents the structure of a hypothetical input-output table with three industrial sectors: extractive, manufacturing, and services. Remember, the sectors should be defined so as to account for every firm in the region. The sectors should also, ideally, be as disaggregated as possible. For these reasons, this represents a very unrealistic example of the size of an actual table. Keeping the size of the model to just three industries, however, makes required computations much simpler. The structure and use of larger tables remains much the same.

One of the most important things to remember when reading an input-output table is that the rows of the table represent sales and the columns of the table represent purchases. Thus, the 700 that appears in the Extractive row and the Manufacturing column indicates that firms in the extractive industry sold $700 worth of goods and services to firms in the manufacturing sector.

TABLE 6-1A Commodity transactions of a regional economy

		Intermediate Demand				
Commodity	Extractive (mil.$)	Manufg (mil.$)	Services (mil.$)	Total (mil.$)	Final Demand (mil.$)	Gross Output (mil.$)
Extractive	100	700	0	800	4,625	5,425
Manufacturing	50	200	0	250	6,400	6,700
Services	75	300	75	450	4,905	5,355
Value added	5,000	5,500	230	10,730	1,800	28,730
Imports	200	0	5,000	5,200	0	5,200
Total inputs	5,425	6,700	5,355	17,480	17,730	33,930

Looked at the other way, we could say that the 700 also represents a $700 purchase by the firms in the manufacturing sector from firms in the extractive sector. The 50 in the Manufacturing row and the Extractive column represents a $50 transaction between manufacturing (the seller) and extractive (the buyer), and so on.

The same industrial sectors identified on the left-hand margin of the table appear along the top of the table. The sales and purchases between these sectors represent sales and purchases of "intermediate" goods and services. These are goods and services produced for the purpose of facilitating further production. Semifinished goods would be an obvious example of intermediate production, but so would the services of lawyers, bankers, transportation agencies (in all cases not involving a final transportation use), and any other sector input or output oriented toward helping other industries with their own production.

The value added row of the table represents another form of sale—the sale of resources of production to each sector. In a theoretical sense, the resources of production include land, labor, capital, and enterprise. In a more practical sense, this row generally includes the income received by local households for whatever contribution they make to the production process.

These resource inputs are not generally considered to be intermediate even though the sale takes place so further production can occur. Rather, they represent final inputs that add to the income of households as opposed to industrial sectors.

Imports represent sales to local industries by industries and resource holders outside of the locality's defined boundaries. Although not shown above, in reality final demand accounts for a large, often a major, share of an area's imports; the smaller and less diversified the area, the larger the share. Remember, imports—and exports too, for that matter—are defined in terms of payments.

Finally, final demand consists of sales for final uses. The usual categories making up final demand include household consumption (by households located in the region), government purchases of goods and services, gross private domestic investment (including inventory changes), and exports (again, defined in terms of the payment made).

The gross output and input values are equal. This is due to the fact that the transactions table really represents a type of cost-accounting sheet for a regional economy—debits equal credits. The elements in the table that force this balance (which is a balance by definition) are profits or losses. This is because the final value of output is made up of all the costs that go into production, with profits and losses making up the difference.

In summary, the transactions table has three identifiable parts: the intermediate transactions component, representing sales and purchases between firms; the final payments plus imports component, representing resource inputs into the firm's production plus inputs from outside the region; and final demand, representing the sale of goods and services for final use. The table balances between inputs and outputs, with profit as the balancing mechanism.

The transactions table contains a great deal of useful information in its own right. The regional balance of trade (exports–imports) can be discerned from this table, as can gross regional product (the dollar value of all final goods and services produced with the economy minus imports). The level of interaction between local industries and between industries in the region and households can be seen in this table. Finally, the relation between local household income and production is depicted in the transactions matrix.

The principal use for this table is found in the construction of the other two tables of the input-output system. As mentioned, the transactions table alone represents a cost-accounting sheet for the region, nothing more or less. It is descriptive rather than analytical, and it does not allow for general equilibrium analysis of the type previously described without further modification. The next step uses the transactions table to construct a table of direct requirements, often called the technical coefficients matrix.

The question answered by the technical coefficients table is, If each local industrial sector sells to other local industrial sectors some total value of intermediate goods and services so that the purchasing sectors can produce their own output, how much do the purchasing sectors require from the other local sectors per dollar of output? For example, Manufacturing purchased $300 worth of intermediate output from Services in order to facilitate its own production of $6,700 worth of intermediate and final outputs. How much did Manufacturing buy from Services per dollar of gross output? The answer is 300/6,700 = $0.45. The same computation can be made for each intermediate sale and purchase in the transactions table. The result of these divisions is shown in Table 6-2A.

TABLE 6-2A Direct input requirements

Commodity	Extractive (mil.$)	Manufg (mil.$)	Services (mil.$)
Extractive	0.018	0.104	0.000
Manufacturing	0.009	0.030	0.000
Services	0.014	0.045	0.014
Subtotal	0.041	0.179	0.014
Value added	0.922	0.821	0.043
Imports	0.037	0.000	0.934
Total inputs	1.000	1.000	1.000

The rows are still read as sales and the columns as purchases. Only now the sales are in terms of cents per dollar, and the purchases have the special interpretation of "input requirements" per dollar of output. We call these input requirements because they represent requirements during the period of analysis in order for each sector to produce its own outputs, scaled down to a "dollar of output" basis.

The technical coefficient matrix represents a recipe for production. To produce one dollar's worth of output, the extractive industry needed a pinch of its own intermediate output, a dash of the intermediate output of the manufacturing sector, and a smidgen of the intermediate products of the services industry. For Manufacturing to produce a dollar's worth of output, it required a pinch from Extractive, a dash from Manufacturing, and a smidgen from Services. And so it goes through all the identified industries for the region.

One of the key assumptions of input-output analysis, as mentioned earlier, is that this recipe does not change, regardless of the level of output. Thus, if the extractive industry were to experience an increase in final sales equal to $10,000, it would require another $180 worth of intermediate products from its own firms, $90 from Manufacturing, and $140 from Services. It should be emphasized that this process starts with a change in the final sales of an industry, or from "exogenous" forces. The coefficients in the inter-industry section of the table represent the "endogenous" component of the table.

It can be seen that this first computed table gives the analyst limited ability for impact analysis. He or she could go through the process of assuming any number of changes in the final sales of the identified industries, multiply these assumed changes by the direct requirements coefficients, and come up with estimates as to the direct effects from these changing final sales on each identified industry in the region. To make sure that this process is understood, one might ask, What is the direct effect on each regional industry from an increase in the exports from the manufacturing sector equal to $10 million? The answer is that Manufacturing would increase by $10 million plus a direct intermediate production effect of $300,000, for a total of $10.3 million; the extractive industry would find its intermediate production increasing by $1.4 million; and the services industry would see its intermediate production increase by $450,000.

But this is not the end of the story. If each industry has to increase its output in order to service the increase in final sales of the manufacturing industry, then each must, in turn, increase its intermediate purchases and sales from and to one another to service this second round of expansion in activity. The second round must then be serviced by a third round of outputs.

TABLE 6-3A Round one of $10 million change in final sales

	Intermediate (mil.$)	Final (mil.$)
Extractive	0.300	0.000
Manufacturing	1.040	10.000
Services	0.450	0.000
Total	1.179	10.000

TABLE 6-4A Round two of $10 million change in final sales

Commodity	Extractive (thou.$)	Manufg. (thou.$)	Services (thou.$)	Total (thou.$)
Extractive	18.720	31.200	0.000	499.200
Manufacturing	9.360	9.000	4.050	106.650
Services	14.560	4.200	6.300	156.100
Total	42.640	325.200	10.350	761.950

Each round is smaller than the previous one due to leakages to imports and to local value added, until the process has completely played itself out. The first of three rounds of such a $10,000 increase in final sales is shown in Table 6-3A.

Note that the only exogenous change is the initial change in final demand assumed for the manufacturing industry. The rest of the sales represent the direct first-round results from those sales on the intermediate output of all industries in the region, including Manufacturing. These are recipe requirements for Manufacturing to produce the hypothesized increased final sales.

Table 6-4A presents second-round totals. Note that Manufacturing requires still more intermediate inputs from its own firms, this time to service the additional $300,000 of output it had to produce to directly allow for the initial $10 million increase in final sales. Similarly, the services industry needs to buy from each of the other industries to enable it to produce the additional $450,000 directly required by Manufacturing. Finally, the extractive industry must have additional inputs to produce its additional $1,040,000 for Manufacturing. The rounds of production in Table 6-4A are *indirect* impacts.

Manufacturing has now increased its sales three times: the $10 million that was initially assumed, the $300,000 needed to directly service that increase in final sales, and the $22,410 to service the $300,000 in the first round. The extractive industry has increased its sales by $1,040,000 to service the final sales

TABLE 6-5A Round three of $10 million change in final sales

Commodity	Extractive (thou.$)	Manufg. (thou.$)	Services (thou.$)	Total (thou.$)
Extractive	0.899	2.331	0.000	3.230
Manufacturing	0.449	0.672	0.225	1.346
Services	0.699	0.351	0.351	1.401
Total	2.047	3.354	0.576	5.977

TABLE 6-6A Direct and indirect input requirements

Commodity	Extractive (mil.$)	Manufg. (mil.$)	Services (mil.$)
Extractive	1.019	0.109	0.001
Manufacturing	0.010	1.032	0.010
Services	0.015	0.049	1.015
Total	1.044	1.190	1.026

change for Manufacturing plus the $49,920 to service that first-round increase, for a total of $1,089,920 to this point. And so it goes.

We will now run through a third round of increased production (Table 6-5A), this time to service the second round.

Each additional round is computed in the manner shown above, and the totals are added to determine the total direct and indirect effects from the initial assumed change in the final sales of one of the regional industries. This process is obviously cumbersome. It would be even more difficult—impossible, probably—to work such an iterative scheme for a larger number of industries or for higher direct coefficient values. Fortunately, the system of simultaneous equations represented by an input-output system can be solved using high-speed computers in a matter of seconds, even for the largest of tables. The solution for the system in this example is given in Table 6-6A.

The diagonal of Table 6-6A shows "ones" plus some other number (for example, 1.032 in row 2, column 2). These ones represent the dollar increase to final sales of the industry for which such an exogenous change is assumed. The numbers appearing after the decimal represent the direct (shown in Table 6-2A) plus indirect effects from each assumed change in final sales. Thus, the $10 million change for the example using Manufacturing turns into $10,320,000 total increase in Manufacturing sales: $10 million to final sales, $300,000 in direct sales, and $20,000 in indirect sales. That $10 million in Manufacturing sales turns into an increase of $1,090,000 in sales by the extractive industry—$1,040,000 of that direct and $50,000 indirect. Finally, the $10 million assumed increase in manufacturing leads to an increase of $490,000 in the sales of services—$450,000 of that direct and $40,000 of that indirect.

The total impact on all of the industries in the region combined is $11.9 million (1.190 × 10 million). The 1.190 is called the demand multiplier for Manufacturing, or the total direct and indirect purchases this sector must make from itself and from the other regional industries in order to produce one dollar's final output. To conduct an impact study, simply multiply an assumed change in final demand for any of the industries by the demand multiplier for that same industry. This indicates the direct and indirect effects on the region resulting from the assumed change. The impacts stem from the fact that industries in a region interact with one another through their purchases from and sales to one another. The greater this level of interaction, the greater the industrial demand multiplier.

Thus, the input-output model represents a detailed accounting of the economic base of a region. It can be used to delineate the export structure of the regional economy and the multipliers that emerge from that structure. It also identifies, in final demand, the relationship between local activity, investment, and export activity in relation to the identified industrial structure. As in most models, its weakness is in its assumptions. But, at the least, the input-output system can be used for simulations and sensitivity analyses for a regional economy.

Notes

1. According to *The American Heritage Dictionary,* 3d ed., version 3.6a, Houghton Mifflin Company, New York, 1993, *predict* means "to state or tell about, or make known in advance, especially on the basis of special knowledge." Somewhat along the same line, *project* means "to calculate, estimate, or predict (something in the future), based on present data or trends."

2. The Standard Industrial Classification (SIC) is the statistical classification underlying all establishment-based federal economic statistics classified by industry. This classification was established by the Office of Management and Budget and is used widely by states, industries, and analysts. The Major Group SIC 45 (Air Transportation) is a two-digit classification (that is, the group number consists of two digits) and includes the following four-digit subcategories: Air Transportation, Scheduled (4512); Air Courier Services (4513); Air Transportation, Nonscheduled (4522); Airports, Flying Fields, and Airport Terminal Services (4581).

3. U.S. Department of Commerce, Regional Economic Measurements Division. *Regional Economic Information System:* Unpublished series, 1969–1990.

4. U.S. Department of Commerce, Office of Business Economics. *Regional Projections to 2040.*

5. The estimated RPC for a commodity is derived empirically from a regression using county-level variables and regression coefficients based on the 1977 Multi-Region Input-Output Accounts (MRIOAs). The regression equation for each commodity, based on commodity shipments for the 50 states and the District of Columbia from the 1977 United States Census of Transportation by the U.S. Department of Commerce, is in the form

$$\text{RPC} = a + b \times \text{wage ratio} + c \times \text{employment} + d \times \text{location quotient} + e \times \text{area ratio}$$

where the wage ratio is total employee compensation divided by total employment of the industry, employment is total industry employment, location quotient is the proportion of total regional employment in the specified industry divided by the corresponding ratio for the United States, and area ratio is the proportion of the total U.S. area accounted for by the region. Lack of one or more variables to account for the effects of regional variations in spatial-economic structures remains a serious deficiency of the current model. Moreover, most of the regression coefficients—MRIOA sectors 1 to 84—apply only to the shippable commodities—agriculture to manufacturing. A single constant value of each state represents the remaining 52 MRIOA sectors. For some states, for example, California, the constant value is 1 for each of the 52 MRIOA sectors. For most states, however, this value is less than 1 for most commodities. Finally, each RPC value has an upper limit that is given by the regional net commodity supply/gross-demand-pool ratio (S/DP) ratio.

7

Social and Economic Accounts

Year-to-year and place-to-place changes in urban regional activities are monitored with the use of various social and economic accounts (SEAs). These include statistical measures of population growth, migration, labor force participation, regional income and product, and government spending and financing. The reported changes are unique to the general economy, but not necessarily to the areas in which they occur. Measures of gross regional product (GRP), for example, have unique local histories that often fail to track corresponding national changes.

During the 35-year period from 1959 to 1994, for example, records show that at least one state experienced a peak or a trough of its own business cycle during every year of that period. Meanwhile, the U.S. economy experienced the peaks and troughs of only five business cycles as measured by year-to-year changes in constant dollar gross domestic product (GDP). The differences between state and national performances occur because of place-to-place differences in the industrial composition that generates the nation's GDP and the competitive position of these industries in regional and global markets.

Tracking Urban Regional Economies

The tracking of urban regional economies depends on a variety of statistical sources, including both survey statistics and time series. We focus primarily on the various statistical time series relating to population and labor force, government spending and financing, and income and product—all with reference to individual states, localities, and regions. For the most part, the data we use are readily accessible from the *Statistical Abstract of the United States* in the form of hard copy, compact disk, or the U.S. Department of Commerce home page on the Internet. We demonstrate various uses of these data with a series of 32 tabular summaries that reveal some of the many differences among states and regions that influence their long-term economic prospects, as well as the nature and character of their day-to-day activities.

Population and Labor Force

Accurate and accessible estimates of local population and labor force are among the first sets of statistics sought by analysts and students of urban regional activities. Many of the statistics for counties and minor civil divisions originate from the U.S. Bureau of the Census, supplemented by special surveys, including annual school censuses conducted by local school districts. Data sources for the *Statistical Abstract of the United States,* for example, are cited for each of the more than 1,500 tables, with most sources being available from the U.S. Bureau of the Census and other offices in the U.S. Department of Commerce.

We can, for example, find and use the data series for the Detroit primary metropolitan statistical area (PMSA) showing the persistence of an economic climate that fosters both high wages and high levels of unemployment. We also have data series showing the shift in economic activity in the Washington, D.C., PMSA caused by the federal government reducing its own workforce while at the same time instituting and implementing legislation that calls for a much expanded private sector workforce.

We also have numbers that show the amount and types of income redistribution that occur with the state-to-state distribution of federal funds. These, in turn, help us understand how some states with an above-average influx of federal funds have more state revenues available relative to their gross state product than do states that have below-average shares of federal funds.

Population Growth and Migration

One of the oft-proclaimed virtues of American federalism is the freedom to move people or products—a freedom that has much to do with this nation's economic growth and prosperity. State and regional population growth is, at least to some extent, fueled by the new job opportunities generated from economic growth. We start with population growth and migration, therefore, as building blocks in the construction of regional economic and social accounts for use in forecasting and projecting a region's most likely alternative futures. We use the state-level estimates of regional growth and change. Later, in Chapter 11, we compare our findings along with the advantages and disadvantages of alternative approaches to area and regional delineation in urban regional data compilation.

Table 7-1 shows an aggregation of state-level statistics on population change and demographic measures of births, deaths, and migration. The use of broad

TABLE 7-1 Components of state population change, by census region, 1990–95

Region	Resident Population				Migration			Relative Change, 1990–95		
	Total90 (mil.)	NetChg. (mil.)	Births (mil.)	Deaths (mil.)	Total (mil.)	Internat'l (mil.)	Domestic (mil.)	NetChg. (pct.)	Births (pct.)	Deaths (pct.)
United States	248.7	14.0	21.3	11.7	4.0	4.0	0.0	0.0	0.0	0.0
Northeast	50.8	0.7	4.0	2.5	–0.8	1.0	–1.8	–4.4	–0.7	0.3
New England (NE)	13.2	0.1	1.0	0.6	–0.3	0.1	–0.4	–4.8	–1.1	0.0
Middle Atlantic (MA)	37.6	0.5	3.0	1.9	–0.5	0.8	–1.4	–4.2	–0.6	0.4
Midwest	59.7	2.1	4.8	2.9	0.2	0.4	–0.3	–2.1	–0.5	0.2
East North Central (ENC)	42.0	1.4	3.4	2.0	0.0	0.4	–0.4	–2.2	–0.4	0.1
West North Central (WNC)	17.7	0.7	1.4	0.9	0.2	0.1	0.1	–1.8	–0.8	0.3
South	85.5	6.4	7.3	4.2	3.0	0.9	2.1	1.9	0.0	0.2
South Atlantic (SA)	43.6	3.4	3.6	2.2	1.8	0.5	1.3	2.2	–0.3	0.3
East South Central (ESC)	15.2	0.9	1.2	0.8	0.4	0.0	0.4	0.2	–0.5	0.5
West South Central (WSC)	26.7	2.1	2.5	1.2	0.8	0.4	0.4	2.3	0.8	–0.2
West	52.8	4.8	5.2	2.1	1.6	1.6	–0.1	3.5	1.3	–0.7
Mountain	13.7	2.0	1.3	0.6	1.2	0.1	1.1	8.9	0.9	–0.6
Pacific	39.1	2.8	3.9	1.5	0.4	1.5	–1.1	1.6	1.5	–0.7

Source: *Statistical Abstract of the United States 1996*, pp. 228, 230.

census regions, of course, masks the important consequences of the differential rates of births, deaths, and migration for individual states. The detailed state-by-state estimates, including the identification of each state with a census region, are available in later tables.

The regional aggregates illustrate the difficulty of accurate estimation of internal flows of people and products. Internal flows cancel—that is, place-to-place in-migrants equal out-migrants, in-shipments equal out-shipments. Only births, deaths, and international migrants are additive, if shown individually. Births outnumber deaths in every census region, although important differences occur even on this aggregate scale.

International migration makes these differences even larger. Births, for example, were 1.1 percentage points below the U.S. average over the 1990–95 period in the six New England states. They were 1.5 percentage points above the U.S. average in the five Pacific states. Also, the Pacific states lead all regions in international migrants. Relative death rates reverse positions, with below-average death rates generally accompanying above-average birth rates. This is the result, in part, of differences in the age structure of a region's population.

The 248.7 million people in the United States in 1990 increased in total number by 14 million over the 1990–95 period. Four million of this increase is attributed to international migrants, largely to the Pacific and Middle Atlantic states. International migrants account for a majority of in-migrants in only one of the 32 states with net in-migration, however, as shown in Table 7-2.

Neither high birth rates nor low death rates correlated with high in-migration. States that experienced above-average rates of population growth had the highest rates of domestic in-migration. This was enhanced by above-average rates of international migration. For the 32 states with net domestic in-migration, the total number of in-migrants was only slightly less than total deaths.

Despite an aging national population, total deaths in all states were only 55 percent of total births. Retirement states, like Florida, North Carolina, and Arkansas, accounted for above-average ratios of deaths to births, along with states experiencing earlier periods of population out-migration. Because of much state-to-state variability in birth and death rates, the relationship between migration and net population change certainly was less than perfect, as shown in part by the shift in ranking from the first column to the fourth column of Table 7-2.

Thus far the numbers suggest that states with high levels of domestic in-migration generally experience high rates of population growth. The reverse of this is much less the case for states like California, Alaska, and Hawaii, which experience the positive influence of international migration on net births, as shown in Table 7-3.

Although the nineteen states with net out-migration over the 1990–95 period had a larger total population than the 32 states with net in-migration, their net population growth was less than two-fifths of the population increase in the 32 states with net in-migration. Except for California, Alaska, and Hawaii, sixteen of the nineteen states are in the Northeast or Midwest census regions. As we show later, however, all is not lost for the nineteen "rust belt" states. Established institutions, along with households and businesses having strong and enduring "roots" in their local communities and attracting the new international migrants, are meeting the new challenges sufficiently to foster turnarounds in their collective economic prospects.

The state-level numbers in Tables 7-2 and 7-3 show a concentration of international migrants in only a few states. More than 75 percent go to six states—California, New York, Texas, Florida, Illinois, and New Jersey. Four of the six states also show the highest levels of net out-migration. In fact, international migrants are nearly 2.5 times as numerous in the states with net out-migration in the 1990–95 period than in the states with net in-migration. Losing the former residents of these states is partially compensated for by the gaining of new residents from the growing international migration. Economic opportunity and established social networks are the hidden attributes of the destination states of the international migrants, who, like domestic migrants, also vote with their feet.

Table 7-4 parallels Table 7-2 in providing a state-by-state documentation of the three demographic variables. These projections are based on the historical record to 1993 rather than 1995 and hence lack the precision of perfect hindsight, and they differ from a true "trend line" projection.

Notice, for example, the loss of Montana, but the gain of Hawaii, among the states with domestic in-migration. Note also the net addition of two more states to the list. Projected international migrants show a slight increase from a simple trend line—from 1.1 million over a five-year period to 7.7 million over a 30-year period. Projected births, as might be expected with an aging population, drop from a level 77 percent above births to a level 57 percent above births.

Table 7-5 parallels Table 7-3 in its focus on state-to-state out-migration. It differs, however, in its content. Hawaii, Maine, Kansas, and New Hampshire are no longer in the loss column, but two new states—Iowa and Utah—show projected population out-migration. Projected total state-to-state out-migration, relative to projected international migration, drops sharply. Nonetheless, the seventeen states with projected out-migration during this period gain less from international migration than the 34 states with project in-migration when compared with their relative standing for the 1990–95 period. Also, the ratio of births to deaths drops from 1.88 to 1.69, which still is above the ratio of births to deaths for the out-migration states. Net population change is consistently

Table 7-2 Ranking components of state population change, by domestic in-migration, 1990–95

State	Region	NetChg. Rank	Resident Population				Migration			Relative Change, 1990–95		
			Total90 (thou.)	NetChg. (thou.)	Births (thou.)	Deaths (thou.)	Total (thou.)	Internat'l (thou.)	Domestic (thou.)	NetChg. (pct.)	Births (pct.)	Deaths (pct.)
1 Florida	SA	3	12,938	1,227	1,017	746	928	256	672	3.8	–0.7	1.1
2 Georgia	SA	4	6,478	723	585	286	399	42	357	5.5	0.5	–0.3
3 Texas	WSC	2	16,986	1,738	1,686	688	693	372	321	4.6	1.4	–0.6
4 Arizona	Mountain	7	3,665	553	365	167	343	50	293	9.4	1.4	–0.1
5 North Carolina	SA	6	6,632	563	540	320	306	23	283	2.8	–0.4	0.1
6 Washington	Pacific	5	4,867	564	413	204	341	64	277	5.9	–0.1	–0.5
7 Colorado	Mountain	8	3,294	452	284	121	274	29	245	8.1	0.1	–1.0
8 Nevada	Mountain	12	1,202	328	118	55	259	19	240	21.6	1.3	–0.1
9 Tennessee	ESC	11	4,877	379	388	253	237	14	223	2.1	–0.6	0.5
10 Oregon	Pacific	14	2,842	298	222	139	215	28	187	4.8	–0.7	0.2
11 Idaho	Mountain	28	1,007	157	91	42	106	7	99	9.9	0.5	–0.5
12 Alabama	ESC	22	4,040	213	327	214	91	7	84	–0.4	–0.5	0.6
13 Arkansas	WSC	30	2,351	133	183	135	83	4	79	0.0	–0.8	1.1
14 Wisconsin	ENC	19	4,892	231	370	230	89	18	71	–0.9	–1.0	0.0
15 Utah	Mountain	20	1,723	229	196	52	82	11	71	7.6	2.8	–1.7
16 Indiana	ENC	16	5,544	259	443	269	80	14	66	–1.0	–0.6	0.2

17 Kentucky	ESC	26	3,687	173	281	189	71	8	63	–1.0	–0.9	0.4
18 Virginia	SA	9	6,189	429	507	267	132	71	61	1.3	–0.4	–0.4
19 New Mexico	Mountain	27	1,515	170	146	60	78	18	60	5.6	1.1	–0.7
20 Missouri	WNC	23	5,117	207	401	275	76	19	57	–1.6	–0.7	0.7
21 Minnesota	WNC	18	4,376	234	346	189	77	24	53	–0.3	–0.6	–0.4
22 Montana	Mountain	36	799	71	60	38	48	2	46	3.2	–1.0	0.1
23 South Carolina	SA	25	3,486	187	291	164	42	7	35	–0.3	–0.2	0.0
24 Oklahoma	WSC	31	3,146	132	247	164	38	11	27	–1.4	–0.7	0.5
25 Mississippi	ESC	33	2,575	122	224	137	27	3	24	–0.9	0.2	0.6
26 Delaware	SA	41	666	51	57	32	24	4	20	2.0	0.0	0.1
27 West Virginia	SA	43	1,793	35	116	105	22	2	20	–3.7	–2.1	1.2
28 South Dakota	WNC	44	696	33	57	35	10	2	8	–0.9	–0.4	0.3
29 Wyoming	Mountain	45	454	27	35	18	7	1	6	0.3	–0.8	–0.7
30 Nebraska	WNC	38	1,578	59	124	78	9	6	3	–1.9	–0.7	0.3
31 Vermont	NE	46	563	22	41	25	6	3	3	–1.7	–1.3	–0.2
32 Iowa	WNC	37	2,777	65	201	145	9	8	1	–3.3	–1.3	0.5
Total			122,755	10,064	10,362	5,842	5,202	1,147	4,055	2.6	–0.1	0.1

Source: *Statistical Abstract of the United States 1996*, pp. 228, 30.

TABLE 7-3 Ranking components of state population change, by domestic out-migration, 1990–95

Rank	State	Region	NetChg. Rank	Resident Population: Total90 (thou.)	NetChg. (thou.)	Births (thou.)	Deaths (thou.)	Migration: Total (thou.)	Internat'l (thou.)	Domestic (thou.)	Relative Change, 1990–95: NetChg. (pct.)	Births (pct.)	Deaths (pct.)
1	California	Pacific	1	29,758	1,831	3,116	1,148	–168	1,380	–1,548	0.5	1.9	–0.8
2	New York	MA	29	17,991	145	1,503	874	–478	575	–1,053	–4.8	–0.2	0.2
3	Illinois	ENC	10	11,431	399	1,010	553	–67	231	–298	–2.2	–.3	0.2
4	New Jersey	MA	21	7,730	215	625	375	–41	194	–235	–2.9	–0.5	0.2
5	Massachusetts	NE	39	6,016	57	456	286	–111	83	–194	–4.7	–1.0	0.1
6	Connecticut	NE	49	3,287	–12	250	151	–112	37	–149	–6.0	–0.9	–0.1
7	Michigan	ENC	17	9,295	254	756	426	–79	60	–139	–2.9	–0.4	–0.1
8	District of Columbia	SA	51	607	–53	57	38	–73	16	–89	–14.4	0.8	1.6
9	Pennsylvania	MA	24	11,883	189	856	656	–12	63	–75	–4.1	–1.3	0.8
10	Louisiana	WSC	32	4,220	122	370	202	–56	12	–68	–2.8	0.2	0.1
11	Ohio	ENC	13	10,847	303	848	533	–17	35	–52	–2.9	–0.7	0.2
12	Rhode Island	NE	50	1,003	–14	75	50	–39	7	–46	–7.0	–1.1	0.3
13	Hawaii	Pacific	35	1,108	79	104	38	–5	33	–38	1.5	0.8	–1.3
14	North Dakota	WNC	48	639	3	46	30	–16	2	–18	–5.2	–1.3	0.0
15	Maine	NE	47	1,228	13	83	60	–12	3	–15	–4.6	–1.8	0.2
16	Maryland	SA	15	4,781	262	405	209	51	65	–14	–0.2	–0.1	–0.3
17	Kansas	WNC	34	2,478	88	198	120	2	12	–10	–2.1	–0.6	0.2
18	Alaska	Pacific	40	550	54	60	12	–2	5	–7	4.2	2.4	–2.5
19	New Hampshire	NE	42	1,109	39	83	46	0	5	–5	–2.1	–1.1	_0.1
	Total			125,961	3,974	10,901	5,807	–1,235	2,818	–4,053	–2.5	0.1	–0.1

Source: *Statistical Abstract of the United States* 1996, pp. 228, 230.

negative, except for the three western states (California, Alaska, and Utah) and Maryland.

The projected in-migration derives, of course, from a particular set of assumptions about the continuation of past trends. A different set of assumptions will lead to the new projected levels differing. The U.S. Bureau of Labor Statistics, in its projections of gross domestic product (its generation by various production factors and its disbursement to various economic units), provides alternative scenarios of future product demands and resource supplies, as shown later in this chapter. The U.S. Bureau of the Census also provides alternative sets of U.S. population projections, but these are not available for individual states, except for this set of population projections.

The real value of these projections is not their fallibility but their rationale and challenges to our understanding of regional and interregional economic activities and processes. Projections of international and interstate migration thus test our understanding of the underlying economic and demographic processes and structures accounting for the differing geographic patterns from one period to the next.

Labor Force Participation and Employment

The next series of tables highlights the preparation and use of state-level statistical series that focus on various labor force and employment issues, including labor force participation. Labor force participation is the proportion of the total population, age 16 and older, in the labor force. Low labor force participation masks potentially high unemployment rates. The statistical series to uncover these sorts of issues are available, for the most part, in the *Statistical Abstract of the United States* and other references cited at the bottom of each table.

We used an abbreviated version of the shift-share model described in Chapter 6 to sort the 50 states and the District of Columbia into two groups according to the total regional-share effect for each state. We use industry employment to calculate individual industry effects, which we aggregate at the state level. We call a regional-share effect a "differential effect." We then sort the labor force data into two groups of states, namely, those with a positive differential effect for the 1993–95 period and those with a negative differential effect.

Table 7-6 lists the most commonly cited labor force attributes for each state with a positive differential effect, starting with the absolute levels of population and labor force in 1995. We discuss this application of the shift-share model in detail in the next section.

From the numbers in Table 7-6, together with estimates of gender population, age 16 years and older, we derive each of the three ratios—the employment-population ratio, the labor force participation rate (by gender), and the total

Table 7-4 Components of state population change, by domestic in-migration, 1990–2020

State	Region	Rank NetChg.	Resident Population Tot2020 (thou.)	Resident Population NetChg. (thou.)	Resident Population Births (thou.)	Resident Population Deaths (thou.)	Migration Total (thou.)	Migration Internat'l (thou.)	Migration Domestic (thou.)	Relative Change, 1990–20 NetChg. (pct.)	Relative Change, 1990–20 Births (pct.)	Relative Change, 1990–20 Deaths (pct.)
1 Florida	SA	3	19,082	6,144	5,868	5,430	5,705	1,977	3,728	17.7	–4.5	11.3
2 North Carolina	SA	6	8,901	2,269	3,012	2,254	1,512	161	1,351	4.5	–4.5	3.3
3 Washington	Pacific	4	7,825	2,958	2,676	1,413	1,695	373	1,322	31.0	5.1	–1.6
4 Georgia	SA	5	9,315	2,837	3,531	2,079	1,386	286	1,100	14.1	4.6	1.4
5 Tennessee	ESC	11	6,358	1,481	2,195	1,662	949	77	872	0.6	–4.9	3.4
6 Oregon	Pacific	13	4,294	1,452	1,471	907	888	179	709	21.4	1.9	1.2
7 Arizona	Mountain	8	5,592	1,927	2,120	1,199	1,006	338	669	22.9	8.0	2.1
8 Virginia	SA	7	8,302	2,113	2,864	1,827	1,076	513	563	4.4	–3.6	–1.1
9 South Carolina	SA	16	4,619	1,133	1,696	1,155	593	45	547	2.8	–1.2	2.5
10 Colorado	Mountain	10	4,788	1,494	1,734	850	610	157	453	15.6	2.8	–4.9
11 Alabama	ESC	15	5,189	1,149	2,020	1,374	503	54	449	–1.3	0.1	3.3
12 Arkansas	WSC	32	2,961	610	1,051	841	400	30	370	–3.8	–5.2	5.1
13 Nevada	Mountain	23	2,119	917	836	411	492	137	355	46.5	19.6	3.5
14 Texas	WSC	2	25,402	8,416	10,766	4,870	2,520	2,190	331	19.8	13.5	–2.0
15 Oklahoma	WSC	27	3,948	802	1,373	978	408	96	311	–4.2	–6.2	0.4
16 Missouri	WNC	21	6,044	927	2,240	1,651	339	83	256	–11.6	–6.1	1.6

17 Idaho	Mountain	33	1,581	574	637	292	228	36	193	27.3	13.4	–1.7
18 New Mexico	Mountain	28	2,291	776	921	415	270	91	179	21.5	10.9	–3.2
19 New Hampshire	NE	40	1,391	282	441	308	150	18	132	–4.3	–10.1	–2.9
20 Kentucky	ESC	34	4,256	569	1,614	1,194	150	42	108	–14.3	–6.1	1.7
21 Mississippi	ESC	35	3,071	496	1,229	859	127	24	103	–10.5	–2.2	2.7
22 Wisconsin	ENC	26	5,766	874	2,087	1,419	206	107	100	–11.9	–7.2	–1.7
23 Montana	Mountain	41	1,053	254	388	233	98	1	97	2.0	–1.3	–1.5
24 Delaware	SA	44	857	191	318	226	99	18	82	–1.1	–2.2	3.3
25 Maine	NE	46	1,381	153	439	370	84	6	78	–17.3	–14.1	–0.5
26 Hawaii	Pacific	29	1,845	737	704	280	312	139	73	36.8	13.7	–5.4
27 Wyoming	Mountain	43	661	207	258	114	62	6	56	15.8	7.0	–5.7
28 Indiana	ENC	25	6,427	883	2,462	1,681	101	73	28	–13.8	–5.5	–0.3
29 West Virginia	SA	51	1,817	24	606	608	26	6	20	–28.4	–16.1	3.2
30 Nebraska	WNC	38	1,863	285	718	468	35	24	11	–11.7	–4.4	–1.0
31 Vermont	NE	47	650	87	233	157	11	1	10	–14.3	–8.5	–2.8
32 Minnesota	WNC	19	5,343	967	1,973	1,159	153	145	8	–7.6	–4.8	–4.2
33 Kansas	WNC	30	3,105	627	1,261	736	102	101	1	–4.4	1.0	–1.0
34 South Dakota	WNC	45	851	155	346	198	7	6	1	–7.5	–0.2	–2.2
Total			168,948	44,770	62,088	39,620	22,302	7,637	14,665	6.3	0.1	1.2

Sources: *Statistical Abstract of the United States 1996*, pp. 228, 230; U.S. Bureau of the Census, Current Population Reports, P25–1111, Population Projections for States, by Age, Sex, Race, and Hispanic Origin: 1993 to 2020, by Paul R. Campbell, U.S. Government Printing Office, Washington, D.C., 1994.

TABLE 7-5 Components of state population change, by domestic out-migration, 1990–2020

State	Region	Rank NetChg.	Resident Population: Total20 (thou.)	NetChg. (thou.)	Births (thou.)	Deaths (thou.)	Migration: Total (thou.)	Internat'l (thou.)	Domestic (thou.)	Relative Change, 1990–20: NetChg. (pct.)	Births (pct.)	Deaths (pct.)
1 New York	MA	20	18,951	960	7,769	5,298	–1,511	3,343	–4,854	–24.4	–6.7	–1.2
2 California	Pacific	1	47,853	18,095	10,087	8,329	6,337	10,377	–4.041	31.1	17.6	–2.7
3 Illinois	ENC	9	13,102	1,671	5,479	3,356	–451	1,060	–1,511	–15.1	–1.9	–1.3
4 Michigan	ENC	18	10,277	982	4,410	2,630	–798	196	–994	–19.2	–2.4	–2.4
5 Massachusetts	NE	39	6,299	283	2,198	1,714	–201	677	–878	–25.0	-13.3	–2.2
6 New Jersey	MA	14	9,014	1,284	3,239	2,411	456	1,192	–736	–13.1	–8.0	0.5
7 Ohio	ENC	24	11,735	888	4,521	3,312	–321	156	–477	–21.6	–8.2	–0.1
8 Connecticut	NE	36	3k593	306	1,300	939	–56	273	–329	–20.4	-10.3	–2.1
9 Pennsylvania	MA	31	12,499	616	4,466	3,978	128	339	–210	–24.6	-12.3	2.8
10 District of Columbia	SA	50	638	31	279	213	–35	112	–147	–24.6	–3.9	4.4
11 Louisiana	WSC	22	5,139	919	2,200	1,278	–3	107	–110	–8.0	2.3	–0.4
12 Iowa	WNC	42	3,007	230	1,124	841	–53	53	–107	–21.5	–9.4	–0.4
13 Rhode Island	NE	49	1,077	74	377	302	–1	95	–96	–22.3	-12.2	–0.5
14 Alaska	Pacific	37	855	305	435	80	–50	32	–82	25.7	29.3	-16.1
15 Utah	Mountain	17	2,730	1,007	1,368	377	16	72	–57	28.7	29.5	–8.8
16 North Dakota	WNC	48	715	76	263	167	–21	6	–27	–17.9	–8.7	–4.7
17 Maryland	SA	12	6,250	1,469	2,447	1,431	453	465	–12	1.0	1.3	–0.7
Total			153,734	29,196	61,964	36,656	3,888	18,556	–14,668	–6.3	–0.1	–1.2

Sources: *Statistical Abstract of the United States 1996,* pp. 228, 230; U.S. Bureau of the Census, Current Population Reports, P25–1111, Population Projections for States, by Age, Sex, Race, and Hispanic Origin: 1993 to 2020, by Paul R. Campbell, U.S. Government Printing Office, Washington, D.C., 1994.

TABLE 7-6 Characteristics of the civilian labor force, by state and positive 1993–95 differential employment change, U.S., 1995

State	Region	Unemp. Rate Rank	Pop. Total (thou.)	Labor Force		Employed		Emp.Pop. Ratio (pct.)	Labor Force Participation Rate		Unemp. Rate (pct.)
				Total (thou.)	Female (thou.)	Total (thou.)	Female (thou.)		Male (pct.)	Female (pct.)	
All states			262,754	132,304	60,944	124,900	57,523	47.5	75.0	58.9	5.6
1 Georgia	SA	30	7,201	3,618	1,685	3,441	1,592	47.8	75.8	59.2	4.9
2 Texas	WSC	13	18,724	9,568	4,215	8,991	3,939	48.0	79.3	60.0	6.0
3 Arizona	Mountain	27	4,218	2,120	968	2,012	914	47.7	76.7	59.6	5.1
4 Nevada	Mountain	20	1,530	801	353	758	333	49.5	77.4	60.5	5.4
5 Colorado	Mountain	42	3,747	2,089	957	2,001	917	53.4	80.4	67.2	4.2
6 Florida	SA	18	14,166	6,830	3,160	5,455	2,982	45.6	70.4	54.5	5.5
7 Utah	Mountain	47	1,951	971	415	936	400	48.0	82.5	61.2	3.6
8 Tennessee	ESC	25	5,256	2,712	1,284	2,571	1,218	48.9	74.4	60.3	5.2
9 Oregon	Pacific	32	3,141	1,650	751	1,570	718	50.0	75.4	60.8	4.8
10 North Carolina	SA	40	7,195	3,636	1,737	3,479	1,647	48.4	74.7	59.8	4.3
11 New Mexico	MA	9	1,685	788	353	738	331	43.8	71.9	55.6	6.3
12 Ohio	ENC	33	11,151	5,584	2,604	5,318	2,481	47.7	74.5	58.3	4.8
13 Mississippi	ESC	12	2,697	1,258	599	1,181	562	43.8	71.4	55.8	6.1
14 Arkansas	WSC	31	2,484	1,223	579	1,163	545	46.8	71.9	58.4	4.9
15 Louisiana	WSC	6	4,342	1,956	905	1,821	844	41.9	69.8	53.6	6.9
16 Idaho	Mountain	21	1,163	598	263	565	252	48.6	78.0	62.1	5.4
17 Michigan	ENC	24	9,549	4,745	2,176	4,491	2,062	47.0	74.5	57.9	5.3
18 New Hampshire	NE	44	1,148	633	296	607	284	52.9	80.2	65.3	4.0
19 South Dakota	WNC	50	729	382	176	371	171	50.9	78.2	65.8	2.9
20 Kentucky	ESC	22	3,860	1,861	864	1,761	820	45.6	70.7	56.0	5.4
21 Montana	Mountain	14	870	436	202	410	192	47.1	73.5	59.6	5.9
22 Iowa	WNC	48	2,842	1,559	737	1,504	712	52.9	78.3	66.6	3.5
23 Indiana	ENC	36	5,803	3,134	1,473	2,988	1,398	51.5	78.8	64.2	4.7
24 Nebraska	WNC	51	1,637	897	431	873	419	53.3	79.9	68.4	2.6
25 Kansas	WNC	39	2,565	1.330	622	1,270	591	49.5	77.8	63.4	4.4
26 Wisconsin	ENC	45	5,123	2,846	1,324	2,741	1,277	53.5	79.4	68.2	3.7
Above average, total			124,777	63,225	29,129	60,016	27,601	48.1	75.6	59.6	5.1

Source: *Statistical Abstract of the United States 1996*, P. 397.

unemployment rate. For the 26 states with a positive differential effect for the 1993–95 period, 48.1 percent of the total population was employed, 75.6 percent of males and 59.6 percent of females age 16 years and older were in the labor force, and 5.1 percent of the total labor force was unemployed.

These three relationships show much variance in the age and gender composition of the resident labor force of each state, along with its participation and unemployment rates. Participation rates for females are as low as 53.6 percent (Louisiana), for males as high as 82.5 percent (Utah). Unemployment rates range from a low of 2.6 percent (Nebraska) to a high of 6.9 percent (Louisiana).

Table 7-7 lists the same numbers as in Table 7-6 for the remaining 24 states and the District of Columbia. Their summary totals are consistently lower, except for unemployment, which is a full percentage point higher. Again, the numbers differ widely, with West Virginia having the lowest employment-population ratio and labor force participation rates and next to the highest (exceeded only by Alaska) unemployment rates.

The two measures of economic activity change (the proportional and differential effects) represent the combined national growth and industry-mix effects and the regional-share effect, respectively. They clearly fail, however, to unequivocally differentiate between economically "weak" and "strong" states, at least as measured by the three employment variables cited earlier. We explore the possibilities of improving the usefulness of this approach, however, following a close look at the derivation of the two measures of period-to-period changes in state-level economic activity.

Table 7-8 summarizes the shift-share findings for two periods, 1993–95 and 1995–96, for the 26 states with positive differential effects. It also shows the 1995–96 differential effect rankings.

For the nation, of course, the state-level differential effect totals cancel one another, the same as state-to-state migration. Clearly, much information about the workings of a nation's economy is lost by aggregation. Questions remain, however, regarding the usefulness of state-level differential effects in sorting states into meaningful categories for further analysis and study. The sharp shift in rankings from 1993–95 to 1995–96 illustrates the extreme volatility of the measure. While nine of the ten top-ranking states in the 1993–95 listing remain top ranking in the 1995–96 period, seven of the next ten drop to an even lower ranking in the 1995–96 period.

The District of Columbia and the 23 states that demonstrate a negative aggregate differential effect for the 1993–95 period include the highest and lowest ranking states for the 1995–96 period, namely, California and New York (Table 7-9). One clue to the radical shift in rankings for the two periods is apparent correlation between the proportional effect and the differential effect for the

TABLE 7-7 Characteristics of the civilian labor force, by state and negative 1993–95 differential employment change, U.S., 1995

State	Region	Unemp. Rate Rank	Pop. Total (thou.)	Labor Force		Employed		Emp.Pop. Ratio (pct.)	Labor Force Participation Rate		Unemp. Rate (pct.)
				Total (thou.)	Female (thou.)	Total (thou.)	Female (thou.)		Male (pct.)	Female (pct.)	
51 New York	MA	10	18,136	8,493	3,931	7,955	3,700	43.9	70.1	52.8	6.3
50 California	Pacific	3	31,589	15,415	6,792	14,206	6,258	45.0	74.8	56.5	7.8
49 Pennsylvania	MA	15	12,072	5,838	2,646	5,495	2,495	45.5	71.5	55.0	5.9
48 New Jersey	MA	7	7,945	4,064	1,868	3,803	1,753	47.9	75.1	58.7	6.4
47 District of Columbia	SA	1	554	283	145	258	133	46.6	70.4	61.4	8.9
46 Connecticut	NE	19	3,275	1,709	812	1,615	767	49.3	76.0	60.7	5.5
45 Maryland	SA	28	5,042	2,723	1,314	2,584	1,242	51.2	77.8	64.5	5.1
44 Illinois	ENC	26	11,830	6,083	2,814	5,770	2,674	48.8	77.1	60.3	5.2
43 Massachusetts	NE	23	6,074	3,168	1,498	2,998	1,425	49.4	75.0	60.8	5.4
42 Washington	Pacific	8	5,431	1,805	1,261	2,626	1,181	48.4	76.5	60.6	6.4
41 Hawaii	Pacific	16	1,187	580	284	546	270	46.0	74.3	61.2	5.9
40 Virginia	SA	38	6,618	3,496	1,657	3,338	1,573	50.4	78.0	63.0	4.5
39 Missouri	WNC	34	5,324	2,832	1,361	2,697	1,293	50.7	76.8	65.0	4.8
38 Rhode Island	NE	5	990	485	235	451	219	45.6	71.7	58.4	7.0
37 South Carolina	SA	29	3,673	1,859	878	1,765	822	48.1	74.1	59.9	5.1
36 Alabama	ESC	11	4,253	2,062	957	1,933	898	45.5	72.4	55.3	6.3
35 Maine	NE	17	1,241	642	305	605	290	48.8	72.3	61.6	5.7
34 Alaska	Pacific	4	604	302	141	280	132	46.4	79.2	66.4	7.3
33 Minnesota	WNC	46	4,610	2,589	1,222	2,493	1,186	54.1	81.3	69.6	3.7
32 Oklahoma	WSC	37	3,278	1,547	719	1,471	687	45.0	71.9	55.7	4.7
31 Vermont	NE	43	585	320	153	306	147	52.3	78.0	65.3	4.2
30 Delaware	SA	41	717	381	180	365	173	50.9	77.2	62.9	4.3
29 West Virginia	SA	2	1,828	790	356	728	331	39.8	63.4	46.3	7.9
28 Wyoming	Mountain	35	480	256	117	244	112	40.8	79.3	64.1	4.8
27 North Dakota	WNC	49	641	334	159	323	154	50.4	77.1	64.9	3.3
Total			137,977	69,056	31,805	64,858	29,915	47.0	74.5	58.3	6.1

Source: *Statistical Abstract of the United States 1996*, p. 397.

TABLE 7-8 Nonfarm employment change, by state and positive 1993–95 differential change, U.S., 1993–96

Rank	State	Region	1995–96 Change	1993 (thou.)	Change, 1993–95 Total (thou.)	Change, 1993–95 Prop. (thou.)	Change, 1993–95 Diff. (thou.)	1995 (thou.)	Change, 1995–96 Total (thou.)	Change, 1995–96 Prop. (thou.)	Change, 1995–96 Diff. (thou.)	1996 (thou.)
	All states			110,525	6,678	6,678	0	117,203	2,351	2,351	0	119,554
1	Georgia	SA	5	3,109	308	184	124	3,402	111	69	42	3,528
2	Texas	WSC	4	7,482	546	443	103	8,023	215	165	50	8,242
3	Arizona	Mountain	2	1,586	197	99	97	1,796	113	42	71	1,896
4	Nevada	Mountain	6	671	118	48	70	786	54	15	39	843
5	Colorado	Mountain	10	1,671	168	104	64	1,834	58	38	20	1,897
6	Florida	SA	3	5,571	429	370	59	5,996	183	120	63	6,183
7	Utah	Mountain	8	810	98	49	49	908	47	20	27	955
8	Tennessee	ESC	41	2,329	175	137	38	2,499	31	45	–14	2,534
9	Oregon	Pacific	9	1,308	109	78	30	1,418	58	32	26	1,475
10	North Carolina	SA	7	3,245	210	184	26	3,460	100	70	30	3,555
11	New Mexico	Mountain	38	636	64	37	26	682	4	15	–11	694
12	Ohio	ENC	48	4,918	314	292	22	5,221	64	97	–33	5,296
13	Mississippi	ESC	37	1,002	73	54	19	1,075	15	22	–7	1,090
14	Arkansas	WSC	28	994	75	56	19	1,069	17	20	–3	1,086
15	Louisiana	WSC	23	1,659	116	98	18	1,772	36	36	0	1,811
16	Idaho	Mountain	16	437	41	26	15	477	15	11	4	492
17	Michigan	ENC	13	4,006	246	234	12	4,274	93	83	10	4,345
18	New Hampshire	NE	14	502	37	30	6	540	21	11	10	560
19	South Dakota	WNC	26	319	25	19	6	344	5	8	–3	349
20	Kentucky	ESC	30	1,548	95	89	6	1,643	28	32	–4	1,671
21	Montana	Mountain	21	326	25	20	5	351	8	7	1	359
22	Iowa	WNC	32	1,279	78	76	3	1,358	23	28	–5	1,380
23	Indiana	ENC	44	2,627	154	151	3	2,787	32	50	–18	2,813
24	Nebraska	WNC	20	767	48	46	2	816	19	18	1	834
25	Kansas	WNC	19	1,133	68	66	2	1,198	27	25	2	1,228
26	Wisconsin	ENC	27	2,413	142	141	2	2,559	47	50	–3	2,602
	Total			52,337	3,957	3,132	825	56,288	1,424	1,129	295	57,718

Source: *Statistical Abstract of the United States 1995, 1996,* and *1997,* pp. 425, 418, and 423, respectively.

TABLE 7-9 Nonfarm employment change, by state and negative 1993–95 differential change, U.S., 1993–96

Rank	State	Region	1995–96 Change	1993 (thou.)	Change, 1993–95 Total (thou.)	Change, 1993–95 Prop. (thou.)	Change, 1993–95 Diff. (thou.)	1995 (thou.)	Change, 1995–96 Total (thou.)	Change, 1995–96 Prop. (thou.)	Change, 1995–96 Diff. (thou.)	1996 (thou.)
51	New York	MA	51	7,752	119	469	−350	7,872	46	174	−128	7,917
50	California	Pacific	1	12,045	389	735	−347	12,422	341	259	82	12,775
49	Pennsylvania	MA	49	5,123	125	315	−190	5,253	60	94	−34	5,308
48	New Jersey	MA	47	3,491	115	215	−100	3,601	34	67	−33	3,640
47	District of Columbia	SA	50	670	−37	39	−66	643	−20	18	−38	3,640
46	Connecticut	NE	40	1,531	33	91	−58	1,562	19	32	−13	1,583
45	Maryland	SA	45	2,102	79	136	−57	2,183	25	45	−20	2,206
44	Illinois	ENC	46	5,331	269	323	−54	5,593	77	107	−30	5,676
43	Massachusetts	NE	17	2,841	133	179	−46	2,976	62	58	4	3,036
42	Washington	Pacific	11	2,253	96	137	−41	2,347	63	50	13	2,412
41	Hawaii	Pacific	42	539	−6	35	−41	533	−4	10	−14	529
40	Virginia	SA	25	2,919	149	178	−28	3,070	62	64	−2	3,130
39	Missouri	WNC	34	2,395	127	146	−19	2,521	43	48	−5	2,564
38	Rhode Island	NE	36	430	11	26	−15	440	1	8	−7	442
37	South Carolina	SA	35	1,570	78	89	−11	1,646	28	34	−6	1,676
36	Alabama	ESC	43	1,717	86	96	−10	1,804	22	37	−15	1,825
35	Maine	NE	39	519	23	31	−9	538	−2	10	−12	540
34	Alaska	Pacific	33	253	9	14	−5	262	1	6	−5	263
33	Minnesota	WNC	15	2,243	131	135	−4	2,379	58	49	9	2,432
32	Oklahoma	WSC	12	1,247	67	70	−3	1,316	40	29	11	1,354
31	Vermont	NE	24	257	13	16	−3	270	5	5	0	275
30	Delaware	SA	18	349	17	21	−2	366	11	9	2	377
29	West Virginia	SA	29	653	35	37	−2	688	10	13	−3	698
28	Wyoming	Mountain	31	210	10	11	−1	219	1	5	−4	221
27	North Dakota	WNC	22	285	17	17	0	302	7	7	0	309
	Total			58,724	2,097	3,563	−1,465	60,806	990	1,238	−248	61,811

Source: *Statistical Abstract of the United States 1995, 1996,* and *1997*, pp. 425, 418, and 423, respectively.

two top-ranking states in the later period. New York also has a proportional effect in the 1995–96 period, but it ranks lowest in its differential effect. We note, however, that sixteen states and the District of Columbia retain their 1993–95 negative rankings into the 1995–96 period. Volatility of industry rankings is exactly what we might expect in a dynamic and robust national economy, with individual states contributing to the nation's economy in varying degree.

The more difficult problem to tackle is the interpretation of the findings and the identification of industry-specific sources of the positive and negative differential effects. This is a problem of measuring and understanding industry structure and its role in accounting for varying degrees of industry volatility. We address this issue in the next two tables.

Industry Structure

We again introduce the one-digit Standard Industrial Classification (SIC) code industry breakdown to identify sources of change in state-level employment from one period to the next attributable to changes in industry structure. In 1995, the private services sector accounted for the largest share of total industry employment at the national level—33.1 million (28.2 percent) of the 117.2 million industry total. Wholesale and retail trade accounts for 27.9 million (23.5 percent) of total employment, followed by government with 19.3 million and manufacturing with 18.5 million. Each of the latter accounts for 16.5 percent and 15.8 percent, respectively, of the total; the remaining four nonagricultural industries account for 18.4 million, or 15.7 percent, of the total.

Table 7-10 presents the distribution of the nation's nonfarm industry employment totals among eight one-digit industry groups for those states with a positive differential effect for the 1993–95 period. These industries account for 56.3 million (48 percent) of total 1995 nonfarm industry employment. They account for more than 48 percent of total employment in four industry groups—mining, construction, manufacturing, and wholesale and retail trade—and less than 48 percent in the remaining four industry groups. They also include eight of the top ten in total employment change over the 1993–96 period. Among the top regional clusters are Mountain and Midwest, with sixteen of the twenty-one states in these two census regions.

The twenty-five states and the District of Columbia listed in Table 7-11 fall under the category of negative differential effects for the 1993–95 period. The Northeast, South, and Pacific regions dominate this category, particularly the Northeast, with its finance, insurance, real estate, and private services. As suggested earlier, this table includes seven of the top twenty states in total employment change over the 1993–96 period, and two of the three largest in total employment in 1995.

TABLE 7-10 Nonfarm industry employment, by state and positive 1993–95 differential change, U.S., 1995

Rank	State	Region	1993–96 Tot.Chg.	Total (thou.)	Mining (thou.)	Construction (thou.)	Manufacturing (thou.)	Transportation and Public Utilities (thou.)	Wholesale and Retail Trade (thou.)	Finance, Insurance, and Real Estate (thou.)	Services (thou.)	Government (thou.)
	All states			117,203	579	5,158	18,468	6,165	27,585	6,830	33,107	19,311
1	Georgia	SA	4	3,417	10	152	588	215	865	174	838	575
2	Texas	WSC	1	8,027	156	409	1,030	475	1,949	437	2,122	1,449
3	Arizona	Mountain	9	1,783	12	117	193	86	449	107	520	299
4	Nevada	Mountain	18	789	12	62	37	41	157	36	347	97
5	Colorado	Mountain	10	1,839	14	103	191	118	458	113	539	303
6	Florida	SA	3	6,000	7	304	482	303	1,548	376	2,056	924
7	Utah	Mountain	25	908	8	54	124	52	220	48	239	163
8	Tennessee	ESC	12	2,503	6	108	542	137	588	110	636	376
9	Oregon	Pacific	20	1,417	1	68	228	71	359	87	364	239
10	North Carolina	SA	8	3,455	3	174	861	164	799	144	760	550
11	New Mexico	Mountain	35	690	16	46	45	31	164	30	193	165
12	Ohio	ENC	5	5,232	13	207	1,101	229	1,276	270	1,387	749
13	Mississippi	ESC	34	1,075	5	45	258	50	228	40	232	217
14	Arkansas	WSC	33	1,069	4	44	259	64	22	42	237	177
15	Louisiana	WSC	23	1,775	47	105	188	108	417	80	473	357
16	Idaho	Mountain	38	477	3	30	71	23	121	24	109	96
17	Michigan	ENC	7	4,252	8	154	975	166	1,000	196	1,114	639
18	New Hampshire	NE	37	539	0	19	102	20	141	29	152	76
19	South Dakota	WNC	42	344	2	14	46	16	88	19	88	71
20	Kentucky	ESC	26	1,643	25	73	314	91	396	65	392	287
21	Montana	Mountain	41	351	6	16	23	21	96	16	96	77
22	Iowa	WNC	31	1,357	2	55	250	61	341	77	341	230
23	Indiana	ENC	16	2,781	6	130	684	140	669	131	632	389
24	Nebraska	WNC	36	815	2	34	112	49	205	52	210	511
25	Kansas	WNC	32	1,201	7	52	192	68	296	58	290	238
26	Wisconsin	ENC	14	2,555	3	100	601	119	587	136	631	378
	Total			56,294	378	2,675	9,497	2,918	13,659	2,897	14,998	9,272

Source: *Statistical Abstract of the United States 1995* and *1996*, pp. 425 and 418, respectively.

TABLE 7.11 Nonfarm industry employment, by state and negative 1993–95 differential change, U.S., 1995

Rank	State	Region	1993–96 Tot.Chg.	Total (thou.)	Mining (thou.)	Construction (thou.)	Manufacturing (thou.)	Transportation and Public Utilities (thou.)	Wholesale and Retail Trade (thou.)	Finance, Insurance, and Real Estate (thou.)	Services (thou.)	Government (thou.)
51	New York	MA	21	7,871	4	251	944	403	1,614	724	2,537	1,394
50	California	Pacific	2	12,434	30	488	1,790	630	2,927	737	3,730	2,102
49	Pennsylvania	MA	17	5,248	20	201	939	272	1,197	303	1,596	720
48	New Jersey	MA	24	3,606	1	124	500	252	851	228	1,081	569
47	District of Columbia	SA	51	643	0	9	13	20	52	30	265	254
46	Connecticut	NE	39	1,564	0	51	280	72	341	133	466	221
45	Maryland	SA	30	1,181	1	128	176	106	530	128	690	422
44	Illinois	ENC	6	5,599	13	216	967	324	1,317	384	1,577	801
43	Massachusetts	NE	13	2,974	1	90	445	128	688	204	1,024	394
42	Washington	Pacific	22	2,349	3	123	332	120	583	122	622	444
41	Hawaii	Pacific	50	533	0	26	17	41	136	37	165	111
40	Virginia	SA	11	3,068	11	168	402	157	700	161	872	597
39	Missouri	WNC	19	2,521	6	111	421	159	603	146	685	390
38	Rhode Island	NE	47	441	–1	14	85	15	98	25	144	61
37	South Carolina	SA	29	1,648	2	87	378	71	384	69	363	294
36	Alabama	ESC	27	1,803	10	87	391	89	411	77	395	343
35	Maine	NE	45	542	0	22	92	22	139	26	148	93
34	Alaska	Pacific	49	262	9	13	17	23	54	12	61	73
33	Minnesota	WNC	15	2,374	7	83	426	117	577	138	645	381
32	Oklahoma	WSC	28	1,314	32	48	170	73	312	65	344	270
31	Vermont	NE	46	270	1	12	45	12	64	12	79	45
30	Delaware	SA	43	366	–1	19	62	16	83	41	96	50
29	West Virginia	SA	40	688	27	33	82	40	159	27	183	137
28	Wyoming	Mountain	48	220	16	14	10	14	52	8	48	58
27	North Dakota	WNC	44	302	3	14	21	19	79	14	81	71
	Total			60,821	195	2,432	9,005	3,195	13,951	3,851	17,897	10,295

Source: *Statistical Abstract of the United States 1995* and *1996*, pp. 425 and 418, respectively.

We turn to the use of another tool described in Chapter 6, namely, excess activity analysis. We intend to quantify differences in each state's distribution of industry employment compared with the nation's distribution. This gives us a measure of excess or deficit employment for each industry. The excess activity approach is a variant of the industry location quotient method, also cited in Chapter 6. Excess employment correlates with exports of goods and services, while deficit employment correlates with imports.

Table 7-12 summarizes the results of using the excess activity model to derive excess and deficit industry employment for all states and the District of Columbia. The table then shows state ranking according to excess employment in two industry groups—manufacturing and trade.

The top ten states in manufacturing account for 80 percent of total excess manufacturing employment, while the top ten states in wholesaling and retailing account for 78 percent of the total. Excess manufacturing employment is mostly in the long-established manufacturing states of the Midwest and the South, while excess trade employment occurs in states with many retirees or large trade centers serving an extended trading region covering several states. Fifteen states account for the remaining excess manufacturing employment, while twenty states account for the remaining excess trade employment. The lesser concentration of excess manufacturing and trade employment in the remaining states is accompanied by a wider distribution of excess employment in the remaining industries.

The tendency for dispersion of excess employment among several industries, with less concentration in any one industry, is partly a function of the level of detail in industry classification. The net excess employment for a one-digit industry classification is generally less than for a two-digit which, in turn, is less than for a three-digit, and so on. If all industries were represented on a three-digit basis, for example, the excess employment would be much larger. The deficit employment also would be larger. Industry aggregation thus masks the degree of interdependence that exists between regions because of industry specialization and interregional trade—the hallmarks of local and regional economies exercising their particular comparative advantages in local and regional markets.

All states and the District of Columbia are ranked according to excess employment in private services and government. As shown in Table 7-13, the top ten states in private services from the listing in all the tables account for 95 percent of the total excess private services employment, while the top ten states in government employment account for 67 percent of the U.S. total. The excess private services employment is mostly in states with above-average incomes, many retirees, and major metropolitan centers specializing in business

TABLE 7-12 Excess and deficit nonfarm employment, by manufacturing and trade ranking, U.S., 1995

Rank	State	Region	Mining (thou.)	Construction (thou.)	Manufacturing (thou.)	Transportation and Public Utilities (thou.)	Wholesale and Retail Trade (thou.)	Finance, Insurance, and Real Estate (thou.)	Services (thou.)	Government (thou.)
	Manufacturing:									
1	North Carolina	SA	–14	22	317	–18	–14	–57	–216	–19
2	Michigan	ENC	–13	–33	305	–58	–1	–52	–87	–62
3	Ohio	ENC	–13	–23	277	–46	45	–35	–91	–113
4	Indiana	ENC	–8	8	246	–6	14	–31	–154	–69
5	Wisconsin	ENC	–10	–12	198	–15	–14	–13	–91	–43
6	Tennessee	ESC	–6	–2	148	5	–1	–36	–71	–36
7	South Carolina	SA	–6	14	118	–16	–4	–27	–103	22
8	Pennsylvania	MA	–6	–30	112	–4	–38	–3	114	–145
9	Alabama	ESC	1	8	107	–6	–13	–28	–114	46
10	Arkansas	WSC	–1	–3	91	8	–10	–20	–65	1
	Top 10		–76	–52	1,918	–156	–36	–302	–878	–418
	Other excess		–45	–79	477	33	123	17	–423	–104
	Total		–121	–131	2,395	–122	86	–285	–1,300	–522
	Wholesale and retail trade:									
1	Florida	SA	–23	40	–463	–13	136	26	361	–65
2	Georgia	SA	–7	2	50	35	61	–25	–127	12
3	Texas	WSC	116	56	–235	53	60	–31	–145	126
4	Ohio	ENC	–13	–23	277	–46	45	–35	–91	–113
5	Washington	Pacific	–9	20	–38	–4	30	–15	–42	57
6	Arizona	Mountain	3	39	–88	–8	29	3	16	5
7	Oregon	Pacific	–6	6	5	–4	25	4	–36	6
8	Colorado	Mountain	5	22	–99	21	25	6	20	0
9	Iowa	WNC	–5	–5	36	–10	22	–2	–42	6
10	Minnesota	WNC	–5	–21	52	–8	18	0	–26	–10
	Top 10		58	134	–504	17	451	–68	–112	25
	Other excess		–1	–27	–351	68	164	–89	–60	296
	Total		57	107	–855	85	615	–158	–172	321

Source: *Statistical Abstract of the United States 1995* and *1996*, pp. 425 and 418, respectively.

TABLE 7-13 Excess and deficit nonfarm employment, by private services and government ranking, U.S., 1995

Rank	State	Region	Mining (thou.)	Construction (thou.)	Manufacturing (thou.)	Transportation and Public Utilities (thou.)	Wholesale and Retail Trade (thou.)	Finance, Insurance, and Real Estate (thou.)	Services (thou.)	Government (thou.)
	Private services:									
1	Florida	SA	−23	40	−463	−13	136	26	361	−65
2	New York	MA	−35	−95	−296	−11	−239	265	314	97
3	California	Pacific	−31	−59	−169	−24	1	12	218	53
4	Massachusetts	NE	−14	−41	−24	−28	−12	31	184	−96
5	Nevada	Mountain	8	27	−87	−1	−29	−10	124	−33
6	Pennsylvania	MA	−6	−30	112	−4	−38	−3	114	−145
7	District of Columbia	SA	−3	−19	−88	−14	−99	−7	83	148
8	Maryland	SA	−10	32	−168	−9	17	1	74	63
9	New Jersey	MA	−17	−35	−68	62	2	18	62	−25
10	Connecticut	NE	−8	−18	34	−10	−27	42	24	−37
	Top 10		−138	−198	−1,218	−51	−288	375	1,558	−39
	Other excess		−2	91	−317	12	38	−7	78	109
	Total		−140	−107	−1,536	−39	−251	368	1,636	70
	Government:									
1	District of Columbia	SA	−3	−19	−88	−14	−99	−7	83	148
2	Texas	WSC	116	56	−235	53	60	−31	−145	126
3	New York	MA	−35	−95	−296	−11	−239	265	314	97
4	Virginia	SA	−4	33	−81	−4	−22	−18	5	91
5	Louisiana	WSC	38	27	−92	15	−1	−23	−28	65
6	Maryland	SA	−10	32	−168	−9	17	1	74	63
7	Washington	Pacific	−9	20	−38	−4	30	−15	−42	57
8	Oklahoma	WSC	26	−10	−37	4	3	−12	−27	53
9	California	Pacific	−31	−59	−169	−24	1	12	218	53
10	New Mexico	Mountain	13	16	−64	−5	2	−10	−2	51
	Top 10		101	−1	−1,268	0	−249	162	450	805
	Other excess		46	88	225	34	177	−198	−771	400
	Total		146	87	−1,044	35	−73	−36	−321	1,205

Source: *Statistical Abstract of the United States 1995* and *1996*, pp. 425 and 418, respectively.

and professional services. The excess government employment is the most widely dispersed of the four industry groups.

The two notable areas of concentration are contrasts in government activity, namely, the federal government in the District of Columbia and the incidence of many local governments in Texas. Six states account for the remaining excess private services employment, while twenty-one states account for the remaining excess government employment. Again, the lesser concentration of excess private services and government employment in the remaining states is accompanied by a wider distribution of excess employment in the remaining industries.

The four sets of industry comparisons provide additional documentation of the degree of industry specialization and interregional trade that emerge from the attempts of many thousands of businesses to assert their individual comparative advantages from day to day and week to week. Over longer periods, the result is the appearance of much overlapping of identical product flows, frequently in opposite directions. The product flows, of course, are neither identical nor available at the time and point of sale. High levels of data aggregation by industry and region thus add to the illusion of high potential, if not actual, levels of local and regional self-sufficiency.

Government Spending and Financing

Government spending and financing is a separate topic well beyond the scope and content of urban regional economics. This area of economic life accounts for nearly two-fifths of the nation's gross domestic product—37.3 percent ($9.643 per capita) in 1994. We confine our discussion to state and local governments which, in 1994, accounted for 44.4 percent of total government revenues from own sources and 47.2 percent of spending, exclusive of intergovernmental transfers. The ranking of government revenues and spending is confined to state governments, which accounted for 25.3 percent of total government revenues from own sources and 25.2 percent of total direct government expenditures.

Driving Forces

The determinants of state and local spending and finances are a combination of the capacities to levy and collect taxes and acquire other revenues and the constituencies seeking ways of gaining access to these revenues to satisfy particular spending priorities. In 1994, state and local governments acquired more than $1.3 trillion in total revenues and spent $70 billion less than this total.

From a long-term perspective, we find state and local government finances carrying an increasingly larger burden of public welfare and health care expenditures. Annual rates of increase in public welfare spending were 2.3 percent above the overall average for all direct expenditures in the 1980–90 period and

6.2 percent above the average in the 1990–94 period (Table 7-14). The difference was even larger for health care spending in the 1980–90 period at 7.9 percent, but this difference dropped to 3.2 percent in the 1990–94 period.

Public welfare spending is much greater than health care spending and increases at a faster rate. Intergovernmental transfers from the federal government offset much of the increases in the two periods.

Spending on hospitals is increasing also, reaching nearly the same rate of increase as overall state and local government spending. Contributing to all of these increases are the insurance trust expenditures for government workers. These have increased from a rate 0.7 percent below the overall spending increases in the 1980–90 period to 4.1 percent above the overall annual increase in the 1990–94 period. The per capita burden increased from $1,993 in 1980 to $4,180 in 1990, and $5,114 in 1994—all in current, not constant, dollars. As a percent of gross domestic product, the totals grew from 16.7 in 1980 to 18.6 in 1990, and 19.8 in 1994—a 7.4 and 5.0 percent annual relative increase in the two periods, respectively. The ranking of individual state revenues and expenditures in 1995 fits into this larger framework of overall government spending.

Thus, the two critically important social issues of our time—public welfare and health care—are driving the increasing costs of financing state and local governments. These increases were greater in the 1980–90 period than in the 1990–94 period—increases covered in large part by intergovernmental transfers from the federal government.

State government revenues establish the fiscal constraints on state government spending. The forecasting of state revenues thus becomes an important part of the state fiscal processes as state governments assume an increasingly larger share of all government financing. State revenue forecasts must account for likely changes in the determinants of state revenues, starting with the gross state product and its linkages to both gross domestic product and the unique conditions that differentiate one state's economy from that of another.

Table 7-15 is the first of two rankings of state government revenues that we identify as fiscal constraints on state government spending. We show, first, the individual states with above-average per capita state revenues. These twenty-three states account for roughly half of all state revenues, including intergovernmental transfers from the federal government. New York and California lead the list of federal transfers with nearly one-fourth of the total. They account for 23.2 percent of all state revenues in 1995. Alaska leads the list of revenues from its own sources by a very wide margin, thanks to its oil-related economic base. Even among the above-average revenue-generating states, 1995 per capita revenue ranges widely—from $12,202 for Alaska to $2,857 for Iowa. State government revenues totaled to $903.8 billion in 1995, of which $215.6 originated

TABLE 7-14 State and local governments, summary of finances, 1980, 1990, and 1994 (in current dollars)

Selected Revenues and Expenditures	Total			Per Capita			Relative Change	
	1980 (bil.$)	1990 (bil.$)	1994 (bil.$)	1980 (dol.)	1990 (dol.)	1994 (dol.)	1980–90 (pct.)	1990–94 (pct.)
Revenues	**451.5**	**1,032.1**	**1,331.4**	**1,993**	**4,150**	**5,114**	**0.0**	**0.0**
From federal government	83.0	136.8	215.4	367	550	828	–3.5	5.5
Public welfare	24.9	60.0	111.3	110	241	428	0.6	10.2
Highways	9.0	14.4	18.3	40	58	70	–3.8	–0.4
Education	14.4	23.2	30.4	64	93	117	–3.7	0.4
Health and hospitals	2.5	5.9	9.4	11	24	36	0.3	5.6
Housing and community development	3.9	9.7	13.1	17	39	50	0.9	1.4
Other and unallocable	28.3	23.7	33.0	125	95	127	–10.4	2.1
From state and local sources	368.5	895.3	1,116.0	1,627	3,600	4,286	0.7	–0.9
General, not intergovernmental	299.3	712.7	885.0	1,321	2,865	3,399	0.4	–1.0
Taxes	223.5	501.6	625.5	986	2,017	2,403	–0.2	–0.9
Property	68.5	155.6	197.1	302	626	757	–0.1	–0.5
Sales and gross receipts	79.9	177.9	223.6	353	715	859	–0.3	–0.7
Individual income	42.1	105.6	128.8	186	425	495	1.0	–1.5
Corporation income	13.3	23.6	28.3	59	95	109	–2.7	–1.9
Other	19.6	38.9	47.6	87	156	183	–1.5	–1.4
Charges and miscellaneous	75.8	211.1	259.5	335	849	997	2.2	–1.3
Utility and liquor stores	25.6	58.6	70.1	1113	236	269	0.0	–2.0
Insurance and trust revenues	43.7	124.0	160.9	193	498	618	2.4	0.2
Direct expenditures	**432.3**	**972.7**	**1,260.6**	**1,980**	**3,911**	**4,842**	**0.0**	**0.0**
Direct general expenditures	367.3	834.8	1,074.0	1,622	3,356	4,125	0.3	–0.2
Education	133.2	288.1	353.3	588	1,159	1,357	–1.2	–1.5
Elementary and secondary	92.9	202.0	247.0	410	812	949	–1.0	–1.5
Higher education	33.9	73.4	90.9	150	295	349	–1.2	–1.2
Highways	33.3	61.1	72.1	147	245	277	–6.1	–2.5
Public welfare	45.6	110.5	179.8	201	444	691	2.3	6.2
Health	8.4	24.2	35.3	37	97	136	7.9	3.2
Hospitals	23.8	50.4	65.1	105	203	250	–1.8	–0.1
Police protection	13.5	30.6	38.6	60	123	148	0.2	–0.7
Sanitation and sewage	13.2	28.5	35.7	58	114	137	–1.3	–0.9
Interest on general debt	14.7	49.7	55.0	65	200	211	13.0	–4.2
Utility and liquor stores	36.2	77.8	91.2	160	313	350	–1.4	–2.7
Insurance trust expenditures	28.8	63.3	95.5	127	255	367	–0.7	4.1
Wages and salaries	164.0	341.0	411.0	723	1,372	1,577	–2.4	–2.0
Gross domestic product (rev. pct. GDP)	2,708	5,546	6,736	16.7	18.6	19.8	7.4	5.0

Source: *Statistical Abstract of the United States* 1997, p. 304.

TABLE 7-15 State government revenues, by state and above-average per capita state general revenues, U.S., 1995

Rank	State	Per Cap. GSP Rank	All Rev-enues (mil.$)	Gen'l Revenue Total (mil.$)	Intergovernmental: Intergov. Total (mil.$)	Intergovernmental: From Federal (mil.$)	Intergovernmental: From St.&Loc. (mil.$)	General Reviews: Charges &Miscel-laneous (mil.$)	Taxes: Taxes Total (mil.$)	Taxes: Sales Taxes General (mil.$)	Taxes: Sales Taxes Select (mil.$)	Taxes: License Taxes (mil.$)	Taxes: Income Taxes Individ. (mil.$)	Taxes: Income Taxes Corporate (dol.)	Per Capita Gen.Rev. (dol.)
	All states, total		903,756	739,016	215,558	202,485	13,073	124,310	399,148	132,236	64,615	26,083	125,610	29,075	2,823
1	Alaska	1	8,288	7,358	980	975	5	4,456	1,922	(X)	101	75	(X)	528	12,202
2	Delaware	2	3,441	3,114	563	554	9	956	1,595	(X)	246	525	562	194	4,398
3	Hawaii	7	5,778	5,099	1,138	1,120	19	1,086	2,874	1,363	432	84	926	47	4,329
4	Wyoming	3	2,240	1,952	814	796	18	471	667	208	64	72	(X)	(X)	4,101
5	New York	6	90,997	69,875	26,918	21,683	5,235	8,662	34,294	6,845	4,929	1,014	17,589	2,815	3,849
6	Connecticut	4	13,718	12,060	2,761	2,755	5	1,825	7,474	2,368	1,341	317	2,474	699	3,682
7	Massachusetts	8	24,101	21,545	5,758	5,297	461	4,186	11,601	2,481	1,255	429	5,974	1,206	3,566
8	New Mexico	35	6,634	5,660	1,377	1,334	43	1,438	2,844	1,219	397	156	592	150	3,420
9	Rhode Island	31	4,156	3,386	1,193	1,141	52	703	1,490	457	312	85	530	82	3,406
10	North Dakota	42	2,448	2,143	692	660	32	492	959	283	265	76	143	70	3,354
11	Minnesota	14	18,329	15,210	3,569	3,394	175	2,313	9,328	2,742	1,428	711	3,664	666	3,330
12	Vermont	38	2,074	1,869	633	631	2	435	801	174	221	65	250	48	3,222
13	New Jersey	5	32,675	25,343	6,051	5,781	271	5,685	13,607	4,133	2,842	721	4,540	1,029	3,207
14	Michigan	22	35,328	30,217	7,518	7,130	388	4,976	17,723	5,866	1,801	853	5,473	2,130	3,183
15	Washington	15	23,576	16,765	3,977	3,923	53	2,593	10,196	6,048	1,588	482	(X)	(X)	3,141
16	Montana	48	3,293	2,644	877	861	16	553	1,214	(X)	249	141	372	76	3,089
17	West Virginia	49	6,629	5,628	2,001	1,987	14	895	2,732	793	667	151	710	219	3,086
18	Oregon	33	12,986	9,438	3,065	3,002	63	2,087	4,286	(X)	547	504	2,798	312	3,057
19	Wisconsin	24	16,826	15,337	3,766	3,545	221	2,542	9,029	2,572	1,276	436	3,933	671	3,017
20	California	11	118,303	94,252	29,982	27,527	2,455	11,001	53,269	17,687	4,933	2,871	18,344	5,748	3,001
21	Maine	45	4,208	3,673	1,173	11,169	4	687	1,813	650	275	115	640	63	2,964
22	Kentucky	37	12,846	11,085	3,048	3,039	9	1,751	6,285	1,681	1,246	373	1,965	341	2,896
23	Iowa	29	9,268	8,087	2,189	2,093	96	1,496	4,403	1,463	604	402	1,617	221	2,857
	Total		458,142	371,740	110,043	100,397	9,646	61,289	200,406	59,033	27,019	10,658	73,096	17,315	3,187

Source: *Statistical Abstract of the United States 1997*, p. 311.

from intergovernmental sources and $399.1 million from taxes—nearly half from sales taxes and nearly a third from individual income taxes.

Large state-to-state differences among the 27 states with below-average per capita revenues arise from the various revenue sources, particularly state sales and income taxes, but also including federal transfers. Per capita differences, however, are less dramatic, as listed in Table 7-16. Although the range from the lowest to the highest—Texas at $2,191 and Louisiana at $2,796—is smaller, the percentage difference is larger, once Alaska is omitted from the comparisons. The comparisons of state government revenues present only part of the story, however, especially for Texas, with its many local governments.

The differences in sources of state revenues and the applicable rates generating these revenues make the task of forecasting future state revenues increasingly difficult in the state budgeting process. Constitutional limitations on deficit financing by state governments add to these difficulties.

Further considerations are new spending priorities for state and local governments. These may range from the funding of commercial sports and business convention facilities to new university research centers and public transit facilities, as well as alternative energy sources of unknown adequacy, quality, and cost. Also, differences in overall state and local governmental structures add to the complexity of in-state revenues and state expenditures, even on a per capita basis. Texas is an obvious example of a state with many governmental units competing for tax revenue.

Public Priorities

Government spending priorities differ from state to state, representing the outcomes of political processes and values unique to each state in varying degree. In 1995, total state government expenditures were $836.9 billion, or nearly $2,800 per capita (Table 7-17). Direct expenditures, largely for education and public welfare, totaled nearly $493 billion (58.8 percent) of all expenditures. Education and public welfare, the major targets of state direct expenditures, accounted for 29.8 and 23.3 percent, respectively, of this total. Other direct expenditures, like health care, highways, and corrections, added another $48 billion (5.7 percent) to this total. Intergovernmental transfers, largely to local governments for public education, were $241 billion (28.8 percent) of the total. The difference—$103.4 billion—between total general expenditures of $733.5 billion and total state government expenditures of $836.9 billion represents a variety of expenditures that come under the headings of utilities, liquor store, and insurance trusts. We again group individual states as above or below average in per capita general revenues.

Table 7-17 lists the top sixteen states in per capita general expenditures and twenty-one of the top twenty-three, starting again with Alaska. Two

TABLE 7-16 State government revenues, by state and below-average per capita state general revenues, U.S., 1995

Rank	State	Per Cap. GSP Rank	All Rev-enues (mil.$)	Gen'l Revenue Total (mil.$)	Intergovernmental: Intergov. Total (mil.$)	Intergovernmental: From Federal (mil.$)	Intergovernmental: From St.&Loc. (mil.$)	General Reviews: Charges &Miscel-laneous (mil.$)	Taxes: Taxes Total (mil.$)	Sales Taxes: General (mil.$)	Sales Taxes: Select (mil.$)	License Taxes (mil.$)	Income Taxes: Individ. (mil.$)	Income Taxes: Corporate (dol.)	Per Capita Gen.Rev. (dol.)
24	Louisiana	34	13,956	12,066	4,670	4,642	28	2,719	4,677	1,490	917	445	1,062	283	2,796
25	Idaho	43	3,845	3,151	810	781	29	608	1,733	576	238	149	600	129	2,779
26	Utah	41	6,325	5,304	1,498	1,453	46	1,130	2,676	1,067	292	99	1,024	147	2,778
27	Maryland	16	16,430	13,696	3,111	3,013	98	2,525	8,061	1,951	1,547	357	3,400	366	2,739
28	Pennsylvania	26	40,015	32,772	9,184	9,117	68	5,326	18,262	5,550	3,103	1,873	4,930	1,785	2,717
29	North Carolina	18	22,091	19,170	5,475	5,045	430	2,269	11,426	2,794	2,067	710	4,699	906	2,711
30	Nevada	9	5,478	3,896	782	736	45	415	2,698	1,438	823	308	(X0	(X)	2,665
31	Mississippi	50	8,301	7,056	2,556	2,448	108	900	3,599	1,692	731	226	683	203	2,643
32	South Carolina	40	12,068	9,609	3,069	2,953	116	1,777	4,763	1,794	676	327	1,656	250	2,638
33	Arkansas	46	7,368	6,467	2,078	2,068	10	997	3,392	1,302	587	197	1,047	192	2,636
34	Nebraska	21	4,615	4,195	1,114	1,091	23	861	2,220	781	406	151	741	124	2,583
35	Indiana	30	16,261	14,848	3,438	3,313	125	3,364	8,046	2,710	866	228	3,257	874	2,580
36	Arizona	36	12,593	10,510	3,027	2,681	346	1,260	6,883	2,772	828	344	1,463	417	2,577
37	Kansas	27	7,374	6,495	1,599	1,570	29	1,130	3,765	1,379	516	199	1,233	261	2,546
38	Ohio	23	41,306	28,263	8,172	7,945	227	4,905	15,186	4,752	2,730	1,333	5,553	713	2,545
39	Colorado	13	11,555	9,191	2,683	2,666	17	1,977	4,531	1,231	682	268	2,102	190	2,510
40	Alabama	44	12,280	10,583	3,240	3,194	46	2,265	5,078	1,365	1,280	481	1,483	236	2,508
41	Virginia	12	18,993	16,251	3,474	3,296	179	3,992	8,784	1,921	1,525	430	4,316	368	2,481
42	Illinois	10	34,689	29,064	8,166	7,770	396	4,308	16,590	4,959	3,471	949	5,312	1,481	2,472
43	South Dakota	32	2,090	1,766	665	658	7	407	694	359	179	88	(X)	40	2,443
44	Georgia	19	20,284	17,123	5,224	5,162	62	2,413	9,487	3,539	921	402	3,842	653	2,426
45	Oklahoma	47	9,160	7,724	2,051	1,982	69	1,257	4,416	1,144	707	575	1,417	167	2,372
46	Missouri	28	15,586	12,516	3,802	3,758	44	1,962	6,752	2,348	928	481	2,535	369	2,371
47	New Hampshire	20	3,270	2,674	925	791	134	830	918	(X)	541	112	38	166	2,356
48	Tennessee	25	12,900	11,522	4,045	3,992	53	1,569	5,908	3,361	1,204	603	102	493	2,226
49	Florida	39	37,359	31,013	7,973	7,642	331	4,475	18,5651	10,657	3,670	1,300	(X)	945	2,222
50	Texas	17	49,422	40,352	12,681	12,322	359	7,382	20,289	10,275	6,162	2,791	(X)	(X)	2,191
	Total		445,614	367,277	105,512	102,089	3,425	63,023	199,399	73,207	37,597	15,426	52,495	11,758	2,523

Source: *Statistical Abstract of the United States 1997*, p. 311.

TABLE 7-17 State government expenditures, by state and above-average per capita general revenues, U.S., 1995

Rank	State	PerCap. Expend. Rank	All Expenditures (mil.$)	General Expenditures: Gen'l Expend. Total (mil.$)	Inter-gov'tal (mil.$)	Direct Total (mil.$)	Direct Expenditures: Education (mil.$)	Public Welfare (mil.$)	Health Hospital (mil.$)	Highways (mil.$)	Correction (mil.$)	Administration (mil.$)	Interest on Debt (mil.$)	PerCap. Gen'l Expend. (dol.)
	All states, total		836,894	733,503	240,978	492,525	249,670	194,845	60,003	57,374	26,069	26,078	24,485	2,796
1	Alaska	1	5,599	5,047	1,096	3,952	1,258	643	176	579	147	356	276	8,377
2	Delaware	3	2,980	2,709	510	2,199	920	414	201	246	110	207	219	3,778
3	Hawaii	2	6,015	5,372	144	5,227	1,645	894	522	282	110	280	299	4,555
4	Wyoming	5	2,045	1,803	676	1,127	626	252	108	287	30	75	48	3,763
5	New York	4	81,372	68,308	25,190	43,118	17,494	25,785	6,024	3,047	2,273	2,799	3,393	3,755
6	Connecticut	8	13,576	11,603	2,409	9,194	2,815	2,910	1,287	750	481	575	846	3,548
7	Massachusetts	6	24,282	21,870	4,740	17,130	4,102	6,178	2,008	1,703	660	1,079	1,573	3,602
8	New Mexico	9	6,363	5,876	1,966	3,901	2,363	879	570	698	161	229	98	3,472
9	Rhode Island	7	4,265	3,544	504	3,041	885	902	335	266	114	182	304	3,574
10	North Dakota	14	2,213	2,048	437	1,611	737	438	103	270	18	65	54	3,193
11	Minnesota	13	16,380	14,808	5,629	9,179	5,558	3,923	1,026	1,036	246	495	275	3,209
12	Vermont	15	2,014	1,854	309	1,546	611	511	58	157	41	112	100	3,171
13	New Jersey	10	32,605	26,438	7,901	18,537	7,406	6,773	1,693	2,119	825	891	1,168	3,326
14	Michigan	11	34,669	30,875	13,590	17,285	13,889	6,179	3,346	1,832	1,185	734	715	3,237
15	Washington	12	21,200	17,555	5,340	12,215	7,273	3,730	1,368	1,537	483	435	492	3,222
16	Montana	21	2,988	2,565	685	1,880	981	463	154	298	50	120	118	2,947
17	West Virginia	19	6,262	5,418	1,255	4,164	1,994	1,576	215	663	77	237	147	2,968
18	Oregon	16	11,030	9,509	2,980	6,530	3,086	2,076	837	982	281	677	367	3,020
19	Wisconsin	22	16,302	14,546	5,723	8,823	5,030	3,160	938	1,115	510	410	492	2,840
20	California	18	109,231	94,007	44,893	49,114	30,027	28,513	7,782	4,821	3,676	3,194	2,666	2,978
21	Maine	20	4,179	3,664	750	2,914	1,046	1,257	218	307	63	141	175	2,958
22	Kentucky	27	11,395	10,191	2,790	7,401	4,062	2,776	592	922	221	402	358	2,642
23	Iowa	24	8,586	7,916	2,587	5,329	3,167	1,650	713	979	177	288	116	2,784
	Total		425,551	367,517	132,104	235,417	116,975	101,882	30,274	24,896	11,939	13,983	14,299	3,150

Source: *Statistical Abstract of the United States 1997*, pp. 312–13.

states—Kentucky and Iowa—have slightly below-average per capita general revenues. The rank correlation is close but not perfect. The twenty-three states listed account for 50 percent of state general expenditures—54.8 percent of intergovernmental expenditures and 47.8 percent of direct expenditures.

Many differences occur between states in the distribution of general expenditures, particularly between intergovernmental and direct expenditures. For example, New York and California are well above average in percentage allocations to public welfare. This is balanced by below-average allocations to education and highways. Each of these two states also has above-average intergovernmental transfers. Michigan is exactly the opposite. Massachusetts, on the other hand, has a below-average allocation to education, but slightly above-average allocations to public welfare, health care, and administration.

Table 7-18 lists total and general expenditures of states with below-average general revenues. Overall, these 27 states allocate a larger total and a larger share of general expenditures to education, highways, and corrections than the above-average states. Texas, North Carolina, and Georgia top the list in above-average allocations to education, but are below average in allocations to public welfare. Pennsylvania and Illinois are highest in allocations to public welfare, but are well below average for education. Florida, on the other hand, is at the bottom of the list for public welfare but above average for highways, corrections, and administration. Virginia is above average for highways, corrections, administration, and education, but low for public welfare. State spending priorities, according to the latest reports, vary tremendously from one state to another.

The distribution of state and local full-time equivalent (FTE) employment provides still another indicator of state spending priorities. In 1995, reported state and local FTE employment reached 16 million. This compares with 19 million total employment—part time and full time—reported in the U.S. Department of Commerce Regional Economic Information System (REIS). State and local education, with FTE employment level of nearly 7.1 million, accounts for 44.3 percent of total FTE employment. Health care and hospitals add another 1.4 million, or 8.9 percent of the total.

Table 7-19 provides a summary of part of the functional distribution of FTE employment for the twenty-three states with above-average per capita general revenues. They account for 6.6 million (41.9 percent) of the overall FTE total, with 2.9 million in education and 480,000 in health care and hospitals. Individual states differ, of course, from the overall averages. Michigan and Washington, for example, are well above average in state higher education, but below average in local elementary and secondary education. California, on the other hand, is above average in local health care and hospitals, but below average in local ele-

TABLE 7-18. State government expenditures, by state and below-average per capita general revenues, U.S., 1995

Rank	State	PerCap. Expend. Rank	All Expenditures (mil.$)	General Expenditures: Gen'l Expend. Total (mil.$)	Inter-gov'tal (mil.$)	Direct Total (mil.$)	Direct Expenditures: Education (mil.$)	Public Welfare (mil.$)	Health Hospital (mil.$)	Highways (mil.$)	Correction (mil.$)	Administration (mil.$)	Interest on Debt (mil.$)	PerCap. Gen'l Expend. (dol.)
24	Louisiana	17	14,461	13,135	2,981	10,154	4,245	3,876	1,576	798	353	322	600	3,028
25	Idaho	30	3,360	2,941	944	1,998	1,251	487	130	366	76	97	93	2,522
26	Utah	25	5,780	5,289	1,447	3,843	2,483	924	422	354	144	252	119	2,701
27	Maryland	29	15,069	12,902	3,074	9,829	3,813	2,956	1,045	1,175	747	653	503	2,561
28	Pennsylvania	23	39,394	33,633	9,031	24,602	9,677	11,111	2,916	2,508	955	1,104	1,234	2,789
29	North Carolina	28	20,437	18,735	6,665	12,069	7,761	4,020	1,584	1,818	820	601	258	2,601
30	Nevada	41	4,581	3,718	1,425	2,294	1,351	612	128	415	136	156	99	2,425
31	Mississippi	34	7,4141	6,708	2,279	4,429	2,495	2,631	542	585	139	180	116	2,488
32	South Carolina	26	11,623	9,889	2,367	7,522	3,537	2,447	1,206	594	348	262	180	2,697
33	Arkansas	38	6,616	6,071	1,586	4,485	2,352	1,519	579	613	161	173	117	2,443
34	Nebraska	33	4,250	4,094	1,144	2,950	1,440	916	444	545	85	119	83	2,497
35	Indiana	35	15,284	14,326	5,115	9,211	5,840	3,207	870	1,398	351	333	285	2,471
36	Arizona	43	11,621	10,072	3,992	6,080	3,601	2,468	545	925	437	333	180	2,340
37	Kansas	32	7,116	6,460	2,206	4,254	2,876	1,212	514	828	191	233	71	2,520
38	Ohio	31	34,990	28,129	9,534	18,595	9,891	7,593	2.322	2,288	1,052	948	816	2,526
39	Colorado	48	9,802	8,434	2,703	5,731	3,691	2,135	371	780	320	290	233	2,250
40	Alabama	37	11,542	10,490	2,620	7,870	4,401	2,291	1,303	924	231	299	194	2,470
41	Virginia	45	17,040	15,664	4,297	11,367	5,979	2,804	1,586	1,983	778	736	511	2,368
42	Illinois	39	32,991	28,845	7,989	20,856	8,505	8,824	2,015	2,651	816	763	1,381	2,446
43	South Dakota	40	1,880	1,774	337	1,438	515	375	107	258	41	83	106	2,432
44	Georgia	36	19,154	17,771	4,850	12,921	7,341	4,590	1,241	1,325	765	397	310	2,465
45	Oklahoma	46	8,990	7,566	2,449	5,118	3,371	1,656	578	726	247	312	162	2,310
46	Missouri	50	12,482	11,400	3,462	7,938	4,285	2,888	929	1,160	265	400	350	2,143
47	New Hampshire	44	3,096	2,718	374	2,344	557	943	160	205	59	134	373	2,367
48	Tennessee	42	13,432	12,578	3,263	9,316	4,303	3,805	1,060	1,220	424	397	186	2,397
49	Florida	47	34,750	32,168	10,950	21,219	10,846	7,046	2,451	2,995	1,599	1.224	873	2,268
50	Texas	49	44,643	40,475	11,797	28,677	16,286	10,635	3,106	3,038	2,590	1,294	752	2,153
	Total		411,339	365,985	108,881	257,110	132,698	92,971	29,730	32,475	14,130	12,095	10,185	2,515

Source: *Statistical Abstract of the United States 1997*, pp. 312–13.

TABLE 7-19 State and local government FTE, by state and above-average per capita state general revenue, 1995

Rank	State	FTE Rank	Overall Emp. Total (thou.)	FTE Total (thou.)	Total Education		Elem. & Sec. Ed.		Higher Education		Public Welfare		Health		Hospitals	
					State (thou.)	Local (thou.)	State (thou.)	Local (thou.)	State (thou.)	Local (thou.)	State (thou.)	Local (thou.)	State (thou.)	Local (thou.)	State (thou.)	Local (thou.)
	Total		19,040	15,988	1,469	5,619	41	5,341	1,331	278	227	265	160	209	496	551
1	Alaska	50	74	40	7.3	11.4	2.9	11.4	3.9	—	1.8	0.1	0.5	0.6	0.3	0.2
2	Delaware	46	51	47	6.8	13.7	—	13.7	6.5	—	2.0	—	1.8	0.2	2.3	—
3	Hawaii	42	112	68	30.4	—	23.2	—	7.1	—	1.1	0.1	2.8	0.2	3.2	—
4	Wyoming	48	58	45	3.3	16.1	—	14.4	3.2	1.7	0.3	—	0.5	0.2	1.4	4.1
5	New York	3	1,420	1,097	49.0	400.8	—	375.3	43.9	25.5	7.5	51.8	8.9	17.2	55.6	61.4
6	Connecticut	31	217	190	18.0	67.0	—	67.0	15.2	—	4.8	1.8	2.6	1.4	12.3	—
7	Massachusetts	14	392	339	23.7	129.5	—	129.5	23.0	—	6.6	2.7	3.9	2.1	11.8	6.1
8	New Mexico	34	163	133	16.9	43.3	—	40.9	15.9	2.4	1.4	0.5	2.9	0.4	5.4	2.6
9	Rhode Island	44	62	54	7.0	18.1	0.4	18.1	5.8	—	1.8	0.1	1.3	0.1	1.5	—
10	North Dakota	47	67	47	7.8	13.6	—	13.6	7.3	—	0.3	0.9	1.3	0.3	1.4	—
11	Minnesota	15	358	335	38.7	109.8	—	106.1	37.2	3.7	2.2	11.0	2.1	3.7	7.2	13.3
12	Vermont	49	45	44	4.6	16.6	—	16.6	4.2	—	1.1	—	0.5	—	0.3	—
13	New Jersey	10	568	493	45.8	178.9	14.8	169.2	27.6	9.8	5.4	11.5	2.5	4.0	17.1	6.4
14	Michigan	8	639	576	63.4	196.2	—	184.2	62.8	12.0	14.6	2.6	2.1	9.7	15.6	13.2
15	Washington	21	438	284	36.8	87.0	—	87.0	36.1	—	7.3	1.2	7.0	4.1	8.1	8.9
16	Montana	41	76	72	7.1	26.7	—	26.5	6.4	0.1	1.3	0.9	0.4	0.6	1.3	0.8
17	West Virginia	37	137	117	13.6	41.0	—	41.0	12.1	—	2.6	—	0.7	1.5	1.5	3.3
18	Oregon	32	234	188	16.2	68.2	—	61.0	15.2	7.2	4.6	0.8	1.5	3.5	7.3	2.3
19	Wisconsin	17	365	319	29.8	114.1	—	105.2	28.5	8.9	1.2	13.3	1.7	4.7	7.6	3.9
20	California	1	2,094	1,545	117.2	553.1	—	490.9	112.6	62.1	3.7	52.0	10.1	33.3	34.5	75.4
21	Maine	38	94	81	6.8	31.1	—	31.1	5.6	—	1.9	0.2	1.1	0.2	1.5	1.1
22	Kentucky	23	282	258	30.3	90.7	—	90.7	26.2	—	4.6	0.5	2.0	4.5	5.5	3.2
23	Iowa	28	226	220	25.9	72.8	—	66.6	24.7	6.2	2.7	1.5	0.3	2.0	8.1	8.9
	Total		8,170	6,591	606	2,300	41	2,160	531	140	81	154	59	95	211	215

Source: *Statistical Abstract of the United States 1997*, p. 324.

mentary and secondary education and state higher education. New York is the lowest ranking in state higher education, but highest in public welfare.

Table 7-20 lists FTE employment for the 27 states with below-average per capita general revenues. These states have the higher levels of total government employment and FTE employment. Texas, for example, has the highest level of excess employment in local elementary and secondary education, but low levels elsewhere. North Carolina is low in local elementary and secondary education but above average in local higher education. These sorts of comparisons help clarify several of the earlier comparisons of per capita general revenues and expenditures.

The annual reports on the state-by-state allocation of federal funding offer additional information on individual state ranking in total spending. They also show how individual states compare with one another in the redistribution of personal and corporate income via the political process. With comparable information on federal tax payments, individual states could assess their balance of payments with the federal government. Of course, difficulties arise in the allocation of corporation income taxes to individual states and their subsequent comparison with the federal funds allocations.

The 1995 federal funds distribution (Table 7-21) starts with a total of $1,320.1 billion, of which $227.5 billion (17.2 percent) is defense related. The nondefense distributions, totaling $1,092.6 billion, cover direct payments to individuals of $691.7 billion, procurement spending of $198 billion, and grants to states and local governments of $214.2 billion. The amount of the total going to salaries and wages—$169 billion in 1995—is shown separately.

According to the previous ranking of states, the twenty-three states with above-average per capita revenue received $544.2 billion (41.2 percent) of total federal funds distribution in 1995. Defense is the smaller part of the total—$80.9 billion (14.9 percent) of the twenty-three states' distribution. The direct payments to individuals of $292.9 billion are 63.2 percent of this total. Overall, the calculated balance for the twenty-three states is a negative $18.2 billion. Alaska leads the list on a per capita basis, followed by New Mexico, Hawaii, and North Dakota. California is the biggest loser, followed by Minnesota, Vermont, and Delaware.

The ranking in Table 7-21 is based on above-average per capita federal funds to states and the District of Columbia. Also listed is the excess or deficit adjusted gross income (AGI, as computed by the Internal Revenue Service), based on the U.S. per capita average, for each state.

A side-by-side comparison of the two columns—the excess or deficit AGI and the excess federal funds distribution—provides a simple test of the income redistribution results of federal funds going to individual states. For the most part, the net excess federal funds accrue to states with a net deficit AGI. The obvious

TABLE 7-20 State and local government FTE, by state and below-average per capita state general revenue, 1995

Rank State	FTE Rank	Overall Emp. Total (thou.)	FTE Total (thou.)	Total Education		Elem. & Sec.Ed.		Higher Education		Public Welfare		Health		Hospitals	
				State (thou.)	Local (thou.)	State (thou.)	Local (thou.)	State (thou.)	Local (thou.)	State (thou.)	Local (thou.)	State (thou.)	Local (thou.)	State (thou.)	Local (thou.)
24 Louisiana	19	350	313	34.1	102.0	—	102.0	30.2	—	6.0	0.7	5.3	0.9	18.7	13.3
25 Idaho	39	93	78	7.6	27.1	—	26.2	7.1	0.9	1.8	0.2	1.2	0.9	1.0	4.3
26 Utah	35	161	132	22.0	39.0	—	39.0	21.1	—	3.0	0.4	1.4	1.1	3.9	1.0
27 Maryland	22	420	274	19.7	104.9	—	96.0	17.7	8.9	7.9	2.3	6.0	3.6	6.9	—
28 Pennsylvania	7	715	579	50.6	208.4	—	200.1	47.6	8.3	11.6	21.7	1.7	5.2	21.0	2.8
29 North Carolina	11	541	477	45.6	162.5	—	149.0	42.6	13.5	1.3	13.4	2.0	13.9	15.6	17.8
30 Nevada	40	92	75	6.1	27.3	—	27.3	6.0	—	1.0	0.4	0.7	0.8	1.7	3.6
31 Mississippi	29	217	206	16.8	69.5	—	63.9	15.3	5.7	3.4	0.3	3.3	0.3	9.0	18.0
32 South Carolina	24	297	252	27.2	80.4	—	80.4	24.3	—	4.9	0.2	8.1	2.0	9.8	14.3
33 Arkansas	33	172	167	18.6	58.2	—	58.2	15.3	—	3.7	0.2	4.0	0.3	4.9	3.8
34 Nebraska	36	151	122	10.8	43.7	—	41.2	10.1	2.5	2.9	1.1	0.7	0.4	4.4	4.3
35 Indiana	13	392	380	46.6	123.0	—	123.0	41.8	—	5.2	1.6	1.7	2.7	11.1	23.0
36 Arizona	25	285	248	24.8	90.8	—	83.6	22.4	7.2	5.3	4.0	2.1	2.7	0.7	4.6
37 Kansas	30	233	203	21.0	70.6	—	64.9	20.3	5.7	1.7	0.8	0.9	3.2	6.6	7.3
38 Ohio	5	740	638	68.0	212.7	—	207.5	65.6	5.2	2.2	22.3	3.8	17.5	17.7	15.1
39 Colorado	26	300	241	32.7	77.4	—	76.0	31.5	1.4	1.0	4.4	1.2	2.4	4.6	7.9
40 Alabama	20	345	286	34.2	86.7	—	86.7	30.3	—	4.3	1.1	6.8	3.9	12.2	20.1
41 Virginia	12	603	428	44.9	149.2	—	149.2	42.2	—	2.3	7.5	5.8	5.3	18.0	3.5
42 Illinois	6	783	633	55.3	232.1	—	212.6	52.4	19.5	13.4	7.1	3.2	6.7	15.2	15.2
43 South Dakota	45	68	48	5.6	16.8	—	16.8	5.1	—	1.0	0.2	0.6	0.2	1.4	0.4
44 Georgia	9	563	517	40.7	181.4	—	180.7	35.0	0.7	7.7	1.4	4.4	8.2	14.4	42.0
45 Oklahoma	27	270	236	25.5	80.0	—	79.6	23.5	0.4	5.8	0.4	3.8	1.0	6.1	9.5
46 Missouri	16	386	324	26.3	118.8	—	113.0	24.4	5.1	7.5	2.1	3.3	3.5	12.4	7.8
47 New Hampshire	43	77	66	5.7	24.7	—	24.7	5.3	—	1.3	2.5	0.8	0.1	1.0	—
48 Tennessee	18	370	315	38.6	100.7	—	100.7	36.5	—	4.6	3.7	3.6	3.3	10.6	12.3
49 Florida	4	907	721	42.2	274.4	—	253.0	39.6	21.4	11.2	5.8	14.7	5.8	16.4	36.8
50 Texas	2	1,416	1,412	91.5	547.9	—	517.3	86.8	30.7	24.0	3.7	10.4	16.1	40.3	43.7
Total		10,946	9,369	863	3,310	0	3,173	800	137	146	110	102	112	286	332

Source: *Statistical Abstract of the United States 1997*, p. 324.

Table 7-21 Federal funds distributions, by state and above-average per capita federal funds, U.S., 1995

Rank	State	PerCap. Gen.Rev. Rank	Total (mil.$)	Defense (mil.$)	Nondefense: Total Non-defense (mil.$)	Nondefense: Direct Pay to Indiv'ls (mil.$)	Nondefense: Procure-ment (mil.$)	Nondefense: Grants to States and Local Govt (mil.$)	Nondefense: Salaries and Wages (mil.$)	Excess (+) or Deficit (–): Adj.Gross Income (mil.$)	Excess (+) or Deficit (–): Federal Funds Total (mil.$)	Excess (+) or Deficit (–): Federal Funds PerCap. (dol.)
	United States		1,320,132	227,525	1,092,607	691,666	197,959	214,239	168,951	0	–35,236	–135
1	District of Columbia	51	21,766	2,308	19,458	2,523	4,103	2,222	11,415	3,746	17,697	31,212
2	New Mexico	8	11,274	1,764	9,510	4,041	3,596	1,714	1,590	–4,914	5,3911	3,257
3	Mississippi	31	14,072	3,091	10,981	7,308	2,295	2,507	1,539	-11,881	6,010	2,251
4	North Dakota	10	3,909	473	3,436	1,704	210	702	566	–1,610	1,433	2,243
5	West Virginia	17	9,550	416	9,134	6,003	445	2,166	788	–7,558	3,761	2,062
6	Montana	16	4,638	330	4,307	2,250	204	906	611	–2,503	1,534	1,792
7	Virginia	41	45,890	19,023	26,867	17,168	11,689	3,180	12,147	10,179	10,5551	1,611
8	Maine	21	6,708	1,502	5,206	3,518	1,023	1,269	773	–2,477	1,955	1,578
9	Hawaii	3	7,603	3,196	4,407	2,898	905	1,088	2,498	1,093	1,791	1,520
10	Missouri	46	31,766	7,797	23,968	14,889	7,455	3,971	3,332	–5,758	7,893	1,495
11	Alabama	40	22,280	4,058	18,222	11,996	3,364	3,209	3,125	-10,078	5,799	1,374
12	Maryland	27	36,576	7,588	28,988	13,600	8,228	3,637	7,414	16,108	6,790	1,358
13	Alaska	1	4,640	1,567	3,072	924	1,007	1,063	1,367	1,497	818	1,357
14	Louisiana	24	21,672	3,541	18,131	10,656	3,037	5,233	2,136	-14,750	5,722	1,326
15	South Dakota	43	3,814	331	3,483	1,848	197	724	541	–1,916	924	1,278
16	Oklahoma	45	15,718	2,796	12,922	8,943	1,142	2,359	2,627	-10,460	4,068	1,249
17	Arkansas	33	11,376	1,072	10,305	7,298	596	2,996	1,020	–8,794	2,999	1,223
18	South Carolina	32	17,097	3,494	13,603	8,839	2,721	2,726	2,503	–7,884	3,526	968
19	Kentucky	22	17,504	2,618	14,886	10,068	1,322	3,096	2,608	-11,232	3,557	929
20	Rhode Island	9	5,473	845	4,628	3,109	499	1,100	641	52	734	738
21	Idaho	25	4,965	401	14,564	2,520	844	778	612	–2,5611	600	529
22	Iowa	23	12,979	564	12,416	7,590	637	2,015	954	–4,509	1,460	516
23	Arizona	36	19,011	3,624	15,388	10,558	2,679	2,996	2,269	–5,257	1,346	330
24	Kansas	37	12,506	2,073	10,433	6,914	1,151	1,666	1,892	–1,913	834	327
25	Nebraska	34	7,439	936	6,504	4,178	561	1,114	971	–1,949	462	284
26	Tennessee	48	25,056	2,272	22,784	13,461	4,478	3,940	2,669	–6,295	1,233	238
27	Utah	26	7,594	1,452	6,141	3,444	1,190	1,209	1,479	–3,627	257	135
28	Pennsylvania	28	61,025	5,348	55,677	39,193	4,525	9,705	5,802	–502	1,265	105
	Total		463,901	84,480	379,421	227,441	70,103	69,291	75,889	–95,752	100,416	1,212

Source: U.S. Bureau of the Census, federal expenditures by state for fiscal year 1994.

exceptions are the District of Columbia, with its large concentration of federal offices, and the neighboring states of Virginia and Maryland. Alaska—a long-time recipient of excess federal funds—also remains on this list.

Altogether, the 27 states with a net deficit AGI received $463.9 billion in federal funds, or 35.1 percent of the 1995 U.S. total, largely in direct payments to individuals. The excess federal funds totaled $100.4 billion, or 7.6 percent of the U.S. total and 21.6 percent of the total federal funds for the 27 states and the District of Columbia. This compares with a net deficit AGI of $95.8 billion.

The ranking of the twenty-three below-average per capita federal funds states tells a somewhat different story from the ranking of the above-average states. Ten of the twenty-three states also have a deficit AGI compared with the U.S. per capita average, including Texas, which had a $27.1 billion AGI deficit in 1995 (Table 7-22). However, the thirteen states with the larger deficit of federal funds have the larger excess AGI.

Altogether, the twenty-three states account for $821 billion, or 62.2 percent of the U.S. total. (Undistributed federal funds account for the remaining 2.7 percent of the U.S. total.) Defense funds totaling $84.5 billion are 55.5 percent of the U.S. total, while procurement spending is 51.1 percent of the U.S. total. Other nondefense spending, like direct payments to individuals at 66.3 percent, brings the overall nondefense spending to 63.6 percent of the U.S. total.

The deficit federal funds total $135.7 billion, or 10.3 percent of the U.S. total and 16.5 percent of the twenty-three-state total. In the aggregate, federal funds reduce the per capita AGI disparities with below-average per capita distributions to above-average per capita AGI states. Also, thirteen of the twenty-three states have above-average per capita general revenues for financing state government expenditures.

Federal funds provide an approach to income redistribution among states based on a form of legislative populism that adheres, at least in appearance, to the principle of equality in per capita income. For example, 28 of the 50 states, fourteen above and fourteen below, fall within 10 percent of the average per capita federal funds distribution. In comparison, twenty-three of the 50 states, eight above and fifteen below, fall within 10 percent of the average per capita AGI. The downward skewed AGI distribution among states has more states with below-average than above-average AGI and with smaller differences between states within the below-average group.

While the data show a reduction in AGI income disparities, a state-by-state accounting of the efficiency of this approach in income redistribution shows a contrary pattern. Some low-income states are also low in federal funds, while some high-income states are also high in federal funds. Other criteria are involved, of course, in the functional distribution of federal funds for specific

TABLE 7-22 Federal funds distributions, by state and below-average per capita federal funds, U.S., 1995

Rank	State	PerCap. Gen.Rev. Rank	Total (mil.$)	Defense (mil.$)	Nondefense: Total Non-defense (mil.$)	Direct Pay to Indiv'ls (mil.$)	Procure-ment (mil.$)	Grants to States and Local Govt (mil.$)	Salaries and Wages (mil.$)	Excess (+) or Deficit (–): Adj.Gross Income (mil.$)	Federal Funds: Total (mil.$)	Federal Funds: PerCap. (dol.)
29	California	20	155,391	36,198	119,192	75,466	30,416	26,219	18,830	–1,355	–1,472	–47
30	Vermont	12	2,411	155	2,256	1,435	109	546	268	–616	–39	–67
31	Wyoming	4	2,344	249	2,095	1,070	121	714	358	–275	–33	–69
32	Georgia	44	32,067	8,463	23,604	15,486	4,799	5,028	5,945	–3,686	–781	–111
33	Florida	49	71,092	12,138	58,954	46,381	8,306	8,018	7,263	–2,467	–2,860	–205
34	North Carolina	29	28,858	5,313	23,546	16,481	1,897	4,862	4,833	–5,459	–1,907	–270
35	Ohio	38	48,023	5,198	42,825	29,488	4,775	8,366	4,467	–3,037	–4,389	–395
36	Oregon	18	13,057	643	12,415	8,275	493	2,355	1,419	–1,175	–1,250	–405
37	Texas	50	79,308	15,538	63,770	41,079	12,842	12,669	9,999	–27,066	–7,608	–413
38	Colorado	39	18,989	4,897	14,092	7,987	4,472	2,102	3,540	5,839	–1,879	–513
39	Massachusetts	7	35,374	6,251	29,123	17,672	6,609	6,261	3,112	17,653	–3,683	–610
40	Washington	15	26,644	5,028	21,615	13,381	4,086	3,924	4,187	7,467	–3,733	–699
41	Wisconsin	19	19,670	1,227	18,443	12,616	1,282	3,450	1,396	1,725	–4,829	–950
42	Indiana	35	22,104	2,470	19,634	13,804	1,674	3,553	2,101	–1,092	–5,720	–994
43	New York	5	90,346	5,533	84,813	51,909	6,142	22,445	7,428	1,512	–18,597	–1,024
44	Minnesota	11	18,797	1,585	17,212	10,113	1,798	3,515	1,617	5,451	–5,219	–1,143
45	Michigan	14	38,975	2,601	36,374	25,535	2,479	7,117	2,887	7,413	–11,876	–1,251
46	Delaware	2	2,950	398	2,552	1,784	167	472	455	1,478	–947	–1,337
47	New Hampshire	47	4,636	607	4,029	2,648	487	956	432	2,451	–1,986	–1,750
48	Illinois	42	49,936	3,149	46,788	31,367	3,222	8,506	5,402	1,971	–22,244	–1,892
49	Nevada	30	6,104	917	5,187	3,409	1,045	797	763	4,073	–3,494	–2,390
50	New Jersey	13	37,328	4,719	32,609	22,685	4,218	6,163	3,739	39,990	–20,453	–2,588
51	Connecticut	6	16,591	3,042	13,548	8,902	2,751	3,028	1,405	20,674	–10,654	–3,253
	Total		820,995	126,319	694,676	458,973	104,190	141,066	91,846	101,468	–135,652	–764

Source: U.S. Bureau of the Census, federal expenditures by state for fiscal year 1994.

purposes and particular agencies of state and local governments and state and local offices and installations of federal agencies.

Income and Product

National income and product are among the leading measures of social and economic well-being. Their measures are reported monthly by the U.S. Department of Commerce and summarized annually in the *Statistical Abstract of the United States* and monthly in the *Survey of Current Business.* Gross domestic product measures the net dollar value of all U.S. industry output. This measure eliminates the double counting of industry output by excluding all intermediate products used as inputs in producing another product. Only the remuneration of the primary inputs of labor, capital, and entrepreneurs measure the value added by any industry. Similar definitions apply to measures of gross area, regional, and state products. The U.S. Department of Commerce currently publishes only the annual gross state product in its statistical abstracts.

Seeing the Big Picture—The National Accounts

Table 7-23 lists the components of the U.S. gross domestic product in current and 1992 dollars for 1970, 1990, and 1995. The GDP estimates relate to the disbursements of goods and services produced in the United States. Personal consumption expenditures make up the largest component of the final demand for goods and services, accounting for more than two-thirds of the total disbursements of GDP. Gross domestic private investment is the most volatile part of GDP, although exports and imports vary even more than private investment over a number of years, for example, 1990–95 compared with 1970–90. The large drop in national defense spending in the 1990s reduced the growth in federal spending.

We view and use various state and regional income series with their conceptual underpinnings rooted in the national product and income accounts. We often shift from gross domestic product to national income and, finally, to personal income, personal disposable income, and personal savings.

From National to State and Local Accounts

The gross state product, like its national counterpart—the gross domestic product—is a central measure of the economic well-being of a state or locality. As we show later, yearly estimates of gross state and local product are now available for any area of one or more counties, within and across state boundaries.

We rank the readily available estimates of gross state product to illustrate its variability and uniqueness from one state and region to the next, starting with the states with above-average per capita GSP. The $6.5 trillion of gross state

TABLE 7-23 Gross domestic product in current and 1992 dollars, U.S., 1970, 1990, and 1995

Item	Total 1970 (bil.$)	Total 1990 (bil.$)	Total 1995 (bil.$)	Relative to Total 1970 (pct.)	Relative to Total 1995 (pct.)	Annual Change 1970–90 (pct.)	Annual Change 1990–95 (pct.)
Gross domestic product (current $)	1,035.6	5,743.8	7,253.8	100.0	100.0	8.9	4.8
Personal consumption expenditures	648.1	3,839.3	4,924.9	62.6	67.9	9.3	5.1
Durable goods	85.0	476.5	606.4	8.2	8.4	9.0	4.9
Nondurable goods	272.0	1,245.3	1,485.9	26.3	20.5	7.9	3.6
Services	291.1	2,117.5	2,832.6	28.1	39.0	10.4	6.0
Gross private domestic investment	150.2	799.7	1,065.3	14.5	14.7	8.7	5.9
Fixed investment	148.1	791.6	1,028.2	14.3	14.2	8.7	5.4
Nonresidential	106.7	575.9	738.5	10.3	10.2	8.8	5.1
Residential	41.4	215.7	289.8	4.0	4.0	8.6	6.1
Change in business inventories	2.2	8.0	37.0	0.2	0.5	6.7	35.8
Net exports of goods and services	1.2	–71.3	–94.7	0.1	–1.3	NA	5.8
Exports	57.0	557.3	807.4	5.5	11.1	12.1	7.7
Imports	55.8	628.6	902.0	5.4	12.4	12.9	7.5
Government consumption expenditures and gross investment	236.1	1,176.1	1,358.3	22.8	18.7	8.4	2.9
Federal	115.9	503.6	516.6	11.2	7.1	7.6	0.5
National defense	90.6	373.1	345.7	8.7	4.8	7.3	–1.5
Nondefense	25.3	130.5	170.9	2.4	2.4	8.5	5.5
State and local	120.2	672.6	841.7	11.6	11.6	9.0	4.6
Gross domestic product (1992 $)	3,397.6	6,136.3	6,742.2	100.0	100.0	3.0	1.9
Personal consumption expenditures	2,197.8	4,132.2	4,577.8	64.7	67.9	3.2	2.1
Gross private domestic investment	426.1	815.0	1,009.4	12.5	15.0	3.3	4.4
Net exports of goods and services	–65.0	–61.9	–114.2	–1.9	–1.7	–0.2	13.0
Exports	158.1	564.4	775.4	4.7	11.5	6.6	6.6
Imports	223.1	626.3	883.0	6.6	13.1	5.3	7.1
Government consumption expenditures and gross investment	866.8	1,250.4	1,260.2	25.5	18.7	1.8	0.2

Source: *Statistical Abstract of the United States, 1997,* p. 443.

output for the 50 states and the District of Columbia originates largely from three of the ten major industry groups, namely, manufacturing; finance, insurance, and real estate; and private services.

The 26 states with positive differential employment growth over the 1993–95 period, which account for 44.2 percent of the overall GSP, have more than their overall share of the GSP originating in only one of the three major groups, namely, manufacturing (Table 7-24). They have more than their overall share of the three other industry groups, namely, mining, construction, and retail trade. With reference to the total employment change over the 1993–96 period cited earlier, only thirteen of the 26 states rank among the top half.

Table 7-24 Gross state product from specified industry, by state and positive 1993–95 differential change, 1994

Rank	State	Region	1993–96 Tot.Chg.	Total (bil.$)	Farming, Forestry, Fisheries (bil.$)	Mining (bil.$)	Con-struction (bil.$)	Manu-facturing (bil.$)	Transporta-tion, Public Utilities (bil.$)	Wholesale Trade (bil.$)	Retail Trade (bil.$)	Finance, Insurance, Real Estate (bil.$)	Services (bil.$)	Govern-ment (bil.$)
	All states, total			6,506.5	114.4	94.0	252.3	1,166.8	583.7	449.0	594.3	1,191.2	1,247.4	813.4
1	Georgia	SA	4	175.0	3.2	0.7	6.3	32.2	21.1	15.9	16.3	26.6	29.8	22.9
2	Texas	WSC	1	461.5	7.6	34.9	18.8	73.6	51.4	34.5	41.6	64.2	79.6	55.3
3	Arizona	Mountain	9	89.5	1.4	1.1	4.8	13.7	8.1	5.5	9.8	15.9	16.9	12.3
4	Nevada	Mountain	18	41.5	0.3	1.2	2.9	1.9	3.3	1.9	4.0	7.6	14.0	4.4
5	Colorado	Mountain	10	95.3	1.7	1.7	4.9	12.1	10.6	6.2	9.8	15.8	19.3	13.2
6	Florida	SA	3	301.8	5.9	0.6	13.7	25.9	28.9	22.1	34.9	63.9	67.5	38.4
7	Utah	Mountain	25	39.7	0.5	1.5	2.0	5.7	3.9	2.5	4.2	5.5	7.7	6.2
8	Tennessee	ESC	12	120.7	1.7	0.2	4.4	29.7	10.3	9.0	13.6	15.2	22.0	14.6
9	Oregon	Pacific	20	70.1	2.2	0.2	3.2	13.7	5.7	5.7	6.6	11.6	12.3	8.9
10	North Carolina	SA	8	177.2	4.2	0.1	6.7	55.7	13.8	11.4	15.9	21.9	24.5	23.0
11	New Mexico	Mountain	35	36.5	0.7	2.8	1.7	5.1	3.6	1.6	3.5	4.8	6.2	6.5
12	Ohio	ENC	5	261.6	3.1	1.2	9.9	71.4	21.8	18.1	25.3	38.6	44.4	27.8
13	Mississippi	ESC	34	48.2	1.5	0.4	1.7	11.4	6.0	2.8	4.9	5.3	7.0	7.2
14	Arkansas	WSC	33	48.3	2.4	0.3	1.7	12.1	6.0	3.0	5.1	5.3	6.7	5.7
15	Louisiana	WSC	23	97.0	1.2	10.7	4.2	16.4	10.8	5.6	8.5	12.4	15.5	11.7
16	Idaho	Mountain	38	23.0	1.5	0.4	1.4	4.3	2.1	1.4	2.4	2.9	3.5	3.1
17	Michigan	ENC	7	227.4	2.3	0.9	8.1	67.4	15.5	16.0	19.5	34.0	39.3	24.4
18	New Hampshire	NE	37	28.1	0.2	0.1	1.0	6.1	2.2	1.7	2.8	6.0	5.3	2.7
19	South Dakota	WNC	42	16.5	11.8	0.3	0.6	1.9	1.3	1.0	1.6	3.3	2.5	2.2
20	Kentucky	ESC	26	83.2	2.3	3.3	3.2	22.7	8.0	4.6	7.5	8.9	11.5	11.2
21	Montana	Mountain	41	16.0	1.0	0.8	0.7	1.2	2.1	1.0	1.7	2.1	2.8	2.6
22	Iowa	WNC	31	65.3	4.8	0.1	2.5	16.5	5.2	4.6	5.8	8.8	9.3	7.7
23	Indiana	ENC	16	131.6	2.4	0.8	6.1	40.4	11.0	8.2	12.4	17.1	19.7	13.5
24	Nebraska	WNC	36	39.6	3.5	0.2	1.6	5.9	4.5	3.1	3.4	5.4	6.2	5.8
25	Kansas	WNC	32	59.0	2.9	0.7	2.3	10.3	7.3	4.4	5.8	7.3	9.3	8.7
26	Wisconsin	ENC	14	119.7	2.9	0.3	5.1	34.4	8.5	7.5	10.9	18.2	18.8	13.1
	Total			2,873.3	63.2	65.5	119.5	591.7	273.0	199.3	277.8	428.6	501.6	353.1
	Industry distribution (pct.)			100.0	2.2	2.3	4.2	20.6	9.5	6.9	9.7	14.9	17.5	12.3

Source: *Statistical Abstract of the United States 1997*, p. 451.

TABLE 7-25 Gross state product from specified industry, by state and negative 1993–95 differential change, 1994

Rank	State	Region	1993–96 Tot.Chg.	Total (bil.$)	Farming, Forestry, Fisheries (bil.$)	Mining (bil.$)	Con-struction (bil.$)	Manu-facturing (bil.$)	Transporta-tion, Public Utilities (bil.$)	Wholesale Trade (bil.$)	Retail Trade (bil.$)	Finance, Insurance, Real Estate (bil.$)	Services (bil.$)	Govern-ment (bil.$)
51	New York	MA	21	544.7	2.5	0.2	15.7	68.9	44.9	34.8	39.1	156.8	120.6	61.2
50	California	Pacific	2	833.9	17.9	4.4	27.5	119.5	60.7	58.3	77.8	186.7	180.4	100.7
49	Pennsylvania	MA	17	279.9	2.9	1.9	11.1	56.4	26.3	17.6	25.9	50.2	57.8	29.8
48	New Jersey	MA	24	242.2	1.3	0.1	8.7	35.7	24.8	22.8	18.6	53.5	50.5	26.2
47	District of Columbia	SA	51	44.7	0.0	0.0	0.4	1.2	2.5	0.6	1.3	6.5	14.6	17.6
46	Connecticut	NE	39	104.3	0.7	0.2	3.4	18.3	7.4	7.1	8.6	27.5	21.3	9.8
45	Maryland	SA	30	125.6	1.2	0.2	6.1	11.1	10.7	8.0	11.5	27.3	27.6	21.9
44	Illinois	ENC	6	317.2	4.8	1.2	13.2	61.2	30.8	26.0	26.9	58.8	62.4	31.9
43	Massachusetts	NE	13	177.3	1.0	0.0	5.6	30.2	12.4	12.9	14.4	40.2	44.0	16.6
42	Washington	Pacific	22	136.3	3.7	0.3	6.7	19.2	11.2	10.6	14.1	24.2	25.8	20.5
41	Hawaii	Pacific	50	34.7	0.5	−0.2	2.0	1.1	3.4	1.4	4.0	8.0	7.1	7.4
40	Virginia	SA	11	170.6	1.8	1.2	7.0	27.9	14.9	9.4	14.5	28.9	31.5	33.5
39	Missouri	WNC	19	121.8	2.3	0.3	5.5	25.9	13.0	9.2	12.2	17.5	22.4	13.5
38	Rhode Island	NE	47	22.7	0.2	−0.1	0.8	4.1	1.7	1.2	2.2	5.1	4.8	2.7
37	South Carolina	SA	29	76.7	1.1	−0.1	3.3	21.6	6.2	4.3	7.9	9.6	10.8	12.0
36	Alabama	ESC	27	84.6	2.0	1.2	3.3	18.8	8.5	5.4	8.7	10.1	13.0	13.6
35	Maine	NE	45	24.6	0.5	−0.1	1.1	4.4	1.8	1.5	3.1	4.4	4.4	3.5
34	Alaska	Pacific	49	22.3	0.3	4.7	1.0	1.0	3.8	0.7	1.5	2.3	2.5	4.5
33	Minnesota	WNC	15	118.7	3.3	0.5	5.0	24.3	9.2	9.8	10.9	20.4	22.2	13.1
32	Oklahoma	WSC	28	63.5	1.9	3.6	1.9	10.8	7.0	3.9	6.5	7.6	10.0	10.3
31	Vermont	NE	46	12.6	0.3	0.1	0.5	2.2	1.2	0.9	1.3	2.1	2.5	1.5
30	Delaware	SA	43	25.2	0.3	0.1	0.8	5.1	1.3	1.0	1.5	9.7	3.2	2.2
29	West Virginia	SA	40	33.5	0.3	3.8	1.6	5.6	4.4	1.8	3.0	3.6	5.0	4.4
28	Wyoming	Mountain	48	15.6	0.4	5.0	0.6	0.6	2.6	0.5	1.0	1.6	1.4	1.9
27	North Dakota	WNC	44	13.0	1.4	0.2	0.6	1.0	1.5	1.2	1.3	1.6	2.1	2.1
	Total			3,633.2	51.2	28.5	132.8	575.1	310.7	249.7	316.5	762.6	745.8	460.3
	Industry distribution (pct.)			100.0	1.4	0.8	3.7	15.8	8.6	6.9	8.7	21.0	20.5	12.7

Source: *Statistical Abstract of the United States 1997*, p. 451.

The twenty-four states and the District of Columbia with negative differential employment change in the 1993–95 period (Table 7-25) account for the larger share of GSP overall. Over the 1993–96 period, moreover, ten of the twenty-four states—only three less than in the positive growth group—rank among the top one-half in total employment change.

Changes in total employment accompany changes in industry output, and hence gross state product, over a period of several years. These changes, in turn, multiply in overall impact, soon affecting state revenue collections and eventually interstate and interregional imports and exports. The high level of data aggregation represented in Tables 7-24 and 7-25 remains, however, a major obstacle in identifying and isolating the attributes of state industry structure associated with year-to-year and longer-term changes in employment and gross state product.

We cited the associated changes in industry employment and output that eventually affect other industries, along with state and local governments and interstate and interregional exports and imports of goods and services. We recognized also changes in human population and labor force characteristics accompanying international and interstate migration. Altogether, these changes affect the productivity of labor and capital and account for their differences over time and from one place to another.

Labor and capital productivity is a key determinant of state and regional wellbeing. We can use the industry product and employment numbers for 1994 to calculate gross product-employment ratios for each industry in a state or any area for which the industries are available and start accounting for their differences.

Reliable and productive tracking of urban regional economies calls for more detailed data on industry and area than is available in the preceding tabular series. We will now explore some alternative approaches to acquiring this greater detail needed for area activity analyses and forecasts. One approach is the construction of regional economic and social accounts from county-level data now readily available from both public and private sources.

Constructing Regional Social and Economic Accounts

The construction and use of a social accounting matrix (SAM) are topical issues for the study of regional income flows. The SAM modeling framework adapts well to the study of natural resource-based regional economies in transition to a new product cycle and the production of new goods and services (Alfsen, Bye, and Lorentsen, 1987). It allows a tracking of income flows originating from the production and sale of goods and service (see References Cited for series of articles in Economic Systems Research, *Journal of the International Input-Output*

Society, Volume 3, Part 3, 1991). Income flows differentiate between those originating within and outside the region.

Basic Accounting Framework

The SAM modeling framework provides an additional benefit to its user: the tracking of factor inputs and factor payments. The existence of income and institutional accounts in SAM allows the tracking of the transitional income effects on specific populations and institutions in the region. Multiplier models that extend their definition beyond the traditional inter-industry multiplier approaches evolve directly from the SAM framework.

Such a framework facilitates the study of factor input requirements—entrepreneurial and venture capital as well as physical infrastructure—in the production of particular goods and services. It also can address problems of preparing individual factor inputs for higher levels of productivity. The factor productivity analysis includes separate assessments of the performance of public and private spending for education and training from preschool to postsecondary and for other regional infrastructure.

The accumulation component of the accounts shown in Figure 7-1 incorporates the payments made by production activities for primary inputs. The consumption portion of the table gives the final purchases by various institutions. Because of the different classifications used in the final payments and demand sectors, the interpretation of the portion of the table where they intersect is somewhat obscure, unless there are intervening concepts to convert the values from one set of definitions to another. These include nonmarket transactions and transfers with assigned values.

Most public policies do not interact directly with the production of firms but affect the nonmarket, institutional relationships within which firms operate. These include industry regulation and tax policies, as well as broad policies and programs affecting the management of public resources such as forests, wildlife,

				Institutions					
Originating Sector	Industries $S_1–S_{11}$	Commodities $S_{12}–S_{20}$	Value Added $S_{21}–S_{24}$	Private $S_{25}–S_{29}$	Govt. $S_{30}–S_{32}$	Nonindustrial S_{33}	Capital $S_{34}–S_{38}$	Rest of World S_{39}	Total S_{40}
Industries	0	T_{12}	0	0	0	0	0	0	T_{19}
Commodities	T_{21}	0	0	T_{24}	T_{25}	0	T_{27}	T_{28}	T_{29}
Value added	T_{31}	0	0	0	0	0	0	0	T_{39}
Institutions, private	0	0	0	T_{44}	T_{45}	0	0	T_{48}	T_{49}
Institutions, govt.	0	T_{52}	T_{53}	T_{54}	T_{55}	T_{56}	T_{57}	T_{58}	T_{59}
Nonindustrial	T_{61}	0	0	T_{64}	T_{65}	0	0	T_{68}	T_{69}
Capital	0	T_{72}	T_{73}	T_{74}	T_{75}	0	T_{77}	T_{78}	T_{79}
Rest of world	T_{81}	0	T_{83}	T_{84}	T_{85}	T_{86}	T_{87}	T_{88}	T_{89}
Total	T_{91}	T_{92}	T_{93}	T_{94}	T_{95}	T_{96}	T_{97}	T_{98}	T_{99}

FIGURE 7-1. Structure of product and income accounts, by originating and receiving sector accounts

fish, rangeland, and water. SAM-based models can be extremely useful in addressing concerns about how these policies and programs affect the flow and distribution of income among institutional groups.

For example, models derived from the general multiplier model shown previously could be used to investigate policy-induced impacts. These impacts stem from such actions as changes in tax and transfer policies, programs concerning public stumpage production, recreation and other user fees, grazing fees, investment policies, and the sale or acquisition of real property.

We follow the paths suggested in the introductory comments in the building of a SAM—an extension of the national income and products accounts (NIPA) to show the institutional origins and destinations of income and product flows within a given economy. First, a brief historical note: The Cambridge Growth Project produced the first SAM in the beginning of the 1960s (Stone, 1961). This work was led by Sir Richard Stone, for which he later received the Nobel Prize in Economics for outstanding contributions to economic science and society.

In the 1970s SAMs were constructed for some developing countries, including Iran, Sri Lanka, and Swaziland. Regional SAM analysis is often associated with developing countries, for example, as a part of a large project in the Muda region of Malaysia (Bell et al., 1976) and another in Sri Lanka (Pyatt and Rose, 1977). An Iowa State University study team constructed a regional SAM for northeast Thailand (Maki, 1980). Regional SAM analysis for developed countries includes the United States (Hanson and Robinson, 1991) and Norway (Alfsen et al., 1987). In the 1960s, Jerald Barnard constructed SAMs for the State of Iowa (Barnard, 1967) and the Province of Manitoba (Barnard et al., 1970). There is also a SAM part of the IMPLAN (*Im*pact Analysis for *Plan*ning) regional modeling system developed by the U.S. Forest Service team at Ft. Collins, Colorado, now maintained by the Minnesota IMPLAN group (MIG).

Most input-output studies emphasize interrelations among the productive components of an economy. Specifically, input-output studies focus on the transaction flows between producing activities (industries) of a region that occur as goods and services move to the consumption and trade sectors. Studies by Stone (1961; 1987) and others (Bulmer-Thomas, 1982; Pyatt and Rose, 1977) have illustrated the usefulness of extending the logic of input-output accounting to the institutional components of economies.

The more general SAM accounting framework differs from typical input-output studies by the added emphasis given to the consumption, income-creation, and accumulation sectors, and their interactions with the production sector. The social accounting matrix explicitly accounts for the disposition of income as it flows throughout an economy, from production activities to factors, institutions, consumption, and eventually back to production.

For a single region, the conceptual problem of constructing a regional SAM is similar to those arising at the national level. Data problems are often severe and derive largely from the fact that a single subnational region is an open economy without its own custom unions to track imports and exports. A multiregion system, however, poses a different range of problems that extend beyond those normally encountered in single region analysis. These problems stem from the extreme difficulties encountered in tracking the many different transactions that occur among buyers and sellers, producers and consumers, private and public institutions, within a given region and between regions. Even though the tracking is quite imperfect, it nonetheless provides a beginning in better understanding the workings of a national economy that depends on people and their institutions within regional economies that are an integral part of a nation's economy.

SAM uses matrix representation. For every row representing the receipts of an account there is a corresponding column showing expenditures of the account. SAM usually is divided into four blocks:

1. Industries and commodities
2. Factors (primary inputs)
3. Institutions (including households, corporations, and governments)
4. Exogenous (accumulation and rest of world) accounts

Each block is further split into components, with aggregation level depending on the availability of data and the scope of the study.

The first block of columns, called production, shows the breakdown of the supply of each commodity among domestic producing sectors and between domestic production and imports. The first block of production rows shows the industries disbursing goods and services to the producing sectors. The pattern of demand for both locally produced commodities and imports relates, therefore, directly to the structure of production and the distribution of income.

The second block of columns, called consumption, is a representation of the conventional input-output tables. In the conventional case, local consumption is exogenous to the demand-driven production system. It represents the purchases of locally produced goods and services and imports by households and governments in the region. The corresponding blocks of rows represent the institutional sources of income available for the purchases of goods and services.

The third block of columns shows the distribution of value-added income to institutional recipients of this income. The value-added rows show the income payments received by households, businesses, and governments in the form of employee compensation, proprietors' income, capital consumption allowances and other value added, and indirect taxes.

The fourth block of columns shows the disbursement to the wage and salary workers in each occupation to the institutional account. The corresponding block of rows shows the receipts of this income from the value-added account.

Institutional entries are in the fifth block of columns, representing the spenders of incomes listed in the row cells. This includes income transfers to the consumption account, selected entries in the institutional account, the capital account, and the rest-of-world account.

The fifth block of rows shows the receipts of institutions. These receipts consist of transfers from the occupational account and transfers within the institutional account, including dividends, interest, and taxes. The government sector also receives indirect taxes from production activities.

Thus all the sources of income for each type of institution are displayed in the institutions block. However, there is no direct mapping from production activities to households so that the activities that particular households depend on for their livelihood are not directly available from the matrix, except by inference through the factor accounts.

The capital account columns show purchases from both the production account and the rest-of-world account. The capital account rows show income receipts from the institutions account and the rest-of-world account.

The rest-of-world account covers all transfers of product and income from the six preceding accounts to the rest of world. The investment figures balance with domestic savings and the rest-of-world account.

The classification of production activities is available at varying levels of industry detail, from one- to four-digit SIC codes and as many as 500 or more different production activities. The activities must include the most important producing unit in the local economy.

Most SAM frameworks, especially if based on current input-output tables, differentiate between commodities and activities. U.S. studies follow the industry-commodity combination in the use matrix (see Chapter 6, Figure 6-1). The total output value of each industry is the sum of the output value of each commodity produced that makes up this output. Thus, each industry has its unique column in a commodity-by-industry conversion (sometimes called the by-products) matrix that shows the proportion of each industry's output derived from its main product and each by-product.

Many SAM-studies use only labor, capital, and government services as factors of production. An auxiliary industry-occupation conversion matrix and earnings per employee matrix convert the value matrix for occupations to a corresponding number of jobs and the earnings per job in each industry and occupational group. The industry-occupation matrix uses services as a bridge for comparing the quantity of labor demanded by local industry in a given local labor market with the quantity of labor available in each occupational category. Its use extends even further in studies of local income distribution.

Whether households, business enterprises, or government agencies, these institutions represent actual decision-making units of the economic system. They are the owners of factors of production, the source of entrepreneurship, and the consumers of commodities produced. For this reason it may be more appropriate to give the institutions block first consideration than to follow the traditional pattern of beginning with the production account and then focusing on industries or factors.

However, no institutional system in a commercial global economy can exist without remunerative productive activity and trade with the rest of world—a condition of both joy and sorrow for the residents of any area in this global village. After households, the second institutional group is the private corporate sector, followed by government. Thus, the individual decision-making units of any area enter explicitly into the overall analysis.

The third institution category is the public sector. From the expenditure side it is interesting to analyze how education sector receipts and expenditures compare with other government receipts and expenditures along with their distribution within the local economy.

A broader view of the economy's structure than just focusing on the production component includes the entire income cycle. That is, if societal institutions like households, governments, and corporations are viewed as important economic agents in accounting for the uniqueness of an area economy, along with productive enterprises, then more than just inter-industry relationships must be included in an economic model. This draws attention to the consumption and accumulation accounts of the input-output table: the final demand and final payments sectors, respectively.

The classifications used for accumulation and consumption accounts lacks an explicit description of important links in the income cycle of a local economy. In particular, relationships between factor earnings and institutional income and consumption are not quickly and readily apparent.

Two important relationships are particularly troublesome. First, the distribution of factor earnings among institutions is aggregated, except for size and skill differentiation of business and labor. Second, nonmarket transfers (such as transfer payments, taxes, and savings behavior) are difficult to trace. They occur in part as intra-institution transfers and in part as inter-institution transfers. As a result, public policies that affect relationships between factor earnings and institutional incomes or inter-institutional distributions of incomes (such as tax rates) are not readily investigated with this set of accounts. Rather, a more complete social accounting system with a detailed classification of individual entries in each of the product and income accounts that also covers successive time periods is needed.

Figure 7-1 illustrates the aggregate structure of a SAM. It shows all income payments made by all economic activities and their distribution among all institutions. This scheme likewise provides additional details concerning the flow of imports and exports to and from an area.

A description of each transaction submatrix in Figure 7-1 may give some help in clarifying the content and uses of the one-period SAM. The individual transaction submatrices are as follows:

1. Industry make matrix, T_{12}, to show the commodity composition of industry disbursements
2. Industry use matrix, T_{21}, to show the commodity composition of industry use of intermediate inputs from local sources
3. Commodities final demand account, T_{24}, to show the purchases of locally produced commodities by resident firms and households
4. Commodities final demand account, T_{25}, to show purchases of locally produced commodities by resident government offices—federal, state, and local
5. Commodities final demand account, T_{27}, to show purchases of locally produced commodities in private capital formation and inventory additions
6. Commodities export trade account, T_{28}, to show purchases of locally produced commodities by nonresident firms, households, and governments, including visitors
7. Value added account, T_{31}, to show purchases of labor and current capital inputs by resident firms
8. Value added account, T_{34}, to show purchases of labor and capital inputs by local households
9. Value added account, T_{35}, to show purchases of labor and capital inputs by nonenterprise local governments
10. Value added account, T_{37}, to show purchases of labor and investment capital inputs by local firms
11. Value added account, T_{38}, to show purchases of capital inputs for export
12. Private institutions transfer account, T_{44}, to show income payments by private institutions to resident private institutions
13. Private institutions transfer account, T_{45}, to show income payments by government institutions to resident private institutions
14. Private institutions export account, T_{48}, to show income payments by nonresident households, firms, and governments to private institutions

15. Government institutions sales account, T_{52}, to show commodity sales by resident government institutions
16. Government institutions transfer account, T_{54}, to show income payments by private institutions to resident government institutions
17. Government institutions transfer account, T_{55}, to show income payments by government institutions to resident government institutions
18. Government institutions transfer account, T_{56}, to show income payments by nonindustrial sector to resident government institutions
19. Government institutions capital account, T_{57}, to show income payments by capital using firms to resident government institutions
20. Government institutions export account, T_{58}, to show income payments by nonresident households, firms, and governments to resident government institutions
21. Nonindustrial sector sales account, T_{61}, to show commodity sales by resident industries
22. Nonindustrial sector transfer account, T_{64}, to show income payments by private institutions to resident nonindustrial sector
23. Nonindustrial sector transfer account, T_{65}, to show income payments by government institutions to resident government institutions
24. Nonindustrial sector export account, T_{68}, to show income payments by nonresident households, firms, and governments to resident nonindustrial sector
25. Private capital sales account, T_{72}, to identify the capital goods portion of the resident commodities sector
26. Private capital final demand account, T_{74}, to show the purchases of private capital by resident firms and households
27. Private capital final demand account, T_{75}, to show purchases of private capital by resident government offices—federal, state, and local
28. Private capital final demand account, T_{77}, to show purchases of private capital for private capital formation and inventory additions
29. Private capital export trade account, T_{78}, to show purchases of private capital by nonresident firms, households, and governments, including visitors
30. Industries import account, T_{81}, to show the purchases of imported goods and services by resident industries
31. Value-added import account, T_{83}, to identify resident purchases of primary inputs from rest of world
32. Private institutions import account, T_{84}, to show transfers from rest of world to resident private institutions

33. Government institutions import account, T_{85}, to show transfers from rest of world to resident government institutions
34. Nonindustrial sector import account, T_{86}, to show transfers from rest of world to resident nonindustrial sectors
35. Private capital sector import account, T_{87}, to show transfers from rest of world to resident private capital formation sector
36. Rest-of-world import account, T_{88}, to show transfers from rest-of-world import account to rest-of-world export account

While input-output analysis consists of a number of different types of tables, the transactions table is the most commonly used. This table shows the flows of commodities from each of a number of producing sectors to all other consuming sectors, both intermediate and final. Intermediate transactions represent the flow of goods and services that are produced and consumed in the process of current production. Final demand represents the ultimate purchases by consumers or government inside or outside of the region for which the table has been constructed. Final inputs include payments to households (as wages and salaries), proprietor income, business taxes, and imports.

The use table is also of interest in that it shows the value of each commodity used by each industry. The term commodity is used here to represent the output of a given industry, since some industries may produce more than one commodity. Input-output models, like computable general equilibrium models, assume the existence of a general equilibrium for each regional economy. For both types of models, this is a restrictive assumption. Probably the more realistic assumption is that of disequilibrium from one period to the next, with general equilibrium being the tendency, but not the reality, at any given moment.

A general multiplier model to capture the distribution of economic consequences throughout the industrial and institutional structure of an economy comes from a social accounting matrix. To begin, the SAM structure shown in Figure 7-2 expands into specific tasks and related data sources for estimating each of the relationships identified in Figure 7-2.

The social accounting matrix thus extends the notion of human migration and commodity flows to the interrelationships among producing, consuming, investing, and saving activities and institutions. It allows the tracking of monetary and ownership linkages within and between economic units and geographic areas.

Natural Resource Analysis and Accounts

Alfsen, Bye, and Lorentsen (1987), of the Central Statistical Bureau of Norway, outlined a classification of natural resources starting with an initial breakdown

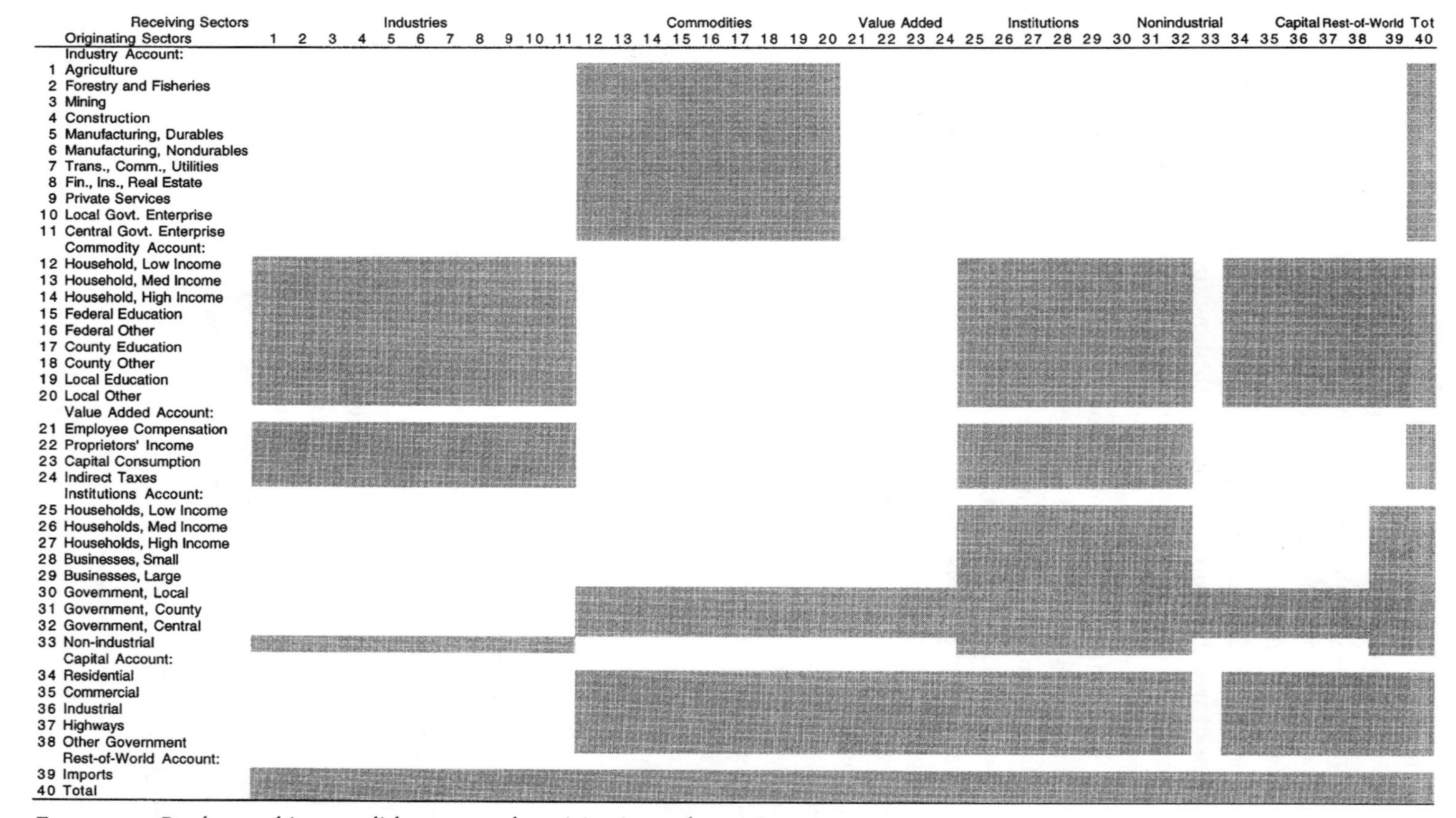

FIGURE 7-2 Product and income disbursements by originating and receiving sectors.

between material and environmental resources. The material resources are tangible resources—mineral and biotic—that are extracted or harvested from nature, and inflowing resources (such as solar radiation, ocean currents, and the hydrologic cycle).

Environmental resources mainly provide services rather than production inputs for the economy (Alfsen et al., 1987, pp. 8–14). In addition, a set of energy accounts includes both material resources (coal, crude oil, and natural gas) and inflowing resources (hydropower and biomass). The forest accounts are separate from the energy accounts since a forest is a renewable biotic resource seldom extracted as a production input in Norway. A set of fish accounts differs from the forest account because of the much shorter reproduction time of fish than forests, the mobility of fish, and their importance as an export commodity.

The proposed set of material resource accounts has three parts: reserves, extraction, and consumption. Resource reserve accounts show the following:

- Resource base reserves, both developed and undeveloped, at beginning and end of period
- Total gross extraction during period
- Adjustments of resources base due to new discoveries and reappraisal of old discoveries; adjustments of resources to new technology, cost of extraction, and related items
- Prices of resources—the extraction, conversion, and trade accounts showing net extraction (gross minus use), net imports (imports minus exports), and changes in stocks during the period

The consumption accounts show domestic use by final use category. The environmental accounts contain information about the state or quality of the resource at the beginning and end of a time period, along with information explaining the change in the state of the resource. Their main purpose is to record in a systematic way the state of environmental resources like air, water, and land. This is difficult because of the lack of accurate and readily maintained indicators of environmental status and the importance of having geographically detailed data series because of the large differences in environmental status within short distances.

The emissions accounts, closely linked to the energy accounts, show emissions of sulfur dioxide from different industry sources and type of emissions, that is, mobile combustion, stationary combustion, and process omissions. The land use accounts show changes in land use for different purposes—for example, buildings, manufacturing facilities, commercial and administrative services, institutions, transportation, and other purposes—and by type of land.

Building a comprehensive set of natural resource accounts for Norway depended on economic expertise from many directions, including energy economics, petroleum economics, environmental economics, and health economics, as well as the counsel of legislators and administrators in the central government. None of the related professional people—economists, biologists, engineers, meteorologists, and physicians (with the possible exception of biologists and economists)—received a basic education in environmental management at the time the natural resource accounts were being implemented. Also, the dependence on direct regulation had resulted in the neglect of long-term incentives for developing alternative and cleaner methods of production. Thus, efforts to effectively integrate the natural resource accounts with the government's economic accounts and budgetary processes—even in a centrally planned economy—introduced additional, almost insurmountable problems of implementation, particularly with the internalization of the external costs of economic growth and development (Alfsen et al., 1987, pp. 42–53).

Area and Community Analysis and Accounts

In contrast to the experience of natural resource accounting in central planning, we refer briefly to several related applications of the economic accounting framework in an impact assessment of a natural resource-based area in extreme northeast Minnesota adjoining Canada and another in central Norway. Their generic classifications include the following:

1. Clear, explicit definition and delineation of the internal and external linkages of each industry product and economic activity with reference to the study region(s)
2. Local decision centers and decision environments for the production, distribution, and use of each product in the study region(s)
3. Economic accounting of the stocks and flows and the spatial-economic incidence of each product and its by-products, including positive and negative income payments generated by the product flows
4. Rural resource-based communities will face several problems of adjustment to changing global market conditions. A Norwegian White Book on regional development, for example, gives some points of view on what can be the greatest challenges. These problem areas also appear in rural communities in the United States, especially in northern states like Minnesota. The problem areas are within a social and economic accounting framework to facilitate the analysis of their consequences for the total economic community.

Refining Tools and Measures of Regional Change

Several types of refinements are available for optimizing the construction and use of regional social and economic accounts. A first-order target for improvement is the input-output model and the steps involved in the formulation and fitting of its various structural relationships. These include (1) changing regional supply, (2) modifying industry production function, (3) editing regional purchase coefficients (RPCs), and (4) controlling for induced effects once better information becomes available (Alward et al., 1989).

Superior local knowledge warrants changing the readymade database values in each category, and the regional purchase coefficients by institution, industry, or commodity. The RPC adjustments for an industry or institution result in the given change being applied to all commodities, by industry or institution. Overlooked, however, is the further differentiation of the local final sales accounts and the industry margins that convert industry output from producer prices to purchaser prices. This process requires detailed and regionally differentiated estimates of final product sales to households, governments, and businesses.

Furthermore, input-output models generally are demand-driven with no supply constraints. While computable general equilibrium models provide for a variety of additional variables that represent various economic trade-offs among inputs and outputs as relative prices change, they still represent changes within a given time period. This presentation aims for the next step in regional modeling—the building of multiperiod dynamic models that can generate regional social and economic accounts for future time periods.

Regional SAM Model

The building of a regional SAM model, as described in this chapter, is at the center of various efforts to enrich and expand existing regional economic models and methods of regional economic forecasting. A SAM makes possible the addressing of distribution issues, that is, the flows of product outputs and income payments to regions differentiated by occupation, income, gender, ethnicity, and other measures of social and economic status.

Such a model and accompanying database also makes possible the differentiation of export-generated commodity shipments and related income flows from simply exogenous or externally originating income flows. For example, income received for exports of timber products to markets outside the region differ from the income received as transfer payments on several counts. While both are "new" dollars, the transfer payments also involve large current income outlays that exceed the transfer payments for most regions and large metropolitan areas. The net transfers are either negative or much smaller than the gross transfers.

The use of input-output multipliers in estimating local and regional impacts of exports accounts for imports by reducing the size of the output multipliers. Transfer payments simply add to available income for local purchases without any inherent multiplier effects except those already accounted for in an area's input-output model.

We approach the task of building a regional system of social and economic accounts with a series of charts that illustrate this task step by step. They highlight the critical data relationships, starting with an abbreviated version of the input-output model presented earlier. Again, we use the standard input-output terminology. The *make matrix* mentioned earlier is the value of each commodity produced by each industry, while the *use matrix* lists the commodities used in each industry's production process. The series of charts lists the *national absorption matrix* of coefficients derived by dividing each element of the use matrix by the respective industry's total output. The series also lists the *national by-product matrix* of coefficients derived by dividing each element of the make table by the make table industry total outlay.

Figure 7-3 displays the underlying data structure of this model with corresponding numerical values derived from the illustrative region's social and economic accounts. The numerical values provide the reference totals for balancing the individual regional accounts, starting with value added. For example, the initial conditions show 6.50 units of value added by Industry 1, which has a total industry outlay of ten units. Thus, the factor absorption rate is 0.65. The remaining 35 percent of the industry outlays are for the three commodity inputs, of which the smallest share—only 5 percent—is for Commodity A. Industry 3 spends 20 percent of its total outlays for Commodity A—the only other industry doing so. Commodity A accounts for 90 percent of Industry 1 output but accounts for only 10 percent of Industry 3 output. For Industry 1, in other words, Commodity A accounts for only 5 percent of its total purchases while accounting for 20 percent of Industry 3 purchases. Also important is the lack of any output for Industry 4. Other data include the distribution of labor income (PoW), nonlabor income (PoR), and transfer payments between institutions, that is, households, government, and capital, as well as estimates of regional purchase coefficients. The coefficient matrices originate from the national data sources, while the estimates of total industry outlay, total value added, gross final demand, and employment originate from state and county data sources.

Step 1 in constructing the regional social accounts is to *adjust the input functions* (Figure 7-4). The important purpose of this step is to reconcile the estimate of commodity input requirements of each industry, as given by the national

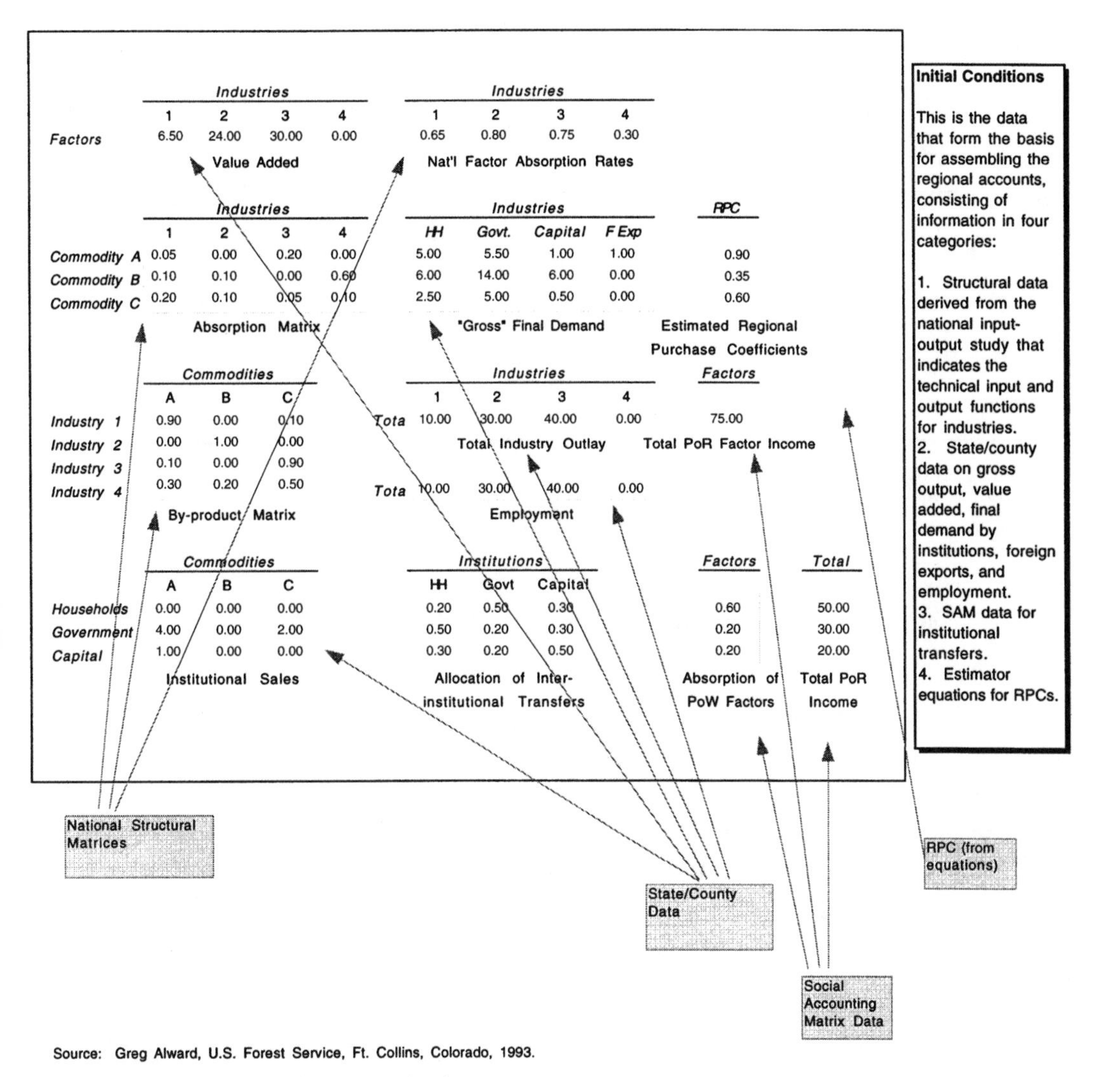

FIGURE 7-3 Initial data and structural information.

absorption matrix, with the regional measures of gross output and its distribution between intermediate and primary inputs. The regional measure of value added takes precedence over the national measure of intermediate demand.

Step 2 is to calculate *total regional commodity demand* (Figure 7-5). We multiply the adjusted regional absorption matrix from Step 1 by the regional industry output to obtain the intermediate input purchases of each industry. We then multiply the matrix of gross final commodity demand by the "summation vector" to obtain total regional final commodity demand. We add the estimates of regional final demand to regional intermediate demand to obtain total commodity demand in the region.

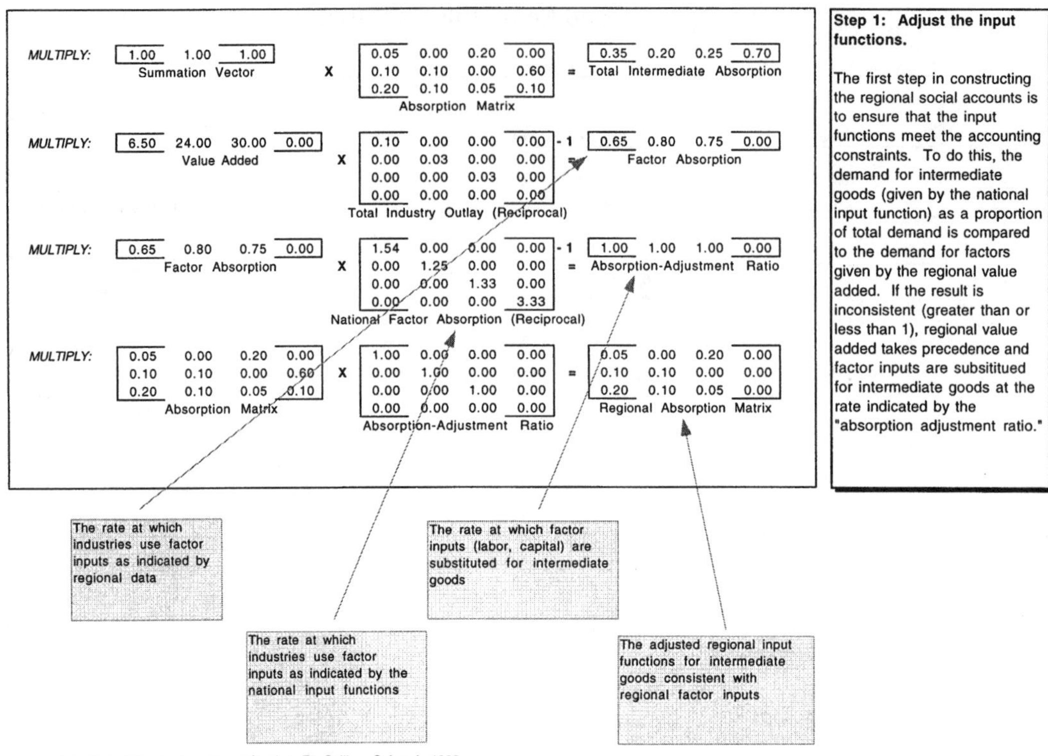

Source: Greg Alward, U.S. Forest Service, Ft. Collins, Colorado,1993

FIGURE 7-4 Adjustment of technical input coefficients.

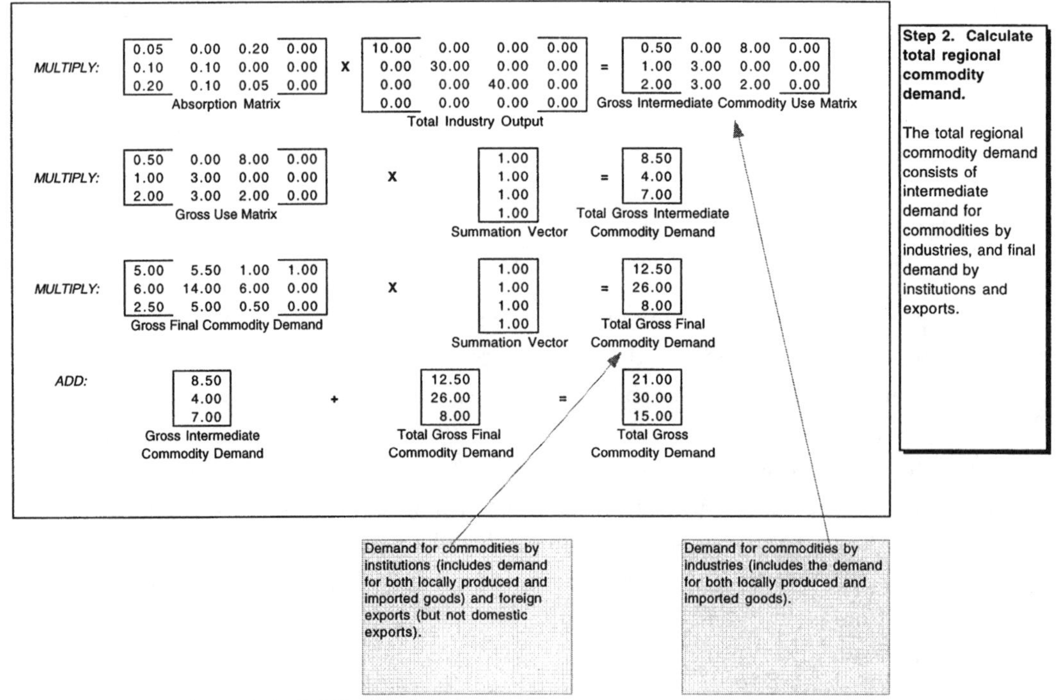

Source: Greg Alward, U.S. Forest Service, Ft. Collins, Colorado, 1993.

FIGURE 7-5 Calculation of total regional commodity demand.

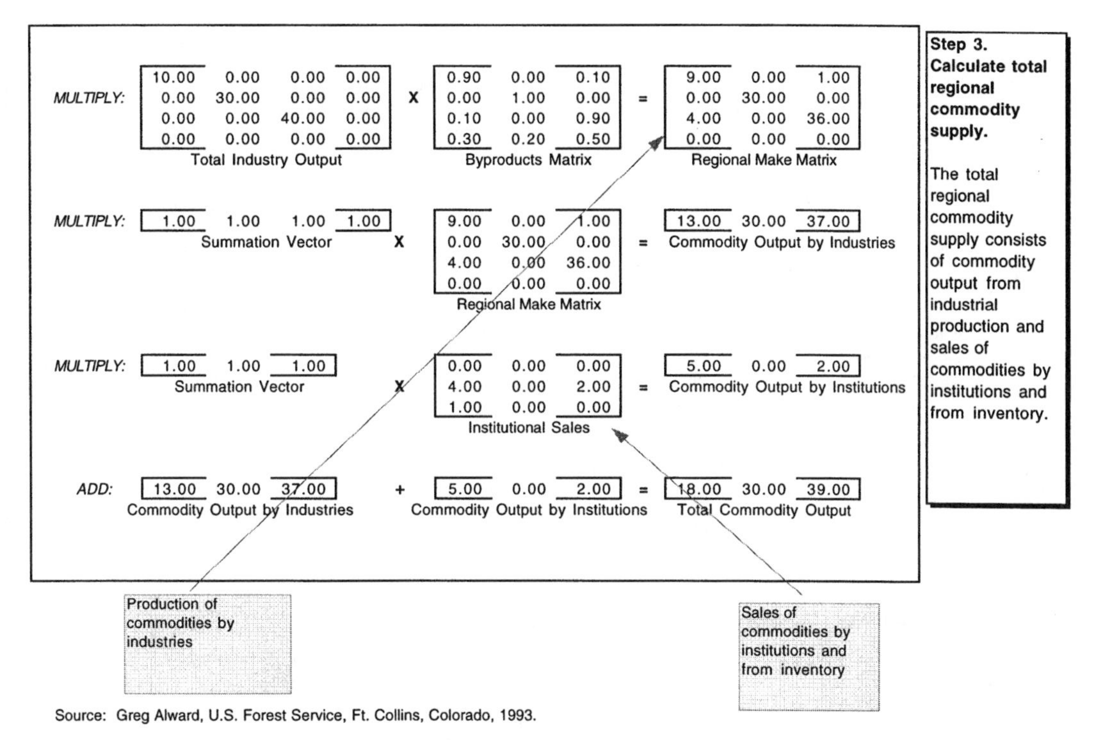

FIGURE 7-6 Calculation of total regional commodity supply.

Step 3 relates to the calculation of *total regional commodity supply* (Figure 7-6). Again, the estimate of total regional industry output enters into the calculation, but in multiplication with the industry by-product ratios from the U.S. by-product matrix. The result is the regional make matrix that shows the commodity production (columns) by each industry (rows). These estimates, together with the estimates of institutional commodity output (commodity sales by government and from inventory depletion), yield the total commodity output for the regional economy.

The U.S. estimates of industry purchases include both domestic production and foreign imports. Thus, the input profile for each industry includes all commodity inputs of that industry. In addition, each industry may produce more than one commodity. The estimates of gross domestic exports relate to both the commodity production and the regional demand for this production.

Step 4 is the derivation of the *regional purchase coefficients* (Figure 7-7). The RPC value times the corresponding value in the regional gross use matrix yields the regional industry use of the locally supplied commodity (Chapter 2, Figure 2.1). Similarly, the *import propensity* for a given commodity times the corresponding value in the regional gross use matrix yields the regional *industry*

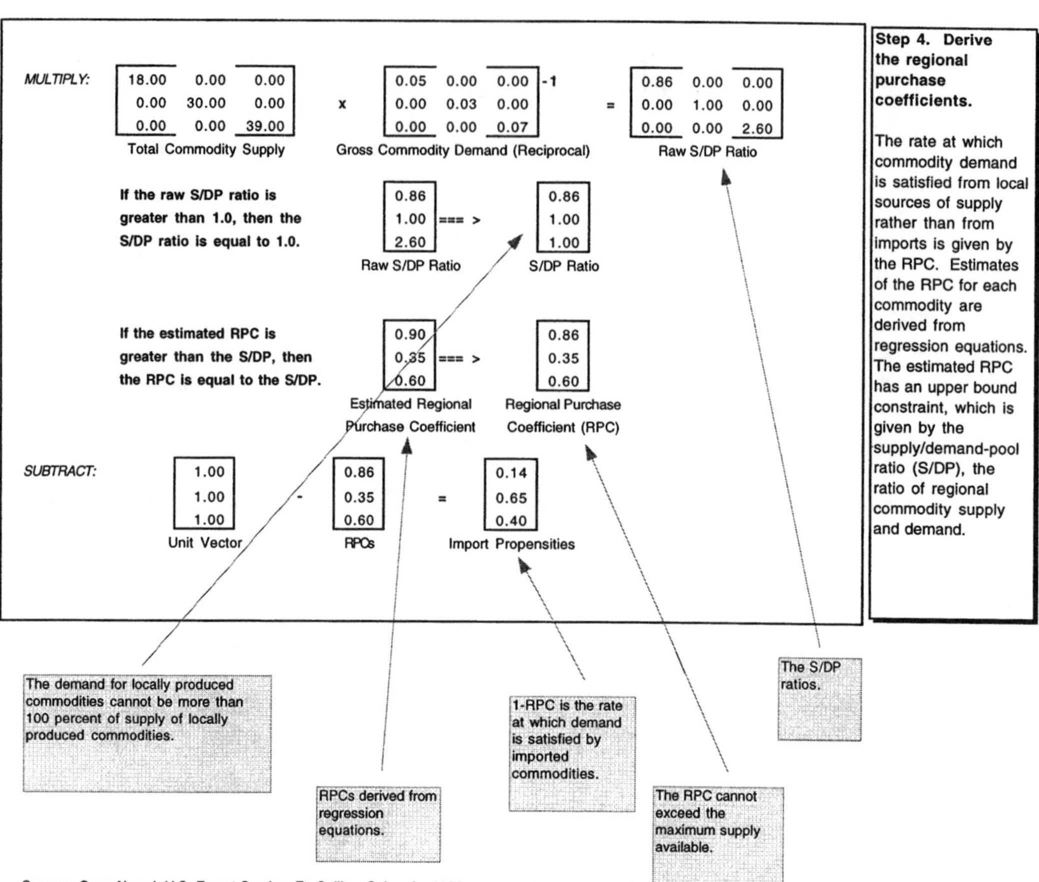

FIGURE 7-7 Calculation of regional purchase coefficients.

domestic imports of each commodity. This procedure applies also in estimating regional institutional use and regional institutional imports, that is, the commodity purchases by the local final demand sectors.

Step 5 is *separating the regional demand for commodities from the import demand* (Figure 7-8). This involves the calculation of regional purchases of commodities produced in the region, using the RPCs presented earlier. It also involves the calculation of imports, using the marginal propensities to import. These import propensities are represented by the total propensity to consume minus the propensity to consume regional production. Similarly, the total purchases of institutions are separated between regional production and imports.

Step 6—the calculation of *domestic commodity exports*—results from subtraction of regional commodity demand from regional gross commodity supply (Figure 7-9). The individual commodity imbalances in the U.S. estimates of

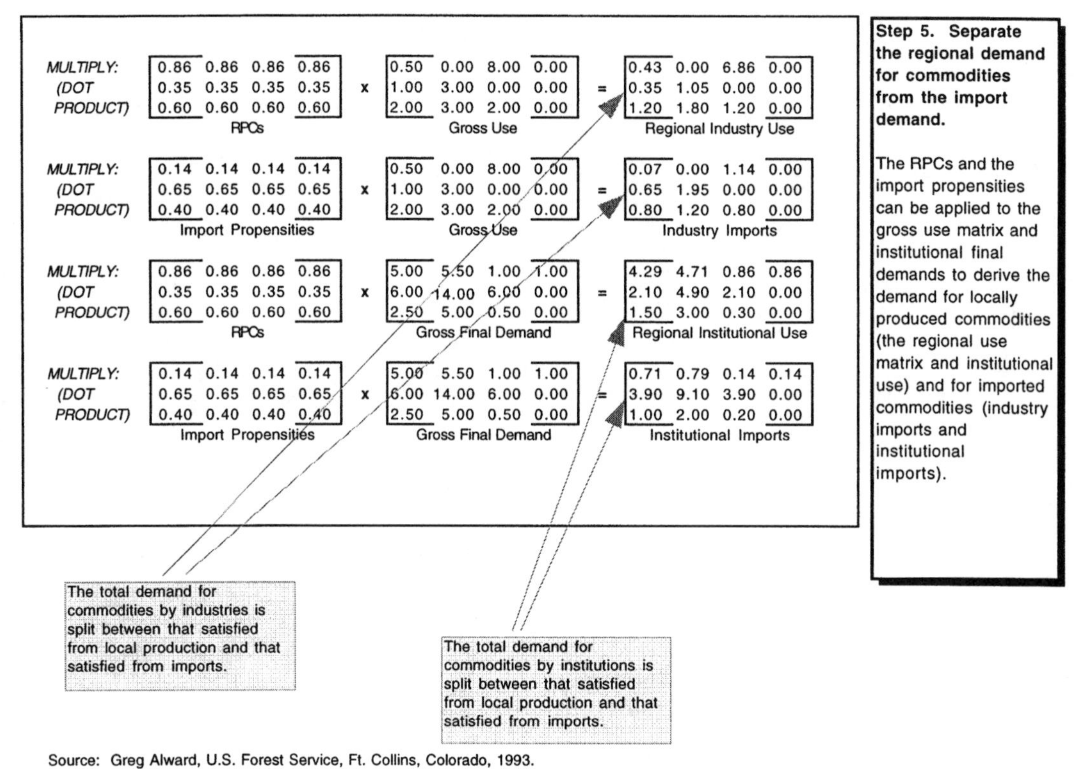

Source: Greg Alward, U.S. Forest Service, Ft. Collins, Colorado, 1993.

FIGURE 7-8 Calculation of imports.

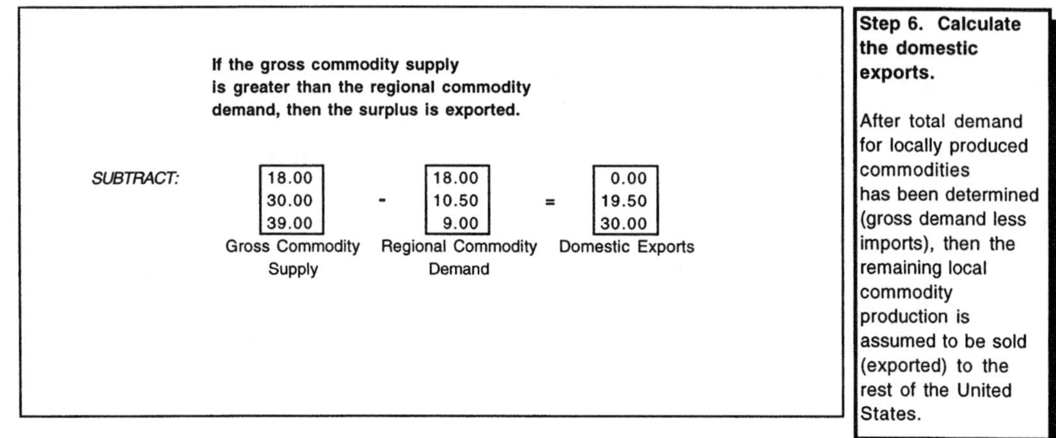

Source: Greg Alward, U.S. Forest Service, Ft. Collins, Colorado, 1993.

FIGURE 7-9 Calculation of domestic commodity exports.

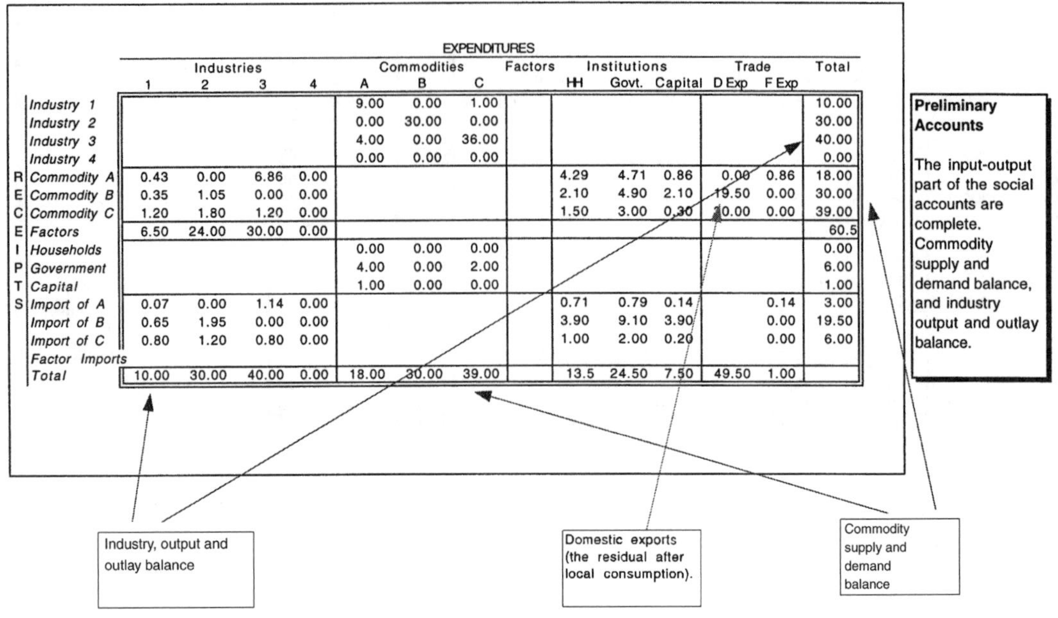

EXPENDITURES

RECEIPTS	Industries 1	Industries 2	Industries 3	Industries 4	Commodities A	Commodities B	Commodities C	Factors	Institutions HH	Institutions Govt.	Institutions Capital	Trade D Exp	Trade F Exp	Total
Industry 1					9.00	0.00	1.00							10.00
Industry 2					0.00	30.00	0.00							30.00
Industry 3					4.00	0.00	36.00							40.00
Industry 4					0.00	0.00	0.00							0.00
Commodity A	0.43	0.00	6.86	0.00					4.29	4.71	0.86	0.00	0.86	18.00
Commodity B	0.35	1.05	0.00	0.00					2.10	4.90	2.10	19.50	0.00	30.00
Commodity C	1.20	1.80	1.20	0.00					1.50	3.00	0.30	[illegible]0.00	0.00	39.00
Factors	6.50	24.00	30.00	0.00										60.5
Households					0.00	0.00	0.00							0.00
Government					4.00	0.00	2.00							6.00
Capital					1.00	0.00	0.00							1.00
Import of A	0.07	0.00	1.14	0.00					0.71	0.79	0.14		0.14	3.00
Import of B	0.65	1.95	0.00	0.00					3.90	9.10	3.90		0.00	19.50
Import of C	0.80	1.20	0.80	0.00					1.00	2.00	0.20		0.00	6.00
Factor Imports														
Total	10.00	30.00	40.00	0.00	18.00	30.00	39.00		13.5	24.50	7.50	49.50	1.00	

Source: Greg Alward, U.S. Forest Service, Ft. Collins, Colorado, 1993.

FIGURE 7-10 Regional input-output accounts, rectangular format.

foreign exports and imports carry through to the individual county or multicounty Micro-IMPLAN models. Domestic exports and imports theoretically balance for the domestic economy as a whole, but not for individual counties or multicounty areas. However, the criteria for allocating the two sets of exports and imports differ greatly. Micro-IMPLAN allocates U.S. *foreign commodity exports* to the regions according to its share of the U.S. commodity production. It also allocates U.S. *foreign commodity imports* to the regions by the same rule. Estimates of a region's total imports and total exports thus derive from a variety of data sources and allocation criteria.

Figure 7-10 shows the outcome of following the six-step procedure—the creation of regional input-output accounts in rectangular format. Industry outlay and output balance in this account. Commodity supply and demand also balance.

While local commodity production provides the basis for allocating foreign exports and imports, uniquely generated local RPCs provide the basis for estimating domestic exports and imports for each county or multicounty area. These estimates of gross domestic imports relate to both the commercial production and the demand for this production in a given region. Model reformulation calls for similar criteria in allocating U.S. foreign imports to individual industries and regions.

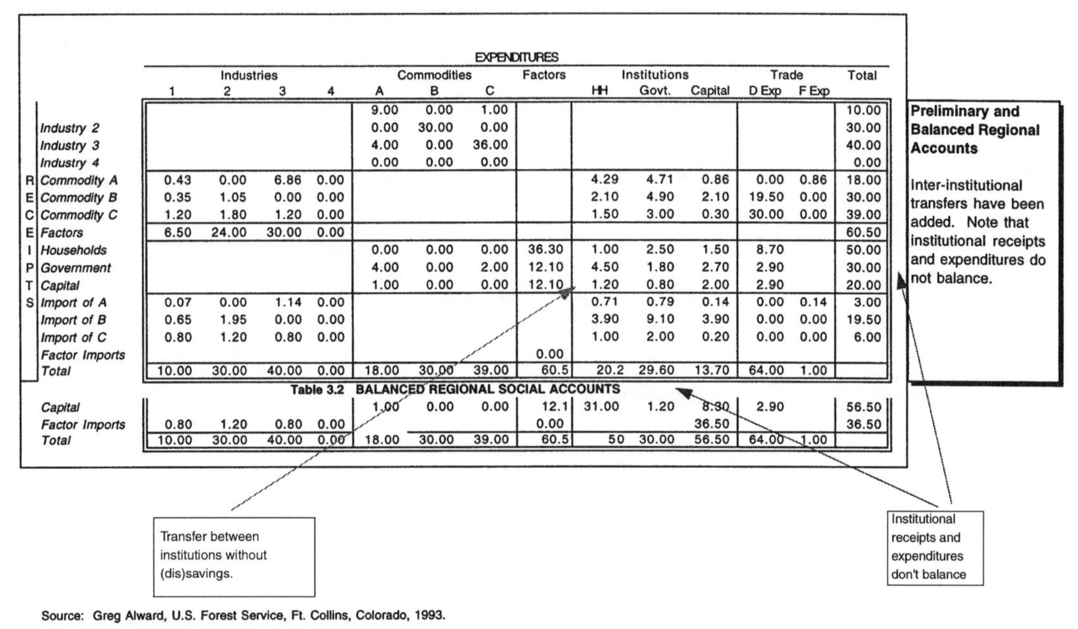

EXPENDITURES

	Industries 1	2	3	4	Commodities A	B	C	Factors	Institutions HH	Govt.	Capital	Trade D Exp	F Exp	Total
					9.00	0.00	1.00							10.00
Industry 2					0.00	30.00	0.00							30.00
Industry 3					4.00	0.00	36.00							40.00
Industry 4					0.00	0.00	0.00							0.00
R *Commodity A*	0.43	0.00	6.86	0.00					4.29	4.71	0.86	0.00	0.86	18.00
E *Commodity B*	0.35	1.05	0.00	0.00					2.10	4.90	2.10	19.50	0.00	30.00
C *Commodity C*	1.20	1.80	1.20	0.00					1.50	3.00	0.30	30.00	0.00	39.00
E *Factors*	6.50	24.00	30.00	0.00										60.50
I *Households*					0.00	0.00	0.00	36.30	1.00	2.50	1.50	8.70		50.00
P *Government*					4.00	0.00	2.00	12.10	4.50	1.80	2.70	2.90		30.00
T *Capital*					1.00	0.00	0.00	12.10	1.20	0.80	2.00	2.90		20.00
S *Import of A*	0.07	0.00	1.14	0.00					0.71	0.79	0.14	0.00	0.14	3.00
Import of B	0.65	1.95	0.00	0.00					3.90	9.10	3.90	0.00	0.00	19.50
Import of C	0.80	1.20	0.80	0.00					1.00	2.00	0.20	0.00	0.00	6.00
Factor Imports								0.00						
Total	10.00	30.00	40.00	0.00	18.00	30.00	39.00	60.5	20.2	29.60	13.70	64.00	1.00	

Table 3.2 BALANCED REGIONAL SOCIAL ACCOUNTS

	1	2	3	4	A	B	C	Factors	HH	Govt.	Capital	D Exp	F Exp	Total
Capital					1.00	0.00	0.00	12.1	31.00	1.20	8.30	2.90		56.50
Factor Imports	0.80	1.20	0.80	0.00				0.00			36.50			36.50
Total	10.00	30.00	40.00	0.00	18.00	30.00	39.00	60.5	50	30.00	56.50	64.00	1.00	

Source: Greg Alward, U.S. Forest Service, Ft. Collins, Colorado, 1993.

FIGURE 7-11 Preliminary regional social accounts.

Figure 7-11 presents a set of *preliminary and balanced regional social accounts*. Inter-institutional transfers are included, but they do not balance at this point.

The industry output and outlay accounts and the commodity supply and demand accounts now balance, but the remaining social accounts—factors, institutions, and trade—do not.

Figure 7-12 presents the complete set of *regional social accounts*. The principal difference between the final two accounts is the introduction of factor imports to cover the imbalance between total exports and total imports. This represents the use of regional savings to pay for the excess of imports over exports.

References Data Systems

The U.S. Departments of Commerce, Labor, and Agriculture maintain the *reference data systems* for Micro-IMPLAN. The U.S. Department of Commerce houses the periodic censuses of population and employment, agriculture, manufacturing, wholesale and retail trade, and selected business services, as well as the annual statistical series on personal income and industry employment and earnings of the employed industry work force. State- and county-level data sources most critical for early fix-up and updating of the current database are

RECEIPTS	EXPENDITURES													
	Industries				Commodities			Factors	Institutions			Trade		Total
	1	2	3	4	A	B	C		HH	Govt.	Capital	D Exp	F Exp	
Industry 1					9.00	0.00	1.00							10.00
Industry 2					0.00	30.00	0.00							30.00
Industry 3					4.00	0.00	36.00							40.00
Industry 4					0.00	0.00	0.00							0.00
Commodity A	0.43	0.00	6.86	0.00					4.29	4.71	0.86	0.00	0.86	18.00
Commodity B	0.35	1.05	0.00	0.00					2.10	4.90	2.10	19.50	0.00	30.00
Commodity C	1.20	1.80	1.20	0.00					1.50	3.00	0.30	30.00	0.00	39.00
Factors	6.50	24.00	30.00	0.00										60.50
Households					0.00	0.00	0.00	36.30	1.00	2.50	1.50	8.70		50.00
Government					4.00	0.00	2.00	12.10	4.50	1.80	2.70	2.90		30.00
Capital					1.00	0.00	0.00	12.10	31.00	1.20	8.30	2.90		56.50
Import of A	0.07	0.00	1.14	0.00					0.71	0.79	0.14	0.00	0.14	3.00
Import of B	0.65	1.95	0.00	0.00					3.90	9.10	3.90	0.00	0.00	19.50
Import of C	0.80	1.20	0.80	0.00					1.00	2.00	0.20	0.00	0.00	6.00
Factor Imports								0.00			36.50			36.50
Total	10.00	30.00	40.00	0.00	18.00	30.00	39.00	60.5	50.00	30.00	56.50	64.00	1.00	

Complete Accounts.

The complete regional social accounts.

Source: Greg Alward, U.S. Forest Service, Ft. Collins, Colorado, 1993.

FIGURE 7-12 Regional social accounts.

the individual state County Business Patterns reports, ES-202 files, on covered industry employment and payroll, and the agriculture censuses.

The social and economic accounts used or incorporated into a regional SAM present a new opportunity for applied regional analysis. The regional SAM shows the flow of income payments and transfers, as well as goods and services, among households, firms, and governments within a region and with the rest of world. It provides a frame of reference for monitoring changes in the distribution of these flows among income-paying and income-receiving sectors from one time period to the next. It makes possible the tracking of a region's production, consumption, and investment and the internal changes in each of these activity areas. It is readily incorporated into existing regional modeling systems, either static (single-period) or dynamic (multiperiod).

PART 3
Applications

Central Cities and Downtown Districts

This is the first of four topical areas selected as applications of urban regional economics. The central cities cited in this chapter are the largest metropolitan centers of nodal regions. As mentioned earlier, the core area is the central place of the modern city region. Its innovative products and production processes, together with a diversity of economic activity and resources, account for its sustained growth and development.

Evolution of Central Cities and Downtown Districts

Urban regional economies are inexorably bound to their central cities and downtown districts. Quick and easy access to global destinations and markets depends on a complex transportation communication system and a large clientele base to support essential, but costly, infrastructure. Only a few of the several hundred metropolitan statistical areas (MSAs) in the United States qualify as major information activity and transfer centers. Add to this the complex communication systems of the downtown districts that offer clients superior opportunities for both face-to-face and distant, real-time communication, and only a very few centers qualify as the central places of metropolitan core economic regions.

Central-City Activities

Michael Porter, in *The Competitive Advantage of Nations* (1990), notes that ". . . proximity increases the concentration of information and thus the likelihood of its being noticed and acted upon . . . At the same time, it tends to limit the spread of information outside because communication takes a form (such as face-to-face contact) which leaks out only slowly." Thomas Hutton and

David Ley (1987), reporting on the findings of two producer surveys in downtown Vancouver, British Columbia, note that "A downtown location usually implies membership in a dense network of business linkages, and the importance of these linkages dictates location centrality within a vertically disintegrated economic complex" (p. 126).

Hutton and Ley document the importance of proximity to business contacts for both the cluster of producer services and the head offices in Vancouver. "Ease of access to customers was of first importance, followed by convenient access to business services, and finally in third rank, access to informal business contacts" (p. 131). "The producer services were the more contact intensive, with 42 percent of the firms stating that more than 90 percent of their business required personal meetings; 22 percent specified that six or more meetings would take place in a typical transaction. Firms which placed a premium on downtown location were rather more likely to emphasize the importance of face-to-face meetings" (p. 132).

The overwhelmingly local sources of inputs for the business services sector mirrored findings of a Greater Seattle-Puget Sound study. "Corporate offices, in contrast, are trading in products where the export market is of far more significance, particularly those closely connected with the staples of mining and forest products" (p. 133). "Both capital intensification and the growing concentration of non-standardized jobs are occurring simultaneously in the downtown corporate complex" (p. 137). According to the authors, despite ". . . a close correlation between changes in provincial gross domestic product, service employment and office development in downtown Vancouver, the relation is asymmetric. Considerable leakage from resource development in the hinterland accrues to Vancouver's corporate service complex, but this complex is largely self-contained for its own inputs. Moreover, the provincial hinterland is a smaller customer of Vancouver's producer services than markets outside the province."

More than 40 percent of firms surveyed expected significant increases in sales outside the province in the 1990s. Producer services include printing, banking, securities, commodities, insurance, real estate, personnel, data processing, accounting, advertising, architecture, planning, geological and engineering consultants, legal services, management consultants, and miscellaneous services.

Business enterprises in the downtown districts of regional cities are the new engines of economic growth in the global information economy. Backward linkages to input suppliers and forward linkages to markets account for the high degree of interdependence of downtown businesses in the metropolitan core region with other businesses in the core and contiguous regions. The downtown economy, with its unique cluster of high-order producer services, is an essential link between many small and independent businesses and the changing product demands of both national and global markets. Without the distributive and other producer services of the downtown economy, the export-producing businesses would lack access to these changing national and global markets and to new production technologies.

Because of the limited number of cities that have downtown districts with a diverse business infrastructure providing high-order banking, finance, insurance, transportation, telecommunications, management, research, and related producer services, property owners in these districts capture an economic rent. This rent correlates with the profitability of business enterprises dependent on superior access to information. Businesses that can afford to pay higher rents locate in central places that provide superior access to product and market information.

Important determinants of business location and development—apart from the location of the business entrepreneur—are costs of labor, energy, land and space, environmental protection, and industry regulation that vary from site to site. In addition, transfer costs—not simply the costs of transportation, but also the costs of the many different producer services required to move a product from its production to final use—are increasingly important. They also include the costs of producing, distributing, evaluating, and using the highly differentiated market and product information essential in completing each business transaction, whether negotiated face-to-face in a downtown office or electronically in a distant commodity exchange. The producer services sector includes commercial printing; radio and television broadcasting; banking; securities exchange; insurance; advertising; computer programming; personnel services; accounting; research laboratories; and architectural, management, and consulting services. These services concentrate in the downtown districts of metropolitan area economies. The producer services sector is an important part of several industry clusters. It has strong forward linkages to intermediate markets.

The export-producing industries in the core metropolitan area and its rural periphery use the producer services concentrated in the downtown district. In turn, the producer services sectors depend initially on the local and regional export-producing industries for their primary market outlets. As the producer-service businesses expand in sales and customers, they acquire market outlets in other metropolitan regions. Airline and telecommunication networks facilitate their interregional linkages.

Central-City Infrastructure

State Fiscal Policy Changes

Historically, state fiscal policy has played an important role in facilitating or inhibiting local economic development. Some of the activities directly or indirectly involving state policy include the following:

1. State-funded highway construction programs that extend the periphery of urbanized areas and reduce congestion and loss of time on the access highways to a downtown or suburban workplace, but which also add to the public costs of urban sprawl.
2. Rationing of municipal water and sewer connections that reduce urban sprawl and the costs of urban growth precipitated by the highway construction.
3. Construction of public facilities in the newly subdivided suburban areas while existing public facilities in the already developed metropolitan core area are being vacated, thus increasing per capita local taxes and public facility expenditures in both areas.
4. Redevelopment programs for the inner city and the first-ring suburbs, supported by downtown development that raises the neighborhood tax base and makes possible transfer payments from the downtown to the neighborhoods—linkage programs that directly tie downtown development to neighborhood needs.
5. Tax transfers to city neighborhoods, which help improve the quality of life in both the downtown and the neighborhoods by reducing the sources and causes of crime.
6. Expenditures toward strengthening the metropolitan downtown economy through targeted assistance programs for new export-producing businesses to replace manufacturing and trade businesses relocating in the suburbs. Such expenditures also bolster use of existing public facilities, thus reducing social costs of urban sprawl and waste of public resources in excess infrastructure that would otherwise occur.
7. Maintenance of the public and private infrastructure of a high-order, world-class metropolitan downtown economy with strong and active linkages to other businesses in the region and state of which the downtown is an integral part.
8. Implementation of studies to assess future air transportation requirements of the increasingly trade-dependent, human-resource-based economy of the state and its metropolitan areas.

Several of the perceived linkages obviously involve state and local governments in contradictory efforts to reduce public costs of economic growth and change. The contradictions in state fiscal policies are increasingly apparent in light of the following:

1. The critical role of public infrastructure in affecting the success of local export-producing business enterprise in global competition.
2. The concentration of producer services—the new agents of economic growth and change—in the downtown districts of regional cities.
3. The symbiotic interdependence of a strong downtown district and healthy surrounding neighborhoods.
4. The increasingly severe constraints on the fiscal capacities of state and local governments to even maintain existing public facilities and services.
5. The expanding real and specific pressures placed upon state government resources to invest in the essential transportation infrastructure of global interchange.
6. The growing awareness of a region's economic base, its critical importance to the region's well-being, and its selective dependence on the public and private infrastructure. More than ever before, every state expenditure now calls for careful scrutiny of opportunity costs. The value of public facilities and services not provided because of an equivalent lack of state funding is a measure of opportunity costs.

If the central city and its metropolitan area also serve as an airline hub in a global air transportation network, it may face additional challenges in the building of a new airport and/or the recycling of its old one. For example, a recently completed task force for such an airline hub presented a series of themes in its final report for airport reuse planning (Metropolitan Airports Commission, 1993). Each of the themes presented to the Reuse Committee by the committee's consulting team emphasized one or more international corporate business activities that involve the resources of state government in their implementation. For example, the airport reuse might include one or more of the following:

1. An international corporate center with a concentration of businesses engaged in international trade, using the site facilities for various administrative, research and development, and manufacturing functions.
2. A world-class biomedical and health care center with direct links to the health care and research facilities in the nearby medical care centers.
3. A major multidisciplinary research complex with both public and private research facilities that share a support infrastructure of university, state and federal staff, and related resources.
4. A major entertainment and outdoor-indoor recreation center with close and easy access to nearby historical and natural resource sites and amenities.

5. A world-class high-technology manufacturing center that builds on an already well-known technology-intensive industry cluster that accounts for an increasingly larger portion of the state's economic base.
6. A multimodal hub transportation center that may include light rail transit, mega-level rail connections to areas outside the region, remote air passenger terminal and/or parking/transfer, and pedestrian circulation. A central hub for goods movement, collection, and distribution (letter, small package, freight) of goods could be integrated into the sites.
7. An aviation reuse facility that would service the noncommercial aviation activities in ways to complement the new airport.

Thus, the nurturing, growth, and maintenance of the "regional city" functions of an actual airline hub may now involve a serious and careful "visioning" of alternative regional and state futures. The visioning process for airport reuse takes into account important linkages of air transportation to the existing corporate headquarters functions in a central city's downtown district.

Documenting Income and Employment Changes

The social accounting matrix (SAM) framework provides summary accounts for documenting the income impacts from market and policy changes for each sector of a region's economy. For example, certain institutional factors account for many of the current difficulties in optimizing public and private investment in city-region infrastructure, particularly transportation. The institutional factors include state and local subsidy of exurban infrastructure and federal tax expenditures (that is, deduction of interest payments on home mortgages) for residential housing, exclusionary use of subdivision and zoning regulations, extraordinarily high transaction and related costs of "doing business" in the area, and blind acceptance (to use Jane Jacobs's words) ". . . of the mercantilist tautology that nations are the salient entities for understanding the structure of economic life" (Jacobs, 1984, p. 50). Thinking of economic development as a national rather than a city process ignores the importance of location and its matrix of unique attributes affecting the viability of business enterprise.

Among the most troubling institutional factors affecting city-region futures is the public subsidy of access roads and highways to the expanding urban periphery of metropolitan areas. The public subsidies invite "leapfrog" residential development that results in costly urban sprawl. Separation of place of residence from place of work also adds to costs of local transportation.

Another troubling feature of urban economic life is misuse of the power given states to establish any land use controls they want, subject to "due process." These powers include zoning regulations, subdivision controls, municipal growth management, and land development fees that protect the exclusionary uses of large lot sizes and open space preserves.

Large lot size and open space requirements favor the construction of high-income housing in open country. Central cities and first-ring suburbs secure the low-income housing, but without the tax base to support the high cost of providing social and economic services for those left behind. Those who have the money to buy large lots and build expensive housing can move away from the problems of the central city and first ring suburbs and at the same time reduce their municipal taxes and mortgage interest payments.

Two serious consequences (apart from its outright unfairness) flow from exclusionary land use controls: the inability of local governments to support areawide concerns and the proliferation of subsidized shopping centers that simply redistribute the total spending of the metropolitan region, but add to its total transportation costs—public and private. No municipality can bribe another, for example, to locate its landfill in the other municipality. Suburban municipalities readily turn down the location of a job-creating facility within their boundaries, especially if it serves the entire metropolitan area. Thus, the exclusionary practices reduce the total jobs available to central city and first-ring suburban residents.

Maintenance and reinforcement of the separation of place of work and place of residence impose large private and social costs. The transportation infrastructure for serving new shopping centers absorbs a large amount of available financial resources of state and local governments. Not only is the total transportation bill higher, but the environmental costs also increase because of the readily available public financing of local transportation infrastructure. For example, one segment of interstate highway connecting suburbs twelve miles west of downtown Minneapolis has a fast lane for vehicles with two or more passengers. This feature cost $420 million. It serves largely the commuter workers in downtown Minneapolis who reside in a very high-income residential area. The city also provides parking space at a very low monthly cost for the fast-lane user.

Another important source of added costs of doing business in the metropolitan core is government, particularly the federal government. Most recent estimates place the cost of government regulation of all industry at $600 billion. Since 1990, American businesses incurred $130 billion more costs as a result largely of compliance with the Fair Labor Standards Act of 1989, the Americans With Disabilities Act of 1990, the Clean Air Act amendments of 1990, and the Civil Rights Act of 1991.[1] Small businesses (those with fewer than 500 employees) created two out of three net new jobs between 1982 and 1990. Since 1989, however, small business profits per worker dropped from $3,500 to $600, while the tax and regulatory burden increased from $900 to $4,300. The added costs place U.S. businesses at an immediate disadvantage in competition with foreign companies that do not carry these costs.

Changing Role of Downtown Districts

The highly specialized and mutually interdependent producer service activities cited earlier cater to the downtown "corporate complexes" of the central cities (Daniels, 1991). These complexes are composed of three parts: headquarters offices, firms servicing headquarters offices, and ancillary services provided to individuals who perform the headquarters and corporate services. These include food, hotel, travel, and retail sales services. A major concern in the analysis of the economic changes taking place is whether or not the national economy is turning into one characterized primarily as a service economy, and if it is, what the implications are.

Downtown Corporate Complex

The 1970s and 1980s saw profound changes in the structure of the U.S. economy. Manufacturing ancillary services' shares of the economy fell from 25 percent in 1969 to 19 percent in 1985 after approximately 60 years of fairly constant employment growth in relation to other sectors. These include trade; finance, insurance, and real estate (FIRE); and services.

Services, at 24 percent of U.S. employment, constitute a larger share of total industry employment than any other group. This process has been given prominence in the "postindustrial" school of thought, which claims that the switch to the services sector is leading toward an inevitable marginalization of the goods-producing sector as the country shifts to an "information economy." Similarly, the "deindustrialization" theory emphasizes this shift to services, but instead of ascribing it to an inevitable evolutionary process, cites the outcome of specific investment decisions (Bluestone and Harrison, 1982). However, the manufacturing share of the total output of goods and services in the United States has fluctuated between 20 and 25 percent since 1948. Thus, while employment has been declining, productivity has been increasing.

One of the most important questions is, what happens to incomes? Does the shift from traditionally well-paid manufacturing imply the transformation of the economy to one of low-paid service workers in low-productivity sectors? Services are often seen as ancillary, providing low-wage jobs with little prospect for advancement. Ancillary services are also seen as being dependent on manufacturing and therefore not capable of generating their own economic growth.

If a bank or manufacturer does its own building maintenance or legal work, these jobs are classified in banking or manufacturing. If they are purchased under contract, they are classified as ancillary services. Many consulting firms, real estate offices, and even financial institutions are too small to maintain their own buildings, operate graphics shops, or manage their own payrolls. Hence, to some extent the shift of employment to services reflects a degree of contracting out. The result is that, by sharing fixed costs, both small manufacturing firms and producer services have a resulting enhanced ability to contribute to the economy.

Research has also identified the importance of specialized producer services in the evolution of a new urban hierarchy of cities. Ancillary services enhance the role of international business activity in sustaining the existence of the large corporate complexes, which are the major element in the ranking of these hierarchies (Noyelle and Stanback, Jr., 1984).

What is not clear is the future security and probable evolution of the corporate complex, especially in downtown districts, in an era of rapid technological and organizational change. This complex is predominantly a feature of the central business districts of major metropolitan areas, but concentrations of producer services also occur in suburban and inner suburban areas. The same advances in transportation and communication that permit the dispersion of production also enable administrative and control functions to be physically separated from the sites of production; therefore, they may not necessarily congregate in the downtown districts. In addition, many routine-type producer services, such as data entry, tend to disperse to the lowest wage sites in the periphery.

Jobs requiring specialized information and frequent decisions function best in central business districts since they require inputs of advice, data, and expertise from a variety of professions and other services that are relatively difficult to obtain in a suburban location. These advanced services—that is, long-range planning and coordination functions associated with corporate headquarters—tend to agglomerate in the downtown of large cities. By concentrating in one place, producer services can support an array of specialized business services necessary for accountants, engineers, management consultants, advertising agencies, and specialized law firms. Moreover, top decision makers and specialists need personal contact to exchange confidential information, partly to avoid its use by competitors.

Corporate decision makers want to be close to centers of government power for maximum influence. Downtowns, in addition, possess advantages for satisfying luxury tastes (restaurants, hotels, and access to culture) that make it easier to attract high-level managers and professionals. And, when available, access to appropriate retail and housing plays a part in the decision of a firm to locate in the downtown district. What's more, the impact of city neighborhoods on firms' decisions to remain or locate in the downtown is a function of the perceived negative effects of crime, infrastructure deterioration, and overall conditions.

A Reference Downtown District

The city of Minneapolis, with a population of nearly 370 thousand, is in a metropolitan core area of more than two million. Minneapolis, in many ways, is characteristic of the 29 metropolitan core areas that also serve as airline hubs in the global air transportation network. A number of these centers are characterized by an orientation to business services in the rapidly expanding intermediate product markets. These include corporate headquarters functions and various producer services, such as banking, insurance, real estate, legal services, educational services, nonprofit organizations, and architectural and engineering services (Hutton and Ley, 1987; Noyelle and Stanback, Jr., 1984; Scott, 1988; Stanback, Jr., and Noyelle, 1982). Also, strong transportation and communication linkages to other major central cities are available locally. The central city that also serves as an air node has become the critical decision information center of the new economic order that powers the engines of domestic regional growth. Minneapolis aptly serves as a reference city for comparisons of the economic activities and performance of other central cities in the global transportation networks.

Downtown Activities

The relative strength of the Twin Cities (St. Paul and Minneapolis) in general, and the Minneapolis downtown in particular, seems to contradict the conventional wisdom about what has been happening in the U.S. economy over the past two decades. That wisdom stressed the retrogression of the snow-belt economy and the strikingly rapid advances of the sun belt. It further stressed both the increasing importance of services, usually defined in terms of retailing and consumer services, and the movement of manufacturing out of the older urban centers to the suburbs and smaller nonurban centers.

Recent findings have emerged that help explain why metropolitan-centered regions like the Twin Cities have maintained a strong position nationally. For one, the northern centers have continued to be home for the control functions of large and medium corporations. Secondly, manufacturing is being transformed, with internal and external services playing increasingly important roles. These producer services (finance, banking, and business services) tend to develop in larger cities. Furthermore, most of these cities remain in the snow belt. Many cities have been able to make substantial progress in compensating for their loss of manufacturing and are gaining strength from the continued growth of producer services.

In an analysis designed to assess the functional relationships between firms in the Minneapolis downtown district, standard industrial classifications were divided into various groupings, depending on which aspect of the relationships was being explored.[2] Initially, 28 groupings or sectors were created and applied to the Dun's list of 4,204 downtown firms within the freeway ring. Because the goal of the analysis was to examine the producer services sector, care was taken to distribute the industries in order to obtain a fine level of detailed data on these firms. The 28 functional sectors were further grouped into three main categories of ancillary services, an amalgam adapted from Stanback, Jr., and Noyelle (1982), and Ley and Hutton (1987). These include (1) the extractive, transformative, and distributive services of agriculture, mining, construction, manufacturing, transportation, wholesale trade, communications, and utilities; (2) producer services, including banking, finance, legal, research, and advertising activities crucial to the production and distribution of goods and services in an advanced capitalist economy; and (3) ancillary services required for the maintenance of the labor force involved in the corporate and service activities, which include social and health services, and the various consumer services such as personal services, hotels, and the like.[3]

Two sets of questions were crafted in a survey to analyze the importance of a downtown location for the economic activities pursued by the corporate complex. One set asked the respondents to rate variables in terms of importance for their decision to locate or remain downtown, and the other asked them to rate their satisfaction with these downtown attributes.[4] Further, those surveyed could be divided into two groups: headquarters and nonheadquarters firms.

There were some differences between the two groups in the ranking of these downtown variables. For example, the idea of centrality ranked first for headquarters firms, with 60 percent citing this as a very important influence for locating downtown. The nonheadquarters firms ranked this number two, with 53 percent of the firms. Of the nonheadquarters firms, 56 percent identified the cost of space as the most important locating factor. Apparently, the nonheadquarters firms are more price-sensitive in terms of rents. Second for the headquarters firms was freeway access, while for nonheadquarters firms freeways were ranked fourth as a consideration.

Apparently more critical were differences between the two groups in ranking the importance of proximity factors. In both cases, proximity to suppliers or backward linkages did not appear to be very strong considerations. Only 7 percent of headquarters and 11 percent of the nonheadquarters firms cited this as the most important factor influencing their decision to locate or remain downtown. In both cases, proximity to customers was stronger, with 39 percent of the headquarters and 47 percent of the nonheadquarters considering this very important.

Almost one quarter of the headquarters firms were getting more than 40 percent of their inputs from downtown and another quarter were getting them from the state outside of the Twin Cities, and another quarter from other parts of the United States. The majority of the nonheadquarters firms were getting their inputs from the rest of the United States, with producer services slightly less so than the other sectors. Of the producer services in this sector, 36 percent reported getting their inputs from downtown.

How much influence do these factors have on the cluster of headquarters and nonheadquarters firms in the downtown district? The 26 factors were cross-tabulated with firms' intentions to relocate in the next eighteen months. Of the nonheadquarters firms, approximately 25 percent intended to relocate within the following eighteen months. Of these, 47 percent were planning to move out of the downtown and 58 percent (6.7 percent of the respondents) were moving elsewhere in the Twin Cites, but not to the central city of St. Paul. One firm intended to relocate to St. Paul.

Headquarters firms were similar, with 24 percent relocating, but with around 40 percent of those moving out of the downtown. Twenty-three percent were moving to a site outside of Minneapolis but still within the Twin Cities metropolitan area. Factors displaying significant differences between those planning to relocate and those not planning to relocate (indicated by low chi square values) were examined more closely. Among nonheadquarters firms, 92 percent of those moving out of the downtown cited cost of space, while 81 percent cited central location in the metropolitan area as being of particular importance.

For headquarters firms, communications facilities (such as PBX, fiber optic and digital termination) were not generally a major concern; 56 percent did not consider this important. Of those firms moving out of the downtown, 57 percent thought communications were not important (Moss and Brion, 1991). Nonheadquarters respondents also failed to cite communications; only 34 percent of those moving out of the downtown considered this important. The ability to conduct business face-to-face was extremely important to both groups; 83 percent of the nonheadquarters firms and 73 percent of the headquarters firms indicated this as a priority.

Again, in looking at the relocation plans of headquarters firms, 66 percent said that proximity to suppliers was not very important. Of those, only 10 percent were moving out of the downtown. Of the nonheadquarters firms, 58 percent said this was not important, and of those, only 13 percent were moving out of the downtown.

Proximity to customers, on the other hand, ranked very high for 665 headquarters firms, with 6.6 percent of those moving out of the downtown. Even more of the nonheadquarters firms, 77 percent, believed this factor was important. Cost of space was important to 81 percent of the headquarters firms, with 9.4 percent of those moving out of downtown. It seemed to impact nonheadquarters firms even more, with 90 percent citing it as important; 12 percent said they were moving out of downtown, and another 13 percent said they were moving to another location in the downtown. Prestige was important for 74 percent of the headquarters firms, and only 7 percent of those were moving out of the downtown. For nonheadquarters firms, 70 percent thought prestige important, while 11 percent of those were moving out of downtown and 15 percent were moving to another location in the downtown.

Access to specialized business contact was identified as a primary advantage by all of the downtown firms surveyed. It remains now to specify the network of linkages with the region as a whole. Of the headquarters firms, 61 percent are selling to other companies in all markets, with only 9.3 percent selling to city and suburbs. However, 35 percent sell more than 40 percent of their final goods to customers in the metropolitan region. In comparison, 48 percent of the nonheadquarters sell to all regions and 19 percent sell only to the city, while 16 percent sell more than 40 percent of their output to firms located exclusively downtown.

The existence of substantial backward linkages within the downtown was identified by both groups. One quarter of the firms reported obtaining more than 40 percent of their inputs from downtown. The headquarters firms were also linked closely to firms outside the state. One quarter of the headquarters firms and 28 percent of the nonheadquarters firms obtained another one quarter of their inputs from out of state. The headquarters firms appeared to be more connected to the state, with one quarter obtaining materials from the state outside of the Twin Cities region. Of the nonheadquarters firms, only 3 percent received inputs from out of state, with 36 percent of these firms getting the inputs from the region outside of the downtown.

Employment Practices

While the rise in producer services has played a major role in employment generation, aggregate data have shown that significant proportions of these jobs have been low paying. Within producer services, a typical employment profile shows a well-paid, predominantly male managerial and professional group and a low-paid, primarily female clerical group. A summary of the Minneapolis survey findings may be instructive:

1. Employment profiles of headquarters and nonheadquarters firms are similar. For example, 18 percent of nonheadquarters firms had more than five managerial staff, and 40 percent had more than five professional technical staff.

2. Part-time employment is used extensively by both types of firms. Of the headquarters firms, 78 percent have one to five part-time workers, with almost 60 percent in the producer services category. The same proportion of nonheadquarters firms rely on part-time workers. In addition, 48 percent of the headquarters firms and 37 percent of the nonheadquarters firms use as many as five temporary employees.
3. New jobs are being created in equal numbers by headquarters and nonheadquarters firms. In the eighteen previous months, 60 percent of the headquarters and 60 percent of the nonheadquarters firms had created new positions, mainly in producer services.
4. Firms hire principally from within the city—48 and 42 percent of the headquarters and nonheadquarters firms, respectively.
5. Growing concern was expressed about a downtown core of high-order services and headquarters being surrounded by declining neighborhoods. With no significant difference between sectors, 95 percent of the respondents consider crime in city neighborhoods bad for their business.
6. Close to 60 percent of respondents consider the city public schools inferior to suburban public schools.
7. Housing availability in the neighborhoods or downtown was not considered in a firm's decision to locate in the downtown, either for rental or ownership.

Mergers and sales of firms within and among industries is one indication of the degree of vertical disintegration, that is, the splitting off of activities into separate industries. Disintegration becomes the glue for future downtown agglomerations. Of the headquarters firms, 16 percent were the result of a merger in the same industry, 6 percent a sale from another firm. Of nonheadquarters firms, 5 percent were a result of a merger in the same industry, and 1 percent were a result of a merger with a firm in another industry. As expected, a large number of nonheadquarters firms were branches, with 55 percent of the branches less than 15 years old, compared to 33 percent of the nonheadquarters firms located downtown. One quarter of both groups had one to five locations other than downtown.

A surprising number of firms had plans to expand the size of their facility. Of the headquarters firms, 39 percent had plans to expand in the following three to five years. Of the nonheadquarters firms, slightly less, 35 percent had expansion plans. In terms of expanded operations, 25 percent of the headquarters firms and 22 percent of the nonheadquarters had plans to expand in the downtown. Of the headquarters firms, 5 percent were planning on expanding elsewhere in the city, another 5 percent elsewhere in the Twin Cities outside St. Paul. The breakdown was similar for the nonheadquarters firms.

Governing the Downtown District

The multiplicity of economic linkages that relate the production, investment, and consumption activities in the central city to the flows of exports and imports are illustrated by diverse functions and activities, such as the following:

1. Intermediate markets for parts manufacturers throughout the region.
2. Visitors to the region's many recreation sites, with much of the region's tourism dollars originating in the central city and metropolitan core area.
3. Government revenues for sharing with low-income communities and counties.
4. Final markets for crafts and other labor-intensive products that have nearby markets for their producers.
5. Employment opportunities, especially for the young people graduating from local institutions of higher education and entering the regional labor markets.
6. Financial and other specialized producer services for facilitating new business start-ups as well as serving established businesses.
7. Interregional trade, with the central city and its region bound together by history, culture, and economics—a new economics of cooperation and interdependence in which city and region help each other by doing what each can do best, and then trading with one another for the benefit of all.
8. Industry value added, in large part, remaining in the local community.
9. World market access that its corporate headquarters role requires, which, in turn, enriches both the central city and the region.
10. Cultural attractions of the metropolitan core area that are accessible to residents of the entire region.

Altogether, these findings show the reference city as two cities—one with a nighttime population, the other with a much larger daytime population. The city infrastructure is built to support the larger population with a financial structure based on the smaller population. The disassociation between the two cities in the distribution of the benefits and burdens of economic growth—illustrated in part by the larger rates of growth in wage and salary payrolls than in taxable valuation—is bound to create increasing fiscal and service pressures in the years to come.

Because local economic effects of state fiscal policy are intertwined with changing global markets and the globally competitive position of local and regional industries, the government role thus pales in comparison with that of the private sector in local economic development. These remarkable and far-reaching changes in industry location and mix and the position of metropolitan areas in the rapidly changing global economy therefore have important implications for state and local governments and their fiscal policies.

One implication of the changing location and mix of industry in metropolitan areas is its high cost in terms of urban sprawl. Another is the possibility of unintended participation of state government in supporting public facility construction in the urbanized periphery of the metropolitan area, while carrying the cost of excess facilities in developed portions of the metropolitan area.

Still another implication of global economic change is the increasing importance of well-placed public infrastructure, much of it supported by state money, in maintaining and improving the competitive advantage of local businesses in global and national markets. Cost savings from reducing urban sprawl thus could be targeted for building a cost-effective business environment for a state's basic industries—the ones that carry the burden of a state's and a nation's economic viability in global competition.

The growth of large cities, particularly the downtown districts of the airline and telecommunications nodes in the global information networks, is premised on the importance of physical proximity in the production, distribution, and use of decision information.

1. Concentration of headquarters offices and related support services in downtown districts of the communication nodes in global information networks is the result of the high cost of building and maintaining a world-class business infrastructure. Such locations are not likely to rise willy-nilly in any state but rather in those core metropolitan regions with an already established infrastructure to serve a concentration of corporate decision centers in close proximity to one another.
2. Business enterprises in the downtown districts of the central cities, like Minneapolis, are the new "engines" of economic growth in the global information economy. Backward linkages to input suppliers and forward linkages to markets account for the high degree of interdependence of downtown businesses in the metropolitan core region with other businesses in the core and contiguous regions.
3. The downtown economy, with its unique cluster of high-order producer services, is an essential link between many small and independent businesses and national and global markets. Without the distributive and other producer services of the downtown economy, the export-producing businesses would lack access to changing national and global markets and to new production technologies.

Finally, the financial predicament of central cities is being experienced by a growing number of smaller cities facing the unavoidable erosion of their downtown tax base. Main Street businesses are losing to new shopping centers supported, in large part, by public subsidy of essential local infrastructure. Thus, long-established central cities, whether in a metropolitan core region or a dominantly rural area, share the common challenge of overcoming the adverse effects of the use of tax dollars to subsidize urban expansion while paying for the rising costs of local government from an eroding local tax base.

A business location in a central city provides access to the large range of producer services and the related public and private infrastructure cited earlier and noted repeatedly in the accompanying references. Such locations are in limited supply. Meanwhile, the demand for the marketable goods and services produced at these locations is increasing—a condition that makes possible the capture of an additional premium at these locations. This premium amounts to a form of location "rent" that is shared by labor in the form of above-average labor earnings. Businesses presumably also capture a share of the location rent in the form of above-average profits.

The manufacturing sector, along with construction, also experienced large negative industry-mix effects in most central cities. Manufacturing and construction are among the most cyclically sensitive sectors of the U.S. economy. They experience below-average industry growth in recession and above-average industry growth in recovery. Manufacturing within central cities is burdened by an additional problem—lack of space for expansion, which is a major reason for the migration of manufacturing to sites outside the city. A space shortage exists at the same time that the city is overstocked in its unused land inventory—land unused because of contamination from its earlier uses as railroad yards, filling stations, and environmentally damaging manufacturing.

First among the top ten in employment share growth in Minneapolis, like other central cities, was the amusement and recreation industry. This industry includes theatrical productions that cater to local and export intermediate product markets as well as commercial sports. The communications and transportation industries were part of the top ten group. The communications industry includes radio and television broadcasting. Again, a majority of the largest increases in employment share were in the producer services sector that is an increasingly important part of the local economic base. In past years, however, Minneapolis served as a trade center for a dominantly agricultural region. The export markets of downtown and neighborhood businesses were delineated by their local trade and service areas. For residentiary activities, of course, the local trade area and the characteristics of their residents are important to local businesses. But for the new economic base, markets outside the local retail trade and service areas are most critical to the city's economic well-being.

Access of the central city base economy to U.S. and global markets is facilitated by its designation as one of the U.S. airline nodes. The airline nodes have similarity of business distributive functions. They also have a multiplicity of modes of transportation—air, rail, highway, and even water, as well as satellite communications. Each airline node also serves as a telecommunications node—an increasingly important function and regional asset in the new information-based global economy. Thus, the central city's airline and telecommunication systems make possible speedy access to distant suppliers and markets. This is critically important to the expansion of producer services and technology-intensive manufacturing in the central city and its economic region.

Restructuring of the reference city base economy in the 1980s paralleled the growth of its airline node as an important business distributive center in the air transportation network. Important recent changes in the economies of Europe, Russia's Siberian frontier, and the Pacific Rim nations, and the airline and telecommunication systems serving these economies, have already affected the flow of exports and imports out of and into the city's economic region. Related present and future local changes include the following:

1. Expansion of trade with the European Economic Community as well as Eastern Europe and Russia and the Pacific Rim countries via the polar routes.
2. Growing dependence of the local business community on improvements in air transportation for both passengers and high value-added products that are critical in maintaining the competitive advantage of local businesses in world markets.
3. Increasing involvement (and importance) of state and local governments in providing the physical facilities and services that enhance the local economic environment for successful business enterprise.

Thus, the strategic location of an airline node in the global air transportation networks provides an initial competitive advantage for sustaining its long-term growth. Much depends, of course, on the resolution of existing and emerging difficulties related to the structural transformation of the national and global economies.

Tracking Downtown District Performance

Urban regional economics, with its focus on the spatial dimensions of regional activity, provides a special understanding of the linkage between regional structure and regional performance. For example, measures of personal and business well-being and the realities they represent vary from one part of a region to another. More likely than not, they are high in the metropolitan core areas, low in the rural periphery. They correlate closely with levels of investment per worker. Where perceived profits are high, with little risk of failure, investment and production follow, along with jobs and people. Such activity quite likely involves a location decision.

Location Economies of Downtown Districts

Theorizing about the location of production and investment has a long history in scholarly literature. These theories have been enriched greatly by the gradual inclusion of all economic units—agricultural and nonagricultural firms, rural and urban households, and public facilities within their reach. The several location paradigms discussed in Chapter 2 are now widely accepted and applied in urban regional economics.

Expanding High-Order Service Clusters

Among the recurring policy themes that relate, in varying degree, to urban regional economics are those focusing on the idea of efficient markets and entrepreneurial governments. They also happen to generate catchy slogans, such as "let the markets do it" and "let's reinvent government." The efficient market theme, however, tends to minimize the importance of knowing about unique regional structures. The markets will take care of that. The reinventing-government theme expressed by Osborne and Gaebler (1993) largely ignores the spatial factor in the organization and delivery of government services in its enunciation of the ten principles of entrepreneurial governance. The authors fail to even acknowledge the reality of location within a regional settlement system and its unique and diverse implications for the successful performance of both markets and governments.

The studies by Hutton, Ley (Hutton and Ley, 1987; Ley and Hutton, 1987), and Scott (1983; 1988) and those reported in *Services and Metropolitan Development: International Perspectives* (Daniels, 1991) document the critical contribution of producer services industries in the downtown district to the growth and prosperity of the entire city region. The export-producing industries in both the core metropolitan area and its rural periphery are closely linked to the producer services concentrated in the downtown district. In turn, the producer services sector depends initially on the local and regional export-producing industries for its primary market outlets. As the producer service businesses expand in sales and customers, they acquire market outlets in other metropolitan regions. Readily available airline and telecommunication services provide quick access to the more distant market outlets.

An apparent anomaly in survey data for downtown Minneapolis is the fact that the fastest-growing industries are part of the base economy. The residentiary sector has lagged in both job and payroll growth in the face of overall population decline. The apparent anomaly has its origins in the growth of a commuting workforce, largely downtown. Net commuters into Minneapolis constituted about 41 percent of the net resident workforce in 1980 and 37 to 38 percent in 1987. Thus, as the number of Minneapolis downtown jobs increased, a significant share of the downtown workforce was made up of commuters.

1. Slightly less than half of the wage and salary workers in Minneapolis are located in the neighborhoods.
2. Payroll per job in the neighborhoods is lagging behind downtown payroll growth.

Even then, by the year 2000, if present trends continue, payroll per job in the neighborhoods will decline to slightly more than 80 percent of its average level in the city.

The higher economic rents earned by downtown property owners are shared in part, as discussed earlier, by the downtown workforce, which is more highly paid than a similar workforce in the suburbs. Much of the downtown workforce resides in the suburbs, where most of the labor earnings are spent.

Spatial Separation of Workforce and Service Facilities

The spatial-economic separation of the workforce of a high-order service facility is illustrated by the findings from a study of the commuting patterns of employees from two Group Health (an important local health care provider) facilities (Minneapolis Community Development Agency, 1991). Employees earning less than $25,000 show place-of-residence clustering around the Group Health location. Zip-code areas farther from Group Health locations have higher concentrations of employees along travel corridors, indicating a possible preference for public transportation. In contrast, employees earning $25,000 to $36,000 and more are more widely dispersed. The residences of these employees are spread over much of the metropolitan area. The study concludes by noting the following:

1. The entry-level employees tend to be clustered in fewer areas than higher-income employees.
2. The entry-level employees tend to live close to travel corridors.
3. The entry-level employees tend to live closer to the headquarters than the other, higher-income employees, who are dispersed over much of the metropolitan area.

The City of Minneapolis study, like other transportation studies relating income levels to commuting patterns within urban regions, shows that lower-income employees tend to reside closer to their place of employment, except where mass transit is available. The findings are not surprising given that the greater the commuting distance the greater the cost, but they still have important implications for neighborhood development. An industrial site has certain negative externalities attached to it: a degree of effluence from energy use, increased heavy truck traffic, less attractive architecture. The positive externalities include job creation for local residents as well as additional property tax base. Generally, only a small proportion of nearby residents find employment in the new nearby work sites. Everything else being equal, a suburban municipality has a willingness to impose higher property taxes and exclude industrial development to the extent that its residents are capable of commuting to other locations for employment. The inner-city and first-ring suburban residents frequently rely on public transportation. On the other hand, inner-city density means that industrial development frequently implies either displacement of existing housing or close proximity to housing or both. Suburban sites are less likely to face displacement and extreme negative "neighborhood effects." Transportation infrastructure tends to support centralization, with systems converging on a central location. This facilitates an inward commuting pattern.

A tragic and ultimately costly dilemma faces City Hall and the Downtown District. Workers from inner-city neighborhoods, who are dependent on public transportation, often are unable to travel to suburban locations for employment. A deficit of affordable housing keeps them in the inner cities and first-ring suburbs. Yet the shortage of affordable housing also creates resistance to neighborhood industrial development. While this separation of place of work from place of residence and the affordable housing dilemma hold important implications for neighborhood development, it also shows a pattern that is repeated many times in the location of other facilities on downtown and near-downtown sites. The infrastructure supporting the workplace of a nonresident commuter thus incurs high costs for the central city.

Downtown property tax revenues quickly reach upper limits established by threats of individual businesses to move to the suburbs with lower economic rents and lower property tax assessments. The city's property tax base may not have the capacity to adequately maintain the essential downtown infrastructure and also meet the increasing demands for neighborhood services. Thus, the benefits of the downtown economy accrue to the entire metropolitan area and region, while many of the added social costs of economic dislocation in the central city neighborhoods accrue to the city's taxpayers.

Managing an Expanding Cultural and Entertainment Complex

One approach to managing an expanding cultural and entertainment complex in downtown districts starts with the preservation of its historic resources. Another approach lies in the business community's affinity for professional sports and involves building new and costly facilities as an essential part of a downtown entertainment complex. In between the two approaches are many others that focus on matters of art and culture, but we overlook these to focus on the two extremes of preserving historic resources and subsidizing professional sports.

Preserving Historic Resources

Bringing together architecture and economics for the preservation of historic resources has double merit. Architecture provides an appreciation of historic structures and their cultural environments. Economics explores the values associated with the historic structures and environments and ways in which these values can be transferred from one use or user to another. Examples include the following:

1. Historic resources preservation in the home countries of many of our distant families and relatives.
2. Role of economics in historic resources preservation in government, as in land use control, and through the creation and use of local markets for the transfer of preservation values.
3. Associations of individual and groups for the purpose of promoting the preservation of historic buildings, sites, and activities.

Use of government in historic resources preservation presents new concerns since government represents a form of "authority"—a word that now has a pejorative meaning for many people. Alan Ehrenhalt, executive editor of *Governing* magazine, in his book, *The Lost City* (1995), says this about authority: "We want to keep the safe streets, the friendly grocers, and the milk and cookies, while blotting out the political bosses, the tyrannical headmasters, the inflexible rules, and the lectures on 100 percent Americanism and the sinfulness of dissent. But there is no easy way to have an orderly world without somebody making the rules by which order is preserved. Every dream we have about recreating community in the absence of authority will turn out to be a pipe dream in the end" (p. 2).

Government, with its particular authority, serves the political processes of our society. Political processes redistribute economic values by taxation, regulation, and condemnation, using the proceeds for income payments to other governments and individuals and for the building and maintaining of our common infrastructure. Foremost among local regulatory means of asserting authority affecting historic preservation values are the land use controls of local government, like building permits. Zoning, which ranges all the way from nuisance and fiscal zoning to design zoning, is also important. For example, suppose a city changed its zoning policy by preserving one part of the city for low-density use and another part for high-density use. Given a system of transferable development rights, the city establishes a preservation zone and development zone, with landowners in the preservation zone having the option to sell their development rights to landowners in the development zone.

Local governments derive their powers to control land use from the states. Zoning is an exercise of the police power of local government if it promotes the public health, safety, and welfare. Court decisions established three criteria for determining the constitutionality of zoning, namely, substantive due process, equal protection, and just compensation.

Local markets for historic resources preservation may provide for the following:

1. Transfer of development rights from low-density to high-density development areas.
2. Visitors to historic buildings, sites, and activities. These range from local tour guides to airports and airlines and related informational services.
3. Management and educational services catering to visitors and resident businesses and households.

Finally, networking educational and research services includes such diverse activities as the following:

1. Demonstrating the contribution of historic resources preservation to a region's quality of life—examples and applications (such as relation of quality of life indicators to residential and office location preferences and site values, including contribution to local government tax base).
2. Forming partnerships with public and private entities with common interests in historic resources preservation.
3. Sponsoring a study of the extended value of historic resources preservation and its benefits and beneficiaries.

Subsidizing Professional Sports

Consumer spending on all entertainment and recreation activities totals about $300 billion in the United States and about $4 billion in the Twin Cities area. Admissions to sports events accounts for about $5 billion of the total for the

TABLE 8-1 Personal consumption expenditures, by type of expenditure, U.S. and Twin Cities, 1990–91

	U.S.		Mpls.-St. Paul MSA	
Type	Total (bil.$)	Relative (pct.)	U.S. Proport'l (mil.$)	BLS Survey (mil.$)
Personal consumption expenditures, total	3,730.0	100	48,612	48,612
Entertainment and recreation	285.7	7.7	3,625	3,949
Books and maps	17.6	0.5	223	294
Magazines, newspapers, and sheet music	23.6	0.6	299	394
Nondurable toys and sporting goods	32.1	0.9	407	359
Wheel goods, durable toys, and sports equipment	31.3	0.8	397	412
Radio, television receivers, records, and musical instruments	50.4	1.4	640	559
Radio and television repair	3.7	0.1	47	49
Flowers, seeds, and potted plants	10.3	0.3	131	136
Admissions to specified spectator amusements	14.0	0.4	179	211
Motion pictures	4.7	0.1	60	70
Legitimate theater	4.5	0.1	57	67
Sports events	4.9	0.1	62	73
Clubs and fraternal organizations	8.4	0.2	107	126
Commercial participant amusements	23.1	0.6	293	346
Pari-mutuel net receipts	3.7	0.1	47	55
Other recreation services	67.4	1.8	855	1,009

Source: U. S. Department of Commerce, Survey of Current Business, Consumer Expenditures Survey, 1990–91.

United States and $80 million for the Twin Cities. Table 8-1 shows two measures of personal consumption expenditures for 1990 and the average for 1990–91. The first is proportional to personal spending in the entire United States for all items. The second is proportional to the spending on individual items based on the Bureau of Labor Statistics (BLS) annual survey of personal consumption expenditures for the Twin Cities MSA. The two sets of estimates compare closely for most items. Admissions to sports events account for 0.1 percent of total personal spending in the United States. This percentage is about 20 percent higher when based on the BLS survey of consumer expenditures for 1990–91.

The Minneapolis-St. Paul MSA accounts for 1.4 percent of the U.S. consumer market. It is the third largest metro area consumer market in the Midwest, exceeded only by Chicago and Detroit. The Cleveland consumer market is slightly smaller than the Twin Cities in total spending, but slightly larger in total consumer units.

Table 8-2 shows the proportion of annual personal spending of consumer units (households) in the United States accounted for by consumer units in the Midwest. Specifically, these data are associated with the eight largest MSAs (which are slightly smaller than the local commuting areas), averaged over the 1990–91 two-year period. The Minneapolis-St. Paul MSA has 1.1 percent of the consumer units. They accounted for 1.3 and 1.4 percent, respectively, of the U.S. total entertainment fees and admissions and total personal spending in the 1990–91 period. Fees and admissions to sports events accounted for about one-third of the total entertainment fees and admissions.

Table 8-3 provides a breakdown of the 1990 input purchases of the professional sports industry in the seven air node and commercial centers. The gross product, for example, represents the contribution to gross domestic product of the professional sports industry. Employee compensation and proprietary income represent the income payments to employees and self-employed workers. Indirect taxes relate to the product-based taxes paid by the providers. They exclude all income, direct taxes, and sales taxes paid by the consumer. The total employment estimate includes both wage and salary and self-employed workers.

All industry sales in the Twin Cities (in this case, the sixteen-county commuting area rather than the entire economic region extending over all or parts of six states) total about $120 billion. Professional sports account for about $110 million of this total. The entire Twin Cities economic region, which extends over five states and has a population of nearly eight million and total industry sales of nearly $300 billion, supports a professional sports industry with about $120 million in total sales. For this region, therefore, like a majority of the 29 U.S. economic regions, professional sports activities concentrate in the larger metropolitan areas. Consumer spending on professional sports in the extended Twin Cities region market accounts for less than $50 million (about 35 percent) of the total industry sales. Most of the remaining sales are to other industries and businesses—local, regional, national, and international.

Leslie Wayne (*New York Times,* July 17, 1996, "Taxpayers Build Stadiums; Owners Cash In") documents the rising public cost of stadiums—about $3.2 billion for the first five years of the 1990s—$100 million more than the cost of libraries and museums. As for job creation, Baltimore's new football stadium cost $127,000 to $331,000 per job, depending on the source (the governor or the legislature) of the estimate. The actual cost per job for Maryland's Sunny Day Development Fund is $6,250.

The real beneficiaries of a new stadium are, of course, the fans, media, and business services (such as advertising), as well as the owners and players, who are among the most highly paid entertainers in the world. Professional baseball

TABLE 8-2 Total personal consumption expenditures and money income of consumer units in selected metropolitan markets as percent of U.S. total, Midwest Region, 1990–91

			Selected Metropolitan Statistical Areas							
Item	All CUs (pct.)	MSA Total (pct.)	Chicago (pct.)	Detroit (pct.)	Mnpls-St. Paul (pct.)	Cleve-land (pct.)	St. Louis (pct.)	Cincin-nati (pct.)	Kansas City (pct.)	Mil-waukee (pct.)
Total annual expenditures	23.3	11.0	3.6	1.9	1.4	1.1	1.0	0.7	0.7	0.6
Entertainment	24.5	11.6	3.6	2.1	1.3	1.1	1.4	0.8	0.7	0.7
Fees and admissions	23.3	12.3	4.3	1.8	1.5	1.1	1.2	0.8	0.8	0.7
Television, radios, and sound equipment	24.8	11.4	3.8	2.1	1.1	1.2	1.0	1.0	0.7	0.5
Pets, toys, and playground equipment	24.9	11.5	4.3	1.6	1.1	1.2	1.1	0.9	0.7	0.7
Other supplies, equipment, and services	24.6	11.2	2.0	2.8	1.3	0.7	2.6	0.5	0.5	0.8
Reading	25.6	11.8	3.9	1.9	1.7	1.1	1.1	0.9	0.7	0.6
Money income before taxes	23.2	11.6	3.8	2.1	1.3	1.1	1.2	0.8	0.8	0.6
Less personal taxes	19.9	9.1	3.6	1.2	1.7	0.7	1.0	0.4	0.3	0.3
Income after taxes	23.5	11.9	3.8	2.1	1.3	1.2	1.2	0.8	0.9	0.6
Number of consumer units	25.2	10.6	3.2	1.9	1.1	1.2	1.1	0.8	0.7	0.6

Source: U.S. Bureau of Labor Statistics, Consumer Expenditures Survey, 1990–91.

TABLE 8-3 Economic indicators for professional sports clusters and all sectors, U.S. and selected air node regions, 1990

Sector and Area Air Node Region	Units	Total 29 regions	Chicago	Detroit	Mpls.-St. Paul	St. Louis	Denver	Kansas City	Cincinnati
Core Labor Market									
Total population	(thou.)	100,886	7,411	5,072	2,581	2,233	1,876	1,680	1,506
Total sales									
Professional sports	(mil.$)	3,037.3	191.8	186.0	109.4	143.7	72.0	121.4	87.5
All sectors	(bil.$)	4,431.3	336.6	242.2	115.3	192.4	86.7	66.4	65.1
Entire Air Node Region									
Total population	(thou.)	247,123	16,337	9,865	7,792	4,751	6,483	7,566	6,747
Total sales									
Professional sports	(mil.$)	4,599	329.4	206.4	117.9	66.2	95.8	141.2	179.7
All sectors	(bil.$)	9,576	690.1	421.6	287.8	181.5	246.4	286.0	258.2
Personal spending									
Professional sports	(mil.$)	1,116.3	100.2	60.4	43.1	22.8	33.1	43.4	28.6
All sectors	(bil.$)	2,643.9	196.9	96.8	84.1	52.5	65.8	85.5	66.0
Gross product									
Professional sports	(mil.$)	2,772	227.6	114.5	74.9	36.9	65.7	79.1	108.1
All sectors	(bil.$)	5,419	377.0	215.2	151.1	95.4	139.1	154.5	137.7
Employee compensation and proprietary income									
Professional sports	(mil.$)	3,491	273.2	146.4	93.5	46.7	85.8	102.4	145.2
All sectors	(bil.$)	3,628	256.8	144.0	105.8	64.2	88.9	101.3	91.7
Indirect taxes									
Professional sports	(mil.$)	290	32.9	10.3	8.6	4.4	3.9	7.7	4.7
All sectors	(bil.$)	436	27.2	15.2	12.0	7.7	13.3	12.8	11.2
Employment									
Professional sports	(no.)	75,973	4,448	3,738	1,687	1,100	1,128	2,416	2,847
All sectors	(thou.)	136,161	9,285	5,055	4,542	2,623	3,764	4,424	3,755

Source: Regional Markets, Inc.; U.S. Bureau of Labor Statistics, Consumer Expenditures Survey, 1990–91.

is part of the first round of consumer spending on all entertainment and recreation activities in the United States, which reached $300 billion in the United States and about $4 billion in the Twin Cities metropolitan area in the 1990s. Most of the consumer spending on professional sports, of course, is in the Twin Cities metropolitan areas. The Minnesota Twins' share of the total consumer spending is about $15 million.

Many sectors of a region's economy bear the benefits and burdens of professional sports. First, and perhaps foremost, competitively structured local metropolitan markets for professional sports ensure that the direct participants, both the consumers and the providers, receive benefits commensurate with the transaction values. Local businesses and households also bear at least some of the locally provided sports benefits and burdens, either directly or indirectly. Finally, state and local governments are important players in their own ways—by collecting the additional taxes and building the additional infrastructure accruing from the sports activities.

Benefits include the income receipts of sports-related resource owners—team players and owners, directly and indirectly sports-impacted businesses, real estate property owners, and local and state governments. The value added by the professional sports activity is a measure of the net local benefits it generates.

The public sector benefits depend, in part, on the actual location and site of the professional sports facility. The highest concentration of economic activity is in the Twin Cities downtown districts. Land values in the Minneapolis downtown district in the January 1995 assessment, for example, ranged from a high of $160 per square foot to $65 and less adjacent to the Target Center and $30 and less adjacent to the Metrodome. Moreover, these values respond to occupancy and vacancy rates in the downtown district which, in turn, respond to the demand for and supply of downtown commercial and industrial space.

Table 8-4 shows the estimated public sector benefits for one of the four major professional sports activities, namely, baseball. Tables showing public sector benefits from football, basketball, and hockey were prepared similarly. These estimates come directly from the final report of the Advisory Task Force on Professional Sports in Minnesota. They show total public sector revenues of $6.1 million. The State of Minnesota would collect about $6.9 million of the total, while the City of Minneapolis would collect $1.2 million. This assumes the use of the wide variety of tax instruments already in use for this purpose from an activity already in existence. The net increase in benefits was not calculated, however.

The tracking of inputs and outputs includes income payments to the direct participants in the professional sports enterprise. The flow of local income payments provides measures of the economic contributions of these participants to

TABLE 8-4 Estimated public sector benefits, Minnesota Twins

Revenue Source	Item	Public Rev .($)
State of Minnesota Income Tax		
Personal income tax		
	Team salaries	2,050,000
	All other related personnel	300,000
Total, income taxes		2,350,000
General State of Minnesota Sales Tax		
State tax collected on tickets		1,216,215
Taxable expenditures by team		97,500
Spectator expenditures		
	Lodging—weekend/weekday	156,471
	Restaurants—weekend/weekday	749,044
State liquor taxes		
		332,205
Visiting team expenditures		
	Lodging	14,452
	Per diem	9,477
Total, sales taxes		2,575,364
Total, State Taxes		4,925,364
General City of Minneapolis Sales Tax		
Spectator expenditures		
	Lodging—weekend/weekday	12,036
	Restaurants—weekend/weekday	57,619
Visiting team expenditures		
	Lodging	1,112
	Per diem	729
Liquor		
	City of Minneapolis	18,456
Total, sales taxes		89,952
Other City of Minneapolis Sales Tax		
	Restaurant	345,713
	Entertainment	78,888
	Hotel	52,592
	City parking revenues	510,300
	Liquor tax	110,735
	Total, other taxes and revenues	1,098,227
Total, City of Minneapolis Taxes and Other Revenues		1,188,178
Total, Public Sector Revenues		6,113,542

Source: Advisory Task Force on Professional Sports Final Report, January 1996.

TABLE 8-5 Calculations and key assumptions for estimating public sector benefits, Minnesota Twins

Item	Rate	Total Revenue ($)	Total Tax ($)
Salaries:			
Team players	0.082	25,000,000	2,050,000
Other personnel	0.05	6,000,000	300,000
Spending:			
(a) Tickets (81 x 21,000 x $11)	0.065	18,711,000	1,216,215
(b) Home team spending	0.065	1,500,000	97,500
(c) Spectator lodging, total		2,407,240	156,471
Weekday (55 x 21,000 x .0156 X 95/2.2)	0.065	778,050	50,573
Weekend (26 x 21,000 x .0691 X 95/2.2)	0.065	1,629,190	105,897
(d) Restaurant, total		11,523,750	749,044
Weekday (55 x 21,000 x $5.25)	0.065	6,063,750	394,144
Weekend (26 x 21,000 x $10)	0.065	5,460,000	354,900
(e) Liquor, total (81 X 21,000 X $2.17)	0.09	3,691,170	332,205
(f) Visiting team lodging (81 x 30 x 1.5 x $61)	0.065	222,345	14,452
(g) Visiting team per diem (81 x 30 x $60)	0.065	145,800	9,477
Total		69,201,305	4,925,364
Assuming additional Minneapolis sales tax of 0.5 percent:			
(a) through (g), except (e)	0.07	15,799,135	1,105,939
(e) (81 x 21,000 x $2.17)	0.095	3,691,170	350,661
Additional .5 percent Minneapolis tax option, total		19,490,305	1,456,601
Minneapolis 3 percent restaurant tax (d)	0.03	11,523,750	345,713
Minneapolis 3 percent entertainment tax (b) + (f)	0.03	2,629,585	78,888
Minneapolis 2 percent hotel tax (b) + (f)	0.02	2,629,585	52,592
Minneapolis $0.30 parking fee (81 x 21,000)	0.3	1,701,000	510,300
Minneapolis 3 percent liquor tax	0.03	3,691,170	110,735
Total, Minneapolis additional		22,175,089	1,098,227

Source: Advisory Task Force on Professional Sports Final Report, January 1996.

local and regional economies. The commercial sports enterprise cluster provides entertainment value to those buying its product—the sports entertainment package. This product must compete with providers of other products purchased by the same parties. Sports entertainment spending includes spending by those attending the sports event, those viewing the broadcast and rebroadcast of all or part of a sports event, those reading the reports of a sports event, sports entertainment revenues, the opposing teams, the sports facility and tenants, and the media. State and local government tax receipts include direct taxes from owners and players and indirect taxes from businesses providing goods and services purchased by games spectators, players, and owners. These do not include the direct taxes or the incremental changes in property taxes from businesses benefiting from the sports-related local purchases.

Table 8-5 provides some closure to the earlier discussion by showing the values that were assumed, much as in the Galloway and Anderson study cited earlier,[1] in the calculation of public sector benefits. Thus, the counting of existing attendance as if it were new attendance is exaggerated even more by the use of assumed values relating to revenue receipts of state and local governments attributed to the revenue-generating activities in a new baseball stadium.

Building a new stadium for the Minnesota Twins, perhaps doubling average daily attendance, would mean large transfers from one entertainment activity to another—all, except for the net new dollars flowing into the state, without any measurable overall increase in state and local revenues. For much of the business community and many ardent baseball fans, however, the Minnesota Twins produce business and entertainment values that exceed their monetary contributions to the enterprise and its products.

Major league baseball is strictly private enterprise. You see them play, you pay—a rule that any team can enforce at the gate. Public financing of a Twins stadium, on the other hand, would disburse the costs among taxpayers throughout the state. The benefits would still belong to the small group of supporting businesses and ardent baseball fans (Table 8-6).

By tracking the flows of private purchases and private income payments, we figure the economic contributions of the professional sports teams to their local and regional economies. Even for the metropolitan area, these contributions are relatively small. We discover this from an economic impact assessment of the Twins enterprise cluster, with and without a new stadium. Such an assessment shows the new dollars coming into Minnesota because of the Twins, the spending of these dollars, and the subsequent income receipts and spending of households and businesses affected by the initial spending. Including, also, the dollars originating within Minnesota would show gross changes, not net changes that exclude transfers from the activities adversely affected by the increases in Twins-related spending.

Twins Metrodome attendance is 1.4 million, or an average of 17.3 thousand per game (given a total of 81 home games). The assumed (not a forecast based on a credible analytical model and data) number for a new ballpark is 2.5 million, or 33 thousand per game—a 79 percent increase. This compares with an average 61 percent increase (ranging from 22 percent for Arlington/Texas to 188 percent for Jacobs Field/Cleveland) for the four new ballparks cited in the Andersen Consulting report. These estimates assume a distribution of attendance by origin that includes both Greater Minnesota and out-of-state origins. Six percent of the fans were from out of state over the 1990–94 period, according to the Final Report of the Advisory Task Force on Professional Sports.

TABLE 8-6 Minnesota Twins attendance and revenue estimates and projected attendance and revenues

Item	Distribution			Total				Per Game				
	Metro Dome		New Ball Park (pct.)	Metro Dome		New Ball Park (mil.)	Increase (mil.)	Metro Dome		New Ball Park (thou.)	Increase (thou.)	Projected Attendance Growth (pct.)
	Using ATF% (pct.)	Arthur Andersen (pct.)		Using ATF% (mil.)	Arthur Andersen (mil.)			Using ATF% (thou.)	Arthur Andersen (thou.)			
Metro area	77	75	70	1.078	1.050	1.750	0.700	13.3	13.0	21.6	8.6	67
Greater Minnesota	17	16	19	0.238	0.230	0.480	0.250	2.9	2.8	5.9	3.1	109
Out of state	6	9	11	0.084	0.120	0.270	0.150	1.0	1.5	3.3	1.9	125
Total Revenues	100	100	100	1.400	1.400	2.500	1.100	17.3	17.3	30.9	13.6	79

Item	Distribution			Total				Per Game				
	Base Period		New Ball Park (pct.)	Base Period		New Ball Park (mil. $)	Increase (mil. $)	Base Period		New Ball Park (thou. $)	Increase (thou. $)	Projected Revenue Growth (pct.)
	MLB Average (pct.)	Metro Dome (pct.)		MLB Average (mil. $)	Metro Dome (mil. $)			MLB Average (thou. $)	Metro Dome (thou. $)			
Gate	37.5	31.6	13.4	23.8	15.4	26.7	11.3	294	190	330	140	74
Stadium	14.2	15.6	15.6	9.0	7.6	13.2	5.6	111	94	163	69	74
Media	43.7	47.9	47.9	27.7	23.4	40.6	17.2	342	288	501	212	74
Other	4.6	5.0	5.0	2.9	2.4	4.2	1.8	36	30	52	22	74
Total	100.0	100.0	100.0	63.4	48.8	84.7	35.9	783	602	1,046	443	74

Source: Minnesota Sports Stadium Advisory Task Force on Professional Sports, Final Report, 1999.

Note: ATF% refers to attendance at 81 home games, 1990–94; MLB Average refers to average team attendance in major league baseball, 1990–94, exclusive of ticket sales (assumed $11 per ticket average for Minnesota Twins).

The report estimates base period Minnesota Twins revenues at $48.8 million. Like the Final Report cited earlier, we estimate an average of $11 per paid attendee, or a season total from game attendance of $15.4 million. The difference between $15.4 million and $48.8 million (that is, $33.4 million) we distribute to the revenue sources according to the major league baseball estimates. We then apply the 74 percent growth rate used in the report to each of the four revenue sources for the future distribution of total revenues. All the in-state values, especially construction, are simply transfers from spending that would occur otherwise. The new dollars refer to the additional dollars, which, if proportional to attendance, would be 6 percent of the increase, or slightly more than $2 million, rather than $35.9 million. Applying a multiplier to account for the indirect effects would still leave a net economic impact below $5 million, with additional state tax revenues being much less.

Simply using the numbers in the Andersen Consulting Report, the increases in new state tax revenue would peak at less than $1 million. This, by the way, is the most optimistic rate of return we can figure—less than one-half of 1 percent on an investment of $200 to $300 million, excluding infrastructure and using the inflated growth rates. The new jobs are fewer than twenty. These findings clearly show that reasons other than economic development impact, such as "every first-class city must have a first-class professional sports team," drive the call for sports stadium subsidies. Meanwhile, professional team owners band together to vigorously and successfully defend their professional sports monopolies.

Calculation of the benefits and costs of a professional sports stadium would be a relatively straightforward exercise were it not for its politics. Because of this, unbiased and technically sound economic analysis is sometimes unavailable. In its place, the daily newspaper reader and the voting public are confronted by the assurances of team owners, media lobbyists, and newspaper editors—all with large vested interests in professional sports—that a $500 million baseball stadium, for example, is "worth the money." Mark Rosentraub warns us in *Major League Losers,* that a "welfare system exists in this country that transfers hundreds of millions of dollars from taxpayers to wealthy investors and their extraordinarily well-paid employees. This welfare system exists—indeed it thrives and continues to grow—because state and local government leaders, dazzled by promises of economic growth from sports, and captivated by a mythology of the importance of professional sports, have failed to do their homework" (p. 3). We start doing our homework with the big picture of professional sports as one of several areas of consumer spending for recreation and entertainment.

Notes

1. Lowell Galloway and Gary Anderson, visiting scholars at the congressional Joint Economic Committee, also note in their recently completed study, Derailing the Small Business Express (1992), that "an increase in costs amounting to 1 percent of sales for a gas station would necessitate, on average, a 1.4 percent decline in wages to avoid layoffs."

2. The Minneapolis Community Development Agency sponsored a mail survey of businesses located in the downtown area of Minneapolis. For the purposes of the study the downtown was defined as the area encircled by the "freeway ring" of Interstate Highway 35 on the east and south, Interstate Highway 94 on the west, and the Mississippi River on the north.

3. The anonymous questionnaire was sent to all nongovernmental firms within the downtown, with no a priori identification of actual headquarters firms. To determine which firms were carrying out corporate headquarters or head-office types of activities, respondents were sorted on the basis of responses. A proxy for head-offices was developed. The first question involved a series of possible strategic decisions, up to a maximum of eight separate activities such as ancillary services decisions on location of the firm and service. The second question asked about actual activities carried out at each downtown location, including production, distribution, and sales. For purposes of identifying headquarters-type operations, those firms that reported that they were carrying out no sales or production at that location and carried out all of the possible strategic decisions were selected. Out of 1,237 respondents to both sets of questions, only 94 firms, or 7.6 percent of respondents to these questions, performed all of the strategic decisions. A full 61 percent of these firms fell into the producer services group, while 29 percent were in the ancillary group, and 10 percent were in the extractive, transformative, and distributive category.

4. The responding firms were categorized with respect to number of employees following the U.S. Department of Commerce County Business Patterns size delineation based on the number of employees at that location. The firms tended to be small, with 59 percent having less than ten employees. The headquarters firms tended to be somewhat older, with 7 percent reporting an age of less than five years and 12 percent of the nonheadquarters firms less than five years. The headquarters firms were mostly privately held corporations (52 percent of respondents); the next most common were partnerships at 13 percent. The nonheadquarters firms were only 44 percent private corporations, with sole proprietorships and subchapter S being the next most common at 16 percent each.

Urban Neighborhoods and Communities

This chapter is an extension of the previous chapter in its focus on the urban community of metropolitan rather than nonmetropolitan areas. The division between metropolitan and nonmetropolitan is somewhat arbitrary. The distinction makes sense, however, in the administration of periodic population censuses as well as various federal government programs and policies. Again, the importance of proximity—place to place and people to people—and the use of it are central concerns in the ensuing discussions. In this chapter, however, all counties in a metropolitan labor market area (LMA) belong to an urban community, and because of proximity to their central city, residents of these counties have easy access to metropolitan area services, new firm foundings, and employment opportunities.

What Is the Urban Community?

Don Eberly (1994, pp. xxvi–xxvii) writes that "the key to community is understanding its voluntary nature. Community cannot be ordered through legislation or summoned forth by verbal admonition—it springs from habits of people's hearts." Amitai Etzioni (1995, p. 25) asks, Which community? What are its boundaries or scope? "It is best to think about communities as nested, each within a more encompassing one. Thus, neighborhoods are parts of more comprehensive suburbs or cities or regional communities. These, in turn, often intersect or are part of larger ethnic or racial or professional communities. And many communities are contextuated by the national society. Ultimately some would aspire to a community which would encapsulate all people." Robert Booth Fowler (Etzioni, 1995, p. 88) argues that "community in American political thought at present engages three types of community: (1) communities of ideas: for example, the participatory democratic and republican models, (2) communities of crisis: for example, the earth community born of the environmental crisis, and (3) communities of memory: for example, religious and traditional ideas of community."

Place, Purpose, and Proximity

We start with an economic definition of an urban neighborhood and community. We define both in terms of place, purpose, and proximity. An urban economic community is the commuting area of a larger central place of metropolitan status. It is a "functional economic area," using Karl Fox's terminology (Fox, 1994). An urban neighborhood is a proximity-based association of residents. In years past, its residents and institutions were sustained with common values of family, school, and church that held to authority and community, sacrificing some choices for stability and order.

Cornelia Butler Flora (Goreham, 1997, p. 113) observes that "almost all definitions of community emphasize the informality and solidarity engendered by relationships and social organization. Communities of interest are composed of interactions among people linked to each other purposively by shared interest and actions. Communities of place are composed of the interactions of individuals who live in a particular locality." The report of the Commission on Global Governance (1995, pp. 43–44) speaks of the "global neighborhood" with its citizens cooperating for many purposes in matters requiring nation-states "to pool their efforts—in other words, calling for neighborhood action." The report goes on to say that "neighborhoods are defined by proximity. Geography, rather than communal ties or shared values, brings neighbors together. People may dislike their neighbors, they may distrust or fear them, and they may even try to ignore or avoid them. But they cannot escape from the effects of sharing space with them." We accept parts of each of these definitions as they apply to urban neighborhoods and communities.

Quantitative Measures

We delineate urban neighborhoods and communities in many ways—by natural boundaries as well as artificially built barriers. We start with the urban community as the commuting area of the local workforce, employed, for the most part, in one or more urban places. It is also the local labor market area. Within the urban community are its proximity-based neighborhoods.

In Chapter 3, we listed the urban communities that serve also as core labor market areas for the proximity-based economic regions. The 29 core metropolitan statistical areas (MSAs, including primary metropolitan statistical areas—PMSAs) in the contiguous 48 states and the District of Columbia contain a total population of 82,678,000. The remaining MSAs account for a total population of 114,834,000. Thus, the two sets of metropolitan areas account for a total MSA population of 197,512,000, or 80 percent of the total population of 247,052,000 in the 29 economic regions.

Delineating the Urban Community

An active labor market area offers the greatest possible internalization of the journey to work and the daily activities of an active resident household. It is the commuting area of the local labor market. It delineates the urban community. The urban community has a downtown district, central city, neighborhoods, suburban business districts, and open country settlements. Residents outside an extended MSA (now defined as a county or multicounty area with a central place or contiguous places with a population less than 50,000) are in a rural or nonmetropolitan labor market area. The essential differences between nonmetropolitan and metropolitan in an economic sense are the diversity and scale of business and industrial activity. In delineating the urban community, we refer to the 221 MSA-centered LMAs, exclusive of the metropolitan core areas of the economic regions cited in Chapter 1.

Metropolitan Areas

A metropolitan LMA provides, among its many other contributions, a favorable environment for the emergence of self-reliant communities that depend upon one another for jobs and services. A majority of all LMAs in the United States have one or more MSA as a core area. In the Minneapolis-St. Paul Economic Region—one of the most rural in the United States, thirteen of the 27 LMAs are centered on an MSA. They are self-reliant in essential community services on an areawide basis and they share in a common labor market. Even the idea, if not the reality, of caring for the impoverished and dislocated households in their extended community is a part of each community's perceived responsibility. Interested local parties have the opportunity to readily exchange information, negotiate, and collaborate within this urban community (Saxenian, 1994, p. 167).

A top priority of local governments in the urban community is financing and building the essential *transportation and communications infrastructure* for the globally competitive businesses within its territorial jurisdiction.

Waste disposal and water supply are increasingly important responsibilities of entrepreneurial government.

The provision of *protective services* for guarding property and human life also taxes the capacities of entrepreneurial governments to manage their costs and effectiveness.

Emerging also is an alternative framework for the study of regional activity systems. This framework and its related set of paradigms start with the internationally competitive business enterprise that leads to the self-organizing industry cluster, entrepreneurial government, and finally, the self-reliant community. This framework puts a premium on a thorough understanding of the regional structures and processes that affect the formulation and implementation of public policy. This particular framework brings together four somewhat separate topical areas into an overall, holistic approach to the understanding and prediction of regional growth and change.

Among the less ubiquitous phenomena of our time is the *globally competitive business enterprise.* This is an export-producing enterprise, one that brings new dollars into the local area. Exports are any goods and services that produce income, originating from outside the area, for local residents and resource owners.

Self-organizing industry clusters are the essential elements of what AnnaLee Saxenian, in *Regional Advantage* (1994), labels a regional "industrial system." The industry clusters nurture close working relationships with one another and depend on the countless numbers of informal exchanges among their workers. These become, as noted by Eric von Hippel, in *The Sources of Innovation* (1988), important sources of product and process innovation for a business enterprise. The individual business clusters are an integral part of a larger complex of business enterprises that form a regional industrial system. This system is large and diverse enough to provide its workers with widely recognized opportunities for quickly moving from one job to another through an efficient and well-functioning local labor market.

Trade is the lifeblood of the self-reliant community. Domestic trade, that is, the trade-generated flow of income across county boundaries, is many times greater than all foreign trade. Even trade between the seven Twin Cities Metropolitan Area counties and the 80 Greater Minnesota counties exceeds Minnesota's total foreign trade. For most businesses, whether its shipments or purchases are domestic or foreign is a moot question since this information is not readily available. Moreover, Minnesota's domestic exports compete with foreign imports, as much as, if not more than, Minnesota's exports compete with another country's exports to a common destination.

The health care industry brings together several industry groups that form a regional industrial system—a *self-organizing industry cluster* focusing on health care. It is self-organizing through its vast network of industry and individual business and professional linkages in production, marketing, procurement, and personnel staffing. In this paradigm of a geography-bounded industry linkage network, customers and suppliers, as well as researchers, drive the product and process innovations that, in turn, help sustain the health care system's importance in a state's economic base.

An *entrepreneurial government* is one that considers its principal task the management of its space economy as a good place in which to do business and to live. Such a government strives in every conceivable way to equitably and effectively accomplish this task. It manages its corporate territorial entity to maximize the satisfactions of its residents.

Urban Neighborhoods

Entrepreneurial government depends on local education and training institutions to achieve exceptional improvements in service delivery and financing. For the central city, these institutions are the crux of its problems and potential. They are problems when they provide inferior services that drive residents to suburban living. They offer potential for improvement when they help educate and train the poor and the dislocated. These are the sorts of conclusions derived from various arguments or suggestions such as the following:

1. The greatest return on investment in education derives from the successful outcomes of *primary and secondary education* programs.
2. Among postsecondary education institutions, *community colleges and technical institutes* relate most closely to the local business community, especially its small- to medium-size businesses.
3. *Four-year colleges* provide the broad educational backgrounds for future managers and owners of successful export-producing businesses.
4. Entrepreneurial government has an important stake in nurturing and supporting its *professional colleges and research universities.*
5. Successful space management by an entrepreneurial government invariably leads to some sort of *economic development*—the deliberate intervention of a government agency in the processes of business growth and change.
6. Acquiring or producing practical *decision information* is an essential first step in economic development.
7. Economic development is often synonymous with various forms of *public financing* of existing and start-up businesses.
8. Acquiring *ownership and control* of a private business by an entrepreneurial government usually occurs as a measure of last resort.

Whether well-documented support exists for each argument or suggestion relates, in part, to our underlying perspectives in approaching these sorts of issues. For example, the effectiveness of state and local governments in dealing with the more complex issues of education and research is being challenged repeatedly, especially where government-sanctioned local monopolies exist for these services.

Optimizing Neighborhood and Community Change

Effective neighborhood planning, when in consort with city planning and various business and educational partnerships, addresses the issue of job creation as well as the delivery of essential services to neighborhoods. In addition, neighborhood planning may include monitoring existing local jobs and services and even advancing priorities for the city's involvement in neighborhood and community development. Residents of the individual neighborhoods may have other priorities, however, than those of the city's planning and development agencies. The experiences of Anacostia, near the nation's capitol (Smith, 1995) and other communities point to the importance of city and neighborhood partnerships in advancing common goals and viewing alternative futures for the neighborhoods. These efforts ultimately become an important part of the strategic planning functions of city government.

Local Economic Development Models

When local people and institutions nurture one another, building strong, stable, and secure places for residents of all ages as well as for businesses, large and small, we call it endogenous growth. The idea of endogenous growth is nothing new. It is central to the success of any self-sustaining economic area, whether the Minneapolis-St. Paul, the Boston, or even the newly rejuvenated Detroit economic area. External stimuli are important only when an area's export-producing businesses can respond successfully to market opportunities existing outside the local area.

A wide range of studies now finds that government subsidies to support ailing export-producing businesses seldom, if ever, succeed in their original purpose. The end results are no different for the recipients of publicly subsidized housing or welfare assistance. What is true of households and businesses is also true of neighborhoods and communities.

We have many examples of neighborhoods and communities seeking help from government, both state and federal, in dealing with local problems stemming from concentrations of poverty, crime, poor schools, and high unemployment. History is replete with them. We have only a few examples, however, of neighborhoods and communities coping with these problems by building from within. Pulitzer Prize–winning journalist Hedrick Smith, for example, profiled the work of community activists in "Across the River," a television documentary about once-dying neighborhoods in the Anacostia area of Washington, D.C.

The Anacostia model for building livable neighborhoods relates in its own way to the underlying values of St. Nick's parish on the southwest side of Chicago as portrayed by Alan Ehrenhalt in *The Lost City: Discovering the Virtues of Community in the Chicago of the 1950s* (1995). Ehrenhalt writes about three different communities—St. Nick's parish; Bronzeville, a community five miles farther east, where the bulk of Chicago's black people lived in the 1950s; and Elmhurst, fifteen miles from Chicago, where middle-class white families were moving "into houses and lives that seemed, even to them, too good to be true" (p. 194). Stable relationships, civil classrooms, and safe streets characterized life in each of the communities. This came at a price, namely, limits on individual choice.

St. Nick's parish was one square mile in area, a place of single-family bungalows with a population of 15,000, "where the walls of one's house did not constitute boundaries, where social life was conducted on the front stoop and in the alley, and where, even inside the house, four or five children in a three-bedroom home made privacy a rare commodity. Television was coming to such neighborhoods in the 1950s, but air-conditioning had not yet arrived, and summer evenings were one long community festival, involving just about everybody on the block, brought to an end only by darkness and the need to go to sleep" (p. 9). The author notes that people who lived in places like St. Nick's parish "went out of their way to make their lives more public still by joining clubs and organizations of every sort" (p. 30).

Ehrenhalt documents life in two other communities in the 1950s besides St. Nick's through personal interviews and many written accounts. He concludes by comparing their prevailing values with those of the generation born in the baby-boom years. Choice is a good thing in life, authority is inherently suspect, sin is not personal but social, and individual human beings are creatures of their society. The primacy of choice "has brought us a world of endless dissatisfaction, in which nothing we choose seems good enough to be permanent and we are unable to resist the pursuit of new selections—in work, in marriage, in front of the television set. This suspicion of authority has meant the erosion of the standards of conduct and civility, visible most clearly in schools where teachers who dare to discipline children receive a profane response" (Preface). The author further suggests that "the repudiation of sin has given us a collection of wrongdoers who insist that they are not responsible for their actions because they have been dealt bad cards in life. When we declare that there are no sinners, we are a step away from deciding that there is no such thing as right and wrong" (Preface).

The city of Curitiba, capital of the state of Paraná in southeastern Brazil, provides another example of building from within. This is a city of more than two million people. Once a center for processing agricultural products, its economic base now serves as a regional powerhouse of industry and commerce. Although experiencing rapid growth, it has less pollution and crime, and higher educational levels than are typical for its region. According to Rabinovitch and Leitman (1996), "Progressive city administrations turned Curitiba into a living laboratory for a style of urban development based on a preference for public transportation over the private automobile, working with the environment instead of against it, appropriate rather than high-technology solutions, and innovation with citizen participation in place of master planning" (p. 47).

Eastside Community Investments, Inc., is another example of a community development organization established to spark economic revitalization in the east-side neighborhoods of Indianapolis. Two development intermediaries are also presented—Community Builders, a large intermediary organization that grew from a small community development group renovating a cluster of blighted townhouses in Boston's South End, and Neighborhood Progress, Inc., a new-generation local intermediary servicing nineteen community groups in fourteen Cleveland neighborhoods (Pierce and Steinbach, 1989). We introduce, also, a failed community development effort, the $18 million "festival" marketplace, Portside, in Toledo. The forging of local partnerships and the forming of local institutions for gradually, but courageously, changing tough inner-city neighborhoods into livable communities requires "building from within."

Community Economic Base and Core Competencies

The *community economic base* is formed by the export-producing activities of local resident populations—businesses, governments, and households. These also include the activities that cater to visitors purchasing locally available goods and services that bring "new dollars" into the community. The total workforce of every self-reliant community has (1) a place-of-work or "job-count" component and (2) a place-of-residence or "person-count" component. The two overlap in part, the exception being the commuters—the place-of-residence worker commuting to another labor market area. The total number of commuters is minimal for a well-delineated labor market area.

Core competencies relate to a capacity to coordinate diverse production skills and integrate multiple streams of technologies, including involvement and deep commitment to working across organizational boundaries. As noted earlier in Chapter 1, identifying a region's core competencies is much like identifying the core competencies of a business. State and local governments that are successful in building affordable transportation and technical education systems that address the critical market and resource access requirements of their export-producing businesses have a "step up" on their competitors in building the regional infrastructure of a globally competitive regional industry cluster.

The metropolitan core area and its air transportation node provide its industry clusters and those linked to the core area with a global reach. These new engines of economic growth (Jacobs, 1984) are powerless, however, without the strong, productive business and governmental linkages that extend well beyond the metropolitan core area. In fact, manufacturing in rural areas was the fastest growing segment of the industry clusters of U.S. metropolitan core areas in the 1980s in large part because of these linkages (Reynolds and Maki, 1990a,b; 1991; Scott, 1988).

Producer services are most abundant and accessible in the metropolitan core area. They are available to export-producing businesses in rural localities, but from a greater physical distance. The rural plant provides one or more of the producer services from within its own premises. Both facilities maintain strong linkages backward in the production process to suppliers as well as forward to customers. Continuing information exchange between the manufacturing enterprise and its suppliers and customers is an important part of each firm's legacy system. Both suppliers and customers are important information sources.

A vibrant, active, self-reliant community has a capacity for mutually reinforcing common interests and values. It has a strong sense of community. Local civic associations, membership organizations, and voluntary work groups help build a sense of community (Putnam, 1993a,b). Social capital is defined as "features of social life—networks, norms, and trust—that enables participants to act together more effectively to pursue shared objectives" (p. 167). Civic engagement refers to "people's connections with the life of their communities, not only with politics" (p. 88). The passing of the "long civic generation" (that is, Americans who came of age during the Depression and World War II and have been far more deeply engaged in the life of their communities than the generations that followed them) appears to be an important proximate cause of the decline of our civic life (Putnam, 1995, p. 34). Education is by far the strongest correlate of civic engagement. Social trust and group membership are highly correlated with one another (pp. 36–37).

Residential mobility exonerates any responsibility for our fading civic engagement (p. 38). Residents of very small towns and rural areas are slightly more trusting and civilly engaged than other Americans. Types of organizations vary by locality, but overall rates of association membership are not very different. Among workers, longer hours are linked to more rather than less civic engagement, so hard work does not prevent civic engagement, nor does poverty or economic inequality (p. 39). Employed women as a group spend more time on organizations than before (p. 40), although these organizations are of different types.

Successful marriage, especially if the family includes children, is statistically associated with greater social trust and civic engagement (p. 41), but divorce may be the consequence, not the cause, of lower social capital, so the disintegration of marriage is probably an accessory to the crime. In the United States, differences in social capital appear uncorrelated with various measures of welfare spending or government size. Erosion of social capital has affected all races, and reversing the civil rights gains of the last 30 years would do nothing to reverse the social capital losses (p. 42). Time and age are ambiguous in their effects on social behavior because of three contrasting phenomena: (1) life-cycle effects, which produce no aggregate effects; (2) period effects on people of all ages who live through a given era, regardless of age, which can produce both individual and aggregate changes; and (3) generation effects that affect all people born at the same time and show up as disparities among age-groups at a single point in time (p. 43). Television is a prime suspect. It destroys social capital by (1) time displacement, (2) effects on the outlook of viewers, and (3) effects on children (pp. 47–48). It is privatizing our leisure time and our lives. The question boils down to, Do we like what it is doing?

We present some elements of one or more paradigms for assessing a region's economic growth prospects and related infrastructure requirements. These concerns are central to a regional policy that offers guidance for individual decision makers of local governments and their resident populations. A paradigm for regional policy is a model of regional structure and activity that provides a basis for understanding and foresight about regional growth and change. It is a framework for well-targeted infrastructure investments, vigorous and sustained capacity building, and continuing efforts in intraregional cooperation.

This framework differentiates the economic activities that occur within a region by their proximity and linkages to the metropolitan core area, which accounts for a corresponding differentiation of their role in regional economic organization. Concentration of high-order producer services and infrastructure characterizes the economy of the metropolitan core area. The spatial differentiation of economic activity and the corresponding differences in the economic base of labor market areas correlate with the spatial differentiation of regional infrastructure. Both the center and the periphery have the basic infrastructure of commodity and people transportation and energy production. The center, however, has the high-order infrastructure of communications, transportation, education, health care, and producer services. The overspill of manufacturing and producer services industries into the surrounding countryside attracts a growing labor force that, in turn, supports a gradually increasing number and variety of consumer-oriented businesses, especially retail trade and personal services. The peripheral labor market areas adapt to their more remote locations by exploiting the gifts of nature that become the principal means of acquiring income from outside sources.

Every downtown district of the world's regional cities exists in close symbiotic relationship with its surrounding neighborhoods. Spillover of crime and other manifestations of social disorganization devastate the safety and security of downtown workers. They eventually precipitate an increasing disassociation between place of work and place of residence while adding to pollution and congestion of downtown access highways. Ultimately, the regional city loses its capacity to function as a viable node in the global air transportation and telecommunication system. Heroic efforts are required to again make such a regional city an advantageous location for new businesses and for the maintenance and growth of existing ones. Again, residents of both areas suffer the consequences of the unintended flight of business services that make the regional city what it is—services that exist in their unique regional city clusters in only a few places.

The Michael Porter volume (1990) cited earlier is replete with well-documented case studies of the importance of business location in relation to a unique infrastructure that provides business enterprise with an initial and sustaining competitive advantage. As noted by Porter in Chapter 4, *The Dynamics of National Advantage*, "The existence of a cluster of several industries that draws upon common outputs, skills, and infrastructure also further stimulates government bodies, educational institutions, firms, and individuals to invest in relevant factor creation or factor-creating mechanisms. Specialized infrastructure is enlarged, and spillovers are generated that upgrade factor quality and increase supply. Sometimes, whole new industries spring up to supply specialized infrastructure to such clusters. Such a mutually reinforcing process is occurring in the United States, where the existence of world-class industries in mainframe computers, minicomputers, microcomputers, software, and logic circuits has sent public and private institutions scrambling to create software training centers and courses" (p. 135). Porter further states that "Proximity increases the concentration of information and thus the likelihood of its being noticed and acted upon. Proximity increases the speed of information flow within the national industry and the rate at which innovations diffuse. At the same time, it tends to limit the spread of information outside because communication takes forms (such as face-to-face contact) which leak out only slowly" (p. 157).

Enterprise Development

Enterprise investment is building from within. Most business foundings are local in origin (Reynolds et al., 1994b; Maki and Reynolds, 1994). Business retention also involves local efforts in addressing the environmental and workplace concerns of adversely affected local businesses, while lack of space hampers business expansion efforts. Business relocation happens most when one locality vies with another for a "footloose" business. Local enterprise investment would focus largely, therefore, on small business development, including venture financing, technical assistance, product or service marketing, and workforce training. The local educational institutions and foundations, like the University of Minnesota's Business Retention and Expansion program and the Northwest Area Foundation, often become involved in the local enterprise investment efforts.

Service Delivery

Service delivery is a high-priority concern of neighborhoods and communities working for stable relationships, civil classrooms, safe streets—goals by Ehrenhalt (1995), listed earlier as being implicit in the lives of people from the three Chicago communities. This includes access to neighborhood schools and the "classrooms without walls" of area colleges and universities. Job training is key to the reduction of neighborhood unemployment levels when complementing the various community efforts in job creation. Access to essential health care services also becomes important to stable neighborhoods and communities. Replacement of high-cost professional service delivery systems with "associations of community"—the family, the neighborhood, the church, and the voluntary organization—is especially well-suited for dealing with the most tragic cases of health care.[1]

To assess the effects of the redevelopment, Levine (1987) compared Baltimore's economic performance throughout the early 1980s with all frost belt cities of similar size. Baltimore was not successful in maintaining its industrial base, and its per capita income lagged behind the other cities compared. Poverty rates increased during the so-called renaissance years, and housing conditions remained among the worst in U.S. cities. He argues that three groups benefited from the corporate center redevelopment: developers and real estate speculators, suburban professionals, and "back-to-the-city" professionals and "out-of-town" tourists.

Levine explains the failure to distribute benefits evenly with several points. The attempt to adjust to national and international structural trends by converting the downtown into an advanced service center created a downtown isolated from the local component of the city's economy. In addition, tourism-related development created mainly low-wage subsistence jobs. And resting a strategy on tourism and conventions results in facing competition from other cities currently pursuing the same strategy. Further, a flaw in public-private partnerships is that no genuine strategic planning was done to maximize the quality of jobs created or meet social needs in the neighborhoods.

Levine states that cities like Baltimore should alter current policies. Instead of offering a grab-bag of investment incentives, it should offer a higher-quality workforce, a superior infrastructure, and better public services. In addition, the redevelopment agenda-setting process needs to be democratized since the logical outcome of public-private partnerships has been a developer's agenda. And finally, cities can adopt policies that better link downtown redevelopment to neighborhood needs, such as specifications or building requirements levied on highly profitable sectors of the city economy.

Community leadership, as shown by the success of Anacostia and other communities, including those in northern Italy cited by Putnam (1993b), is a key to building both viable neighborhoods and communities. Local leadership may emerge, but only momentarily without continuing recruitment to enlarge the leadership base, training in leadership opportunities and resources, and networking of current leaders into the larger community. Consider the following illustrative examples:

1. Neighbors working with juvenile offenders.
2. Schools moving unemployed youth into new work opportunities with the help of mentors, internships, off-campus learning programs, computer classes, and art and dance programs.
3. Neighborhood-university partnerships focusing on physical reconstruction of blighted neighborhoods through resident education.
4. Neighborhood-city partnerships focusing on long-term planning goals and related activities for local neighborhoods.
5. Neighborhood-business partnerships helping the recovery of those afflicted by drugs in the workplace.
6. Community development organizations creating local jobs and rebuilding neighborhood economic base through community revitalization.

Each activity would be networked into the community development activities. Central to these activities is a well-informed resident population, constructively engaged in building strong and stable neighborhoods and communities.

The preceding examples of community development efforts document the importance of using a variety of approaches for achieving livable neighborhoods and communities. These range from industrial park development to enterprise investment and job preparation through classroom instruction and on-site training. Moreover, the residents of each neighborhood and community have their own sense of "livability." Ehrenhalt (1995) characterizes life in the three Chicago communities by stable relationships, civil classrooms, safe streets. These came at a price, namely, limits on individual choice.

City planning addresses the issue of job creation as well as the delivery of essential services in Chicago's east-side neighborhoods. In addition, city planning may include monitoring existing local jobs and services and even advancing priorities for the city's involvement in neighborhood and community development. Residents of the individual neighborhoods may have other priorities, however, than those of the city's planning and development agencies. The experiences of Anacostia and other communities point to the importance of city and neighborhood partnerships in advancing common goals and viewing alternative futures for the neighborhoods, and in helping individual neighborhoods achieve them with their combined resources. These efforts ultimately become an important part of the strategic planning functions of city government.

Tracking Neighborhood and Community Performance

Many documentaries, in-depth newspaper reports, and scholarly papers give credence to the idea that neighborhoods and communities best grow from within. Actual case studies, however, are few. We summarize several of these studies because of the important insights they yield for an urban community's neighborhood revitalization efforts.

Medical Devices Industry: A Case Study

The medical devices industry is part of the "controlling and scientific instruments" industry group. Medtronic, Inc.; St. Jude Medical, Inc.; Sci-Med, TPI, and Del Pak are a few of the manufacturers in this group. Total U.S. medical industry employment was 289,000 in 1990—an increase of 100,000 since 1977. The 1990 County Business Patterns database from the U.S. Department of Commerce shows the distribution of establishments and total employment, by establishment size class, for the top 40 labor market areas that include twenty of the 29 air transportation nodes (Table 9-1).

The particularly high concentrations in the Los Angeles, Boston, New York, and Minneapolis-St. Paul LMAs account for 27 percent of total employment in the U.S. medical devices industry, while the twenty air transportation nodes account for 55 percent of the total industry employment. The top 40 LMAs (of a total of 382 for the United States) account for slightly more than 84 percent of the total. Moreover, most of the manufacturing facilities concentrate in one

TABLE 9-1 Distribution of establishments and employment in medical devices industry, 1990

LMA Area	State	Region	Estab. LMA	Employ. (pct.)	< 50 (pct.)	50–499 (pct.)	> 500 (pct.)	(pct.)
West								
Los Angeles*	CA	3	25	16.4	11.6	16.9	11.8	10.4
San Diego	CA	3	26	3.6	3.6	3.2	3.5	3.7
San Jose	CA	4	29	2.7	2.7	2.4	2.6	2.8
Denver*	CO	2	19	1.7	2.5	1.2	2.6	2.6
San Francisco*	CA	4	28	3.9	2.4	4.4	1.9	2.6
Seattle-Tacoma*	WA	1	11	0.9	1.8	1.3	0.5	3.0
Salt Lake City*	UT	5	44	1.2	1.7	1.2	0.9	2.6
Portland	OR	1	9	0.8	0.8	0.8	0.9	0.7
Santa Barbara	CA	3	27	0.6	0.6	0.5	1.2	0.0
Total West				31.7	27.6	31.7	25.9	28.3
North								
Boston*	MA	16	190	5.9	6.0	6.1	5.8	6.3
New York*	NY-NJ	12	167	9.8	5.2	9.2	7.2	2.6
Minneapolis-St. Paul*	MN-WI	10	156	4.6	4.9	4.8	4.4	5.3
Chicago*	IL	20	231	7.8	4.3	9.1	5.9	1.9
Entire state	CT	12	187	3.8	3.5	4.7	4.0	2.8
Trenton	NJ	15	180	4.0	3.1	4.9	3.3	2.6
Rochester	NY	13	175	0.4	2.3	0.3	0.4	4.4
Philadelphia*	PA	17	179	5.4	2.3	6.2	3.9	0.0
South Bend	IN-MI	20	261	0.8	1.9	0.5	0.8	3.2
St. Louis*	MO-IL	9	141	1.9	1.8	1.9	2.1	1.4
Providence	RI-MA	16	191	1.8	1.7	1.2	2.9	0.7
Pittsburgh*	PA-WV	19	216	1.1	1.5	0.8	1.3	1.9
Milwaukee	WI	20	277	0.6	1.5	0.3	1.4	1.9
Albany metro area	NY	13	170	0.4	1.3	0.3	1.3	1.4
Cleveland	OH	19	224	2.1	1.2	2.9	1.4	0.7
Allentown	PA	15	183	0.5	1.1	0.6	0.9	1.4
Syracuse*	NY	13	173	0.3	0.8	0.5	0.5	1.2
Baltimore*	MD-DE	14	200	0.6	0.8	0.3	0.2	1.4
Cincinnati*	OH-KY	21	257	0.6	0.7	0.9	1.4	0.0
Kansas City*	MO-KS	6	62	0.8	0.6	1.0	1.3	0.0
Total North				53.4	46.4	56.5	50.4	40.8
South								
Miami*	FL	25	319	1.3	2.1	1.4	1.0	3.2
Atlanta*	GA	24	300	0.8	1.4	1.3	0.9	1.9
Raleigh*	NC	28	346	0.5	0.9	0.5	0.4	1.4
San Angelo area	TX	7	121	0.0	0.9	0.0	0.0	1.9
Memphis area	TN-MS-AR	11	164	0.4	0.8	0.2	0.5	1.2
Jacksonville metro	FL-GA	26	314	0.0	0.9	0.0	0.0	1.9
Tampa metro	FL	26	325	0.9	0.8	0.6	1.8	0.0
Greenville area	SC	27	309	0.3	0.7	0.2	0.7	0.7
Knoxville metro	TN-KY	29	360	0.6	0.6	0.3	0.7	0.7
Dallas metro*	TX	7	107	0.6	0.6	0.6	0.6	0.7
Greenborough	NC-VA	28	333	0.4	0.6	0.2	1.3	0.0
Total South				5.9	10.2	5.4	7.8	13.5
Other LMAs				9.0	15.8	6.4	16.0	17.4
All LMAs				100.0	100.0	100.0	100.0	100.0

Source: U.S. Department of Commerce, County Business Patterns. 1990.
*Air transportation node.

TABLE 9-2 Total employment of specified industry, selected labor market areas, 1990

Industry Title	SIC Code	Mpls-St. Paul (pct.)	Boston (pct.)	New York (pct.)	Chicago (pct.)	Los Angeles (pct.)
Doctors and dentists	801–804	2.5	1.9	1.7	1.4	2.1
Hospitals	806	2.4	4.2	3.8	3.4	2.3
Electronic computers	3571	1.6	1.2	0.0	0.0	0.2
Nursing and protective care	805	1.4	1.6	1.0	0.8	0.8
Computer and data processing services	737	1.1	1.9	0.9	1.1	0.9
Engineering services	871	1.0	1.5	0.7	0.9	1.2
Managerial and consulting services	874	0.8	1.5	1.0	1.2	1.2
Accounting services	872, 899	0.7	0.8	1.0	1.0	1.4
Research and development	873	0.6	1.4	0.8	1.3	0.8
Other medical and health services	807–809, 074	0.5	0.7	0.9	0.4	0.5
Mechanical measuring devices	3823, 3824, 3829	0.4	0.4	0.1	0.1	0.1
Automatic temperature controls	3822	0.4	0.1	0.0	0.1	0.0
Surgical appliances and supplies	3842	0.3	0.0	0.1	0.1	0.1
Surgical and medical equipment	3841	0.2	0.3	0.0	0.2	0.1
Computer storage devices	3572	0.2	0.0	0.0	0.0	0.1
Computer peripheral equipment	3577	0.2	0.2	0.0	0.0	0.1
Electromedical apparatus	3845	0.1	0.1	0.0	0.0	0.0
Search and navigation equipment	3812	0.1	0.6	0.4	0.1	1.0
Analytic instruments	3826	0.1	0.2	0.0	0.0	0.0
Computer terminals	3575	0.0	0.0	0.0	0.0	0.0
Top 20, total		14.7	18.7	12.2	12.0	12.9
Other industry, total		85.3	81.3	87.8	88.0	87.1
All industry, total		100.0	100.0	100.0	100.0	100.0
All industry, total (thousands)		1,738	2,611	6,715	4,331	8,032

Source: University of Minnesota IMPLAN regional modeling system.

or two counties in the core metropolitan areas. The size distributions of these establishments vary greatly from one LMA to the next.

Table 9-2 summaries show the total employment in each of the top twenty industries in the Minneapolis-St. Paul LMA compared with four other high-ranking areas—Boston, New York, Chicago, and Los Angeles. Three of the five medical devices industry clusters in the Minneapolis-St. Paul LMA are in the top twenty in total employment. With the exception of surgical and medical equipment manufacturing in the Boston LMA, the entire sector ranks highest in the Minneapolis-St. Paul LMA. However, this LMA has the lowest total employment for the top twenty. This points to the high levels of medical devices industry concentration in the Minneapolis-St. Paul LMA. It also shows the juxtaposition of the health care, computer hardware, and computer software industries as an essential infrastructure for the emergence and viability of the medical devices industry.

Net exports—that is, gross commodity exports minus gross commodity imports—provide an additional measure of the competitive position of the export-producing sector. High levels of net exports may result from high levels of specialized commodity exports and/or from a wide variety of exports. In the top twenty industry ranking for the Minneapolis-St. Paul LMA, three of the five active medical devices industries rank third, seventh, and ninth in net exports, with a total value in excess of $1 billion in 1990 (Table 9-3).

Value added in each of the five medical devices industries ranks among the top twenty. Four of the five rank in the top eleven along with other scientific and controlling instruments and computer-related manufacturing. While high value-added industries generally rank high in all regions, and low value-added industries generally rank low in all regions, many exceptions occur, as shown in Table 9-4. However, one Minneapolis-St. Paul LMA industry may have the lowest value added, for example, electromedical apparatus, while another, for example, dental equipment, may have the highest. An added twist is the accuracy of the estimates biased toward mean values for each industry. We adjust the estimates based on U.S. industry ratios for state-to-state and county-to-county variations in these ratios, subject to the state and county "control" totals. These are the aggregate values collected and published by state and federal agencies for individual states and counties.

The tabular presentations point to the distinctly different industrial structures in the five LMAs. The medical devices and related industries in the Minneapolis-St. Paul and Boston LMAs are strongly export-oriented. They have a disproportionate share of total employment in large establishments. The New York and Chicago LMAs serve largely a local market. They have a disproportionate share of total employment in small establishments. The Los Angeles LMA has attributes of both a residentiary and an export-oriented economy. However, each LMA with a large export-producing medical devices industry component has a roughly similar infrastructure of health care and producer services providers and computer hardware manufacturers.

The U.S. Office of Technology Assessment lists the surgical and medical instruments and apparatus group of 27 firms as the largest employers in the

TABLE 9-3 Net exports of specified industry, selected labor market areas, U.S., 1990

Industry Title	SIC Code	Mpls-St. Paul (mil.$)	Boston (mil.$)	New York (mil.$)	Chicago (mil.$)	Los Angeles (mil.$)
Electronic computers	3571	1,057	1,171	–247	–427	–686
Doctors and dentists	801–804	886	–40	161	–744	2,771
Surgical appliances and supplies	3842	438	–101	41	–23	344
Computer and data processing services	737	413	2,227	–102	705	540
Mechanical measuring devices	3823, 3824, 3829	401	661	435	130	–161
Automatic temperature controls	3822	370	89	–3	197	137
Electromedical apparatus	3845	334	460	78	–55	245
Surgical and medical equipment	3841	226	604	251	475	316
Engineering services	871	213	1,438	–221	682	2,595
Nursing and protective care	805	125	487	504	–291	–523
Computer storage devices	3572	120	–216	–42	–46	262
Analytic instruments	3826	119	356	–100	22	–209
Other medical and health services	807–809, 074	116	201	137	–536	–192
Search and navigation equipment	3812	79	966	110	170	6,640
Managerial and consulting services	874	45	1,486	378	1,097	1,473
Accounting services	872, 899	27	302	–86	1,160	3,318
Typewriters and office machines	3579	23	–12`	254	–47	–102
Calculating and accounting machines	3578	23	16	–6	–18	–52
Research and development	873	19	1,110	–117	1,787	1,465
Laboratory apparatus	3821	16	–49	19	73	–345
Top 20, Minneapolis-St. Paul Total		5,050	11,154	1,444	4,313	17,836

Source: University of Minnesota IMPLAN regional modeling system.

Twin Cities medical devices industry. All except one are engaged in international markets. The next largest is ophthalmic goods, with three of the seven firms in this group, accounting for 60 percent of total employment, reporting only local market sales. The third ranking employer, with fifteen firms, all but one reporting international markets, is the orthopedic, prosthetic, and surgical supplies and apparatus group. Two firms, both reporting international markets, make up the smallest group—dental equipment and supplies. Of the 51 firms in the Twin Cities medical devices industry, 26 have 50 to 500 employees, while twenty-one have fewer than 50 employees and four have more than

TABLE 9-4 Total value added per worker of specified industry, selected labor market areas, U.S., 1990

Industry Title	SIC Code	Mpls-St. Paul (thou.$)	Boston (thou.$)	New York (thou.$)	Chicago (thou.$)	Los Angeles (thou.$)
Laboratory apparatus	3821	116.8	110.7	128.7	132.8	108.0
Computer storage devices	3572	95.3	150.1	130.5	77.7	123.7
Electromedical apparatus	3845	90.3	109.9	98.0	110.4	96.0
Analytic instruments	3826	82.5	85.5	69.7	68.9	84.5
Search and navigation equipment	3812	75.3	76.6	82.6	71.7	88.4
Electronic computers	3571	69.7	69.6	61.4	57.1	66.8
Surgical appliances and supplies	3842	67.1	70.6	70.3	63.0	60.2
Mechanical measuring devices	3823, 3824, 3829	60.5	68.3	58.9	59.9	59.2
Computer peripheral equipment	3577	60.5	89.3	71.3	77.2	74.0
Dental equipment	3843	60.5	50.3	47.3	51.2	46.2
Surgical and medical equipment	3841	51.9	57.9	67.2	60.2	52.1
Doctors and dentists	801–804	50.7	47.5	51.9	57.3	57.9
Computer and data processing services	737	50.0	61.0	62.8	58.1	54.9
Optical instruments	3827	49.2	70.8	52.0	52.9	50.7
Instruments to measure electrical signals	3825	48.4	57.1	52.7	41.2	52.8
Automatic temperature controls	3822	43.8	32.5	22.2	38.3	36.6
Computer terminals	3575	42.9	50.0	37.1	31.3	37.9
Accounting services	872, 899	40.8	55.7	81.3	56.0	45.4
Ophthalmic goods	3851	32.5	40.5	29.4	36.5	48.7
Typewriters and office machines	3579	29.9	34.7	33.6	31.0	36.7

Source: University of Minnesota IMPLAN regional modeling system.

500. This listing differs from the estimates of total employment based on the 1990 County Business Patterns.

The listing for this industry group includes six four-digit Standard Industrial Classification (SIC) code industries from industrial and commercial machinery and computer equipment manufacturing (SIC 35), thirteen four-digit industries from scientific and controlling instruments manufacturing (SIC 38), one three-digit industry from business services (SIC 73), ten three-digit industries from health services, including veterinary services (SIC 80 and SIC 074), and four three-digit industries from engineering and management services (SIC 87). The six four-digit industries in the medical devices group purchase intermediate commodity inputs from these other, local industries. They also purchase primary inputs. Intermediate inputs are both locally produced and imported. Primary inputs include labor, capital, and entrepreneurship. Their business activity generates income payments that go to the resource owners in the form of employee compensation, self-employed income, indirect business taxes, and gross profits. Resource owners receive the largest share of the final payments originating from the medical devices industry group. Commodity disbursements from these industries and their primary and intermediate input purchases provide aggregate measures of industry structure.

Product market access for small businesses and start-up firms depends, in large part, on established distribution systems. Several of the medical devices firms are large enough to have their own networks of domestic and international clients and customers. They serve as industry leaders in product and process innovation and as models for other firms to emulate. Product innovation drives the industry, but its focus shifts with changes in the methods of delivery and financing of health care. The medical devices industry is thus part of a group of industry clusters that includes the much larger health care industry (SIC 80) as its principal product market.

The *product market* for the Twin Cities medical devices industry was largely local in the beginning. The Mayo Clinic, the University of Minnesota Hospital, and several other large local hospitals generated a growing demand for medical devices. The presence of a local market was a critical factor, along with the medical science research that was taking place in these hospital complexes, in the concentration of this industry in this region. The industry expanded dramatically in the 1960s and 1970s, becoming a leading exporter of medical devices to other areas of the United States and to international markets. These exports constitute an important part of the technology-intensive manufacturing exports from the Twin Cities and hence are an important part of the Twin Cities' economic base.

Medtronic, Inc., a Twin Cities manufacturer of bradycardiac pacemakers, exemplifies a globally competitive business enterprise in the health care sector. It was formed in 1949 by two young men—one a graduate student in electrical engineering, the other an employee of a local lumber business. In the 1950s, the two owners, along with a medical doctor in health sciences at the University of Minnesota, set out to develop an improved pacemaker system to help stimulate the heart activity of medical patients. Despite its near demise in the mid-1960s, it survived and is now the world's leading manufacturer of bradycardiac pacemakers. The Twin Cities has several internationally competitive medical devices manufacturing businesses among its leading exporters to domestic and foreign markets.

Factor market access is critical to any technology-intensive business because of the skill requirements of its workforce. Access depends on the networking of professional and technical workers in informal information exchanges. This is part of an active local labor market for skilled workers. It makes possible the retention of these workers in the Twin Cities, for example, despite the demise and closure of one or more formerly successful medical devices firms.

Factor input mix refers to the proportions of total industry outlays accounted for by intermediate inputs, that is, purchases from other firms, and primary inputs, which include employee compensation as well as capital consumption allowances, incurred tax and dividend liabilities, and retained earnings. Not included among these outlays are the intangibles that result from the many informal networks for information exchange among the workers in these firms. Nor are the contributions of the University of Minnesota Hospital and other hospitals and medical clinics in the area as sources of product and process innovation in the medical devices industry necessarily included in the industry outlays.

The importance of the *local labor market* was alluded to in earlier discussions. This is especially important for the local *health care system*—the complex of industry clusters that includes the medical devices industry—during the forthcoming period of likely industry retrenchment in jobs and additional services.

A first step in using the local labor market is matching jobs to available skills. This is now extremely difficult, if not impossible, without reworking the existing occupational classification system—the *Dictionary of Occupational Titles* (DOT). One important task for the health care system is, therefore, the negotiation of a new segment of the DOT that defines occupational classes by the level of skill attainment, a task currently being addressed by the U.S. Department of Labor.

A further step in matching jobs to available skills is *tracking job vacancies and layoffs* in the health care system. This is an early warning system for health care providers as well as state employment, training, and education agencies.

Another step in matching jobs to available skills is *monitoring and projecting industry activity*. This task is doubly difficult for a health care system facing potential retrenchment and new sets of operating rules forthcoming as a result of national health care reform. The new rules represent an alternative future for projecting the local health care system.

Skill-directed technical education matches skills with available jobs, present and prospective. Technical education administrators and curriculum planners track current economic indicators and forecasts of the changing demands for technical college graduates. These educators at least must track the business cycle and individual product cycles to understand their effects, among other factors, on the local health care system's job market.

National health care reform now challenges the health care system to focus its *educational and training curricula* on *skill-based performance* criteria. The skills pertain to a wide range of required competencies for managing the individual businesses and other organizations in a diverse industrial system. These include the delivery of medical services in hospitals and clinics, managing the facilities providing these services, and manufacturing related medical and health care supplies and equipment that go well beyond the medical devices industry.

University-Industry Collaboration

The purpose here is to assess the role and importance of university-industry collaboration in the medical devices industry. This includes forming, developing, and maintaining the industry cluster as well as adding to the viability and value of the regional health care complex of which the medical devices industry is a part. An industry cluster is a group of interacting business, academic, and government enterprises engaged in the development, production, and marketing of a particular product. It includes the core businesses, the collaborating regional university departments and government agencies, the input supplying businesses, and the organizations purchasing and using the product. The medical devices industry cluster is part of the regional health care complex—a much broader grouping of self-reinforcing industry clusters that develop, produce, and deliver health care services.

The underlying argument stems from Porter's *Competitive Advantage of Nations* (1990) and Saxenian's *Regional Advantage* (1994). They cite the critical role of industry clusters and local labor markets in providing regional environments for successful business enterprise. Porter, arguing for a new theory of national competitive advantage, suggests that "Competitive advantage is created and sustained through a highly localized process" (p. 19). Saxenian observed that "Far from being isolated from what lies outside them, firms are embedded in a social and institutional setting that shapes, and is shaped by, their strategies and structures. The concept of an industrial system illuminates the historically evolved relationships between the internal organization of firms and their connections to one another and to the social structures and institutions of their particular localities" (p. 7). New research findings confirm earlier insights that the most strategic relationships among firms in network systems "are often local because of the importance of timeliness and face-to-face communication for rapid product development" (p. 5). Individual business clusters are an integral part of a larger complex of business enterprises that form a regional industrial system. Following the logic of Porter and Saxenian, we continue to focus on the medical devices industry as part of an industry cluster embedded in the health care industry structure and the daily economic activities of a local labor market area.

The health care industry brings together several industry groups that form a regional industrial system—a self-reinforcing industry cluster focusing on health care.[2] Saxenian (1994, p. 7) suggests we think of a region's industrial system as having three dimensions: local institutions and culture, industrial structure, and corporate organization.[3] She notes further that "spatial clustering alone does not create mutually beneficial interdependencies. An industrial system may be geographically agglomerated and yet have limited capacity for adaptation. This is overwhelmingly a function of organizational structure, not of technology or firm size" (p. 161). Moreover, the clustering is self-reinforcing through its vast network of industry and individual business and professional stages in production, marketing, procurement, and staffing. In this paradigm of a geography-bounded industry linkage network, customers and suppliers, as well as researchers, drive the product and process innovations that, in turn, help sustain the health care system's importance in the state's economic base.

The medical devices industry—a group of manufacturing firms that accounts for about 8 percent of the health care system in Minnesota—serves as an example of a collection of internationally competitive business enterprises that form an industry cluster. With other medical and health care activities and their clients, customers, and suppliers, they form a regional health care system.

The medical devices sector also has close ties with the computer hardware and software sectors and the university-medical complex in several of the labor market areas. In the Minneapolis-St. Paul LMA, for example, the computer-related manufacturing accounts for slightly more than 6 percent of the corresponding industry group in the United States. However, one of its industry components, electronic computers, accounts for more than 9 percent of its total, while another, computer storage devices, accounts for less than 4 percent of its total in the United States. The business services industry component in computer and data-processing services sector, on the other hand, accounts for only 1.7 to 1.8 percent of its U.S. total in the three areas. The computer services industries thus fail to follow the patterns of high concentration established by the computer and medical devices industries in the local areas. However, this industry is among the fastest growing in the metropolitan area.

Also important for the future of the medical devices industry in the Minneapolis-St. Paul LMA are the close ties with the health care providers that are an important initial market for its products. It also has important linkages, present and potential, with the University of Minnesota Health Sciences complex, which offers both research and teaching resources to the industry for its support and benefit. From an industry perspective, collaboration with a regional research university provides parts of two additional platforms for improving industry performance beyond the 5 to 10 percent cost reductions achieved by existing industry research and development efforts. This collaboration may assist in establishing new platforms for the commercial utilization of advanced technologies. It may also assist in advanced development work to validate certain technologies prior to putting them into new applications, for example, trying to understand that fluid dynamics works in a certain way, then looking at how technologies take the components and make them work together. The intellectual property owned by the researcher is not directly marketed. But the new products indirectly market the intellectual property along with the work of the research-and-development staff designing the software, for example, that is part of the proprietary product of the company.

Table 9-5 shows the aggregate measures of interdependency among the selected local industries associated with the medical devices industry cluster from the supply side. These are the input purchases of the five four-digit industries in the Minneapolis-St. Paul LMA's medical devices industry group. The medical devices industry has two sets of backward linkages, namely, to intermediate input supply sources, both local and nonlocal, and to primary input supply sources in the locality of the individual businesses. The income payments to each of two input categories are split about equally. Both linkages are critical to the industry and its viability. The purchases of the various high-order services, ranging from engineering services to research and development, are indicators of the potential for university-industry collaboration.

Table 9-6 provides measures of the backward and forward linkages of the industry cluster (of which the medical devices industry is a part). "Industry" refers to the backward linkages of the entire industry cluster purchases of primary and intermediate inputs. "Commodity" refers to the forward linkages of the industry cluster to the purchases of households; businesses; local, state, and federal governments; and export markets. One industry may produce more than one commodity. Industry gross output provides the ranking of the individual industry outside of the medical devices industries that are listed first. In this ranking, the electronic computer manufacturing industry tops all other industries in most criteria, except value added and personal consumption expenditures. For these criteria, doctors and dentists in the health care services sector top the list in value added and hospitals in personal consumption expenditures.

The total commodity output of the selected industries in the Minneapolis-St. Paul LMA was $18.7 billion—about 14 percent of the $121 billion in the LMA's total commodity output in 1990. The selected commodity exports were more than $5.7 billion, or 14.5 percent of $39.5 billion in total area exports. The selected industries were much smaller importers than exporters, accounting for only $2.2 billion of the $29 billion in overall industry imports. These industries thus contributed $3.5 billion in "excess" exports—the difference between total industry exports and imports.

TABLE 9-5 Purchases of specified production inputs, Minneapolis-St. Paul labor market area, 1990

Industry Title	SIC Code	3841 (mil.$)	3842 (mil.$)	3843 (mil.$)	3845 (mil.$)	3850 (mil.$)	Total (mil.$)
Electronic computers	3571	0	0	0	0	0	0
Computer storage devices	3572	0	0	0	0	0	0
Computer terminals	3575	0	0	0	0	0	0
Computer peripheral equipment	3577	0	0	0	0	0	0
Mechanical measuring devices	3823, 3824, 3829	0.1	0.1	0	0.1	0	0.3
Analytic instruments	3826	0	0	0	0	3	3
Optical instruments	3827	0	0	0	0	0.4	0.4
Surgical and medical equipment	3841	1.3	35.1	0	0	0	36.4
Surgical appliances and supplies	3842	20.2	42.8	0	0	0	63
Dental equipment	3843	0	0	0	0	0	0
X-ray apparatus	3844	0	0	0	0.4	0	0.4
Electromedical apparatus	3845	0	0	0	2.5	0	2.5
Ophthalmic goods	3850	0.1	0.1	0	0	0.8	0.9
Computer and data processing services	7370	1.1	1.8	0	0.9	0.1	3.9
Doctors and dentists	801–804	0	0	0	0	0	0
Nursing and protective care	805	0	0	0	0	0	0
Hospitals	806	0	0	0	0	0	0
Other medical and health services	807–809, 074	0	0	0	0	0	0
Engineering services	871	0.1	1.5	0	0.1	0	1.7
Accounting services	872, 899	1.6	1.8	0	0.4	0.2	4
Managerial and consulting services	874	2.8	3.1	0	0.7	0.3	7
Research and development	873	1.7	2	0	0.5	0.2	4.3
Subtotal		6.3	8.4	0	1.7	0.6	17
Other intermediate inputs (including foreign imports)		106.6	226.8	0.8	128.3	9.4	471.8
Total domestic imported intermediate inputs		97.4	129.2	1.7	97.6	5.0	330.9
Value added		204.0	388.5	2.6	223.1	23.4	841.6
Total industry outlay		414.2	752.9	5.2	450.7	38.3	1,661.3

Source: University of Minnesota IMPLAN regional modeling system.

Measuring the economic impact of a regional research university typically includes estimates of the payroll and expenditure effects of university staff and students on the local community. These, however, may be largely income transfers from one part of a region to another or from one socioeconomic group to another. The focus of this study is the impact of a university-industry partnership on the region's economic development, specifically, on the viability of its

TABLE 9-6 Ranking of selected local industry structure indicators, Minneapolis-St. Paul labor market area, 1990

		Industry Outlays					Commodity Output			
Industry Title	SIC Code	Total (mil.$)	Val.Add. (mil.$)	Int.Inputs. (mil.$)	Int.Imp. (mil.$)	Total (mil.$)	Fin.Dem. (mil.$)	PCE (mil.$)	Govt. (mil.$)	Priv.Inv. (mil.$)
Surgical and medical equipment	3841	414	204	113	97	421	298	1	4	38
Surgical appliances and supplies	3842	753	389	235	129	685	562	25	14	69
Dental equipment	3843	5	3	1	2	6	4	0	0	0
X-ray apparatus	3844					3	0	0	0	0
Electromedical apparatus	3845	451	223	130	98	438	422	0	1	25
Ophthalmic goods	3851	38	23	10	5	37	30	19	1	0
Electronic computers	3571	3,613	1,890	1,235	487	3,555	2,831	145	4	1,417
Doctors and dentists	801–804	2,987	2,204	654	128	2,987	2,987	2,092	8	0
Hospitals	806	2,426	1,042	973	411	2,427	2,427	2,400	18	0
Computer and data processing services	737	1,398	964	354	80	1,412	505	0	92	0
Engineering services	871	849	453	377	20	949	269	0	3	53
Computer storage devices	3572	807	358	322	127	779	546	0	3	362
Accounting services	872, 899	773	520	239	15	782	89	4	58	0
Managerial and consulting services	874	665	405	214	47	665	98	0	53	0
Mechanical measuring devices	3823, 3824	637	405	116	116	617	570	2	2	91
Other medical and health services	807–809, 074	481	222	170	88	481	448	446	2	0

Nursing and protective care	805	472	310	120	42	472	472	347	0	0
Research and development	873	420	250	139	30	420	68	0	43	0
Automatic temperature controls	3822	417	288	64	64	390	390	0	0	0
Computer peripheral equipment	3577	411	189	159	63	418	255	0	1	200
Search and navigation equipment	3812	236	143	39	54	212	191	4	0	2
Analytic instruments	3826	225	107	79	39	213	180	1	1	41
Computer terminals	3575	65	36	21	9	69	69	0	0	0
Instruments to measure electrical signals	3825	62	36	13	13	79	58	0	0	10
Laboratory apparatus	3821	36	20	7	9	32	21	0	0	1
Typewriters and office machines	3579	31	8	13	10	88	62	5	10	10
Optical instruments	3827	19	9	7	3	25	22	1	1	4
Calculating and accounting machines	3578	1	0	0	0	32	27	2	1	1
Total		18,692	10,702	5,804	2,186	18,692	13,899	5,494	317	2,322
Other industry		100,434	59,371	23,233	17,830	101,903	77,846	28,639	8,485	6,867
Grand total		119,126	70,073	29,037	20,016	120,595	91,745	34,133	8,802	9,189

Source: University of Minnesota IMPLAN regional modeling system.

technology-intensive industries, of which medical devices manufacturing is a prime example. This focus for the study calls for the one-on-one executive interview.

The quantitative presentation provides a factual context for the preparation of the one-on-one executive interviews. We address the role of the regional research university in the economic development of its region in partnership with the technology-intensive firms that would benefit from intensive and continuing interaction with the university researchers in concept building and testing. This may be especially relevant in the new advanced technologies that represent high risks for even well-established companies with solid successes in the past to support their search for additional financing. For small, newly established companies, these risks may be insurmountable without the university's commitment to the research and teaching that these companies can depend upon in their efforts to successfully compete or collaborate with well-established companies. The latter usually are large enough to support their own research and training programs. For small businesses, however, the regional research university can become an invaluable source of new business and industrial technologies as well as education and training programs for improving the competencies and productivity of their small businesses workforce. Even for large companies, the regional research university is potentially, if not currently, an important source of new entrants into their workforce. Large companies also can directly support the testing of new concepts that more effectively manage the private costs and benefits of various advanced technologies of long-term concern to them.

Notes

1. The industry itself includes several subgroups: surgical and medical instruments and apparatus (SIC 3841); orthopedic, prosthetic, and surgical appliances and supplies (SIC 3842); dental equipment and supplies (SIC 3843); X-ray apparatus and tubes and related irradiation apparatus (SIC 3844); electromedical and electrotherapeutic apparatus (SIC 3845); and ophthalmic goods (SIC 3851).

2. Kenneth Labich, in his article "The Geography of an Emerging America" (*Fortune*, June 27, 1994, pp. 88–940), notes the high concentrations of medical research with computer hardware manufacturing and computer software services. The medical devices industry also fits this pattern of industry clustering.

3. Saxenian notes further that recent research reveals the interrelations of these three dimensions. For example, economists now recognize that "innovation is a product of interactions among customers and suppliers, a firm's internal operating units, and the wider social and institutional environment" (p. 174).

10

Rural Areas and Communities

We often view rural areas as transitional or peripheral. Transitional rural areas adjoin the metropolitan core area and extend to the outer commuting limits of the core workplaces, and even slightly beyond. Within the 60-mile and sometimes even 100-mile radius of the core area, farm subdivision is a common practice because of the high demand for part-time, hobby, and garden-type residential farms. Off-farm employment of one or more family members supplements farming as an income source. Off-farm employment opportunities occur in manufacturing plants locating or expanding in the transitional rural areas, and in trade and service establishments of growing rural service centers. The LMAs farthest removed from the core lack convenient low-cost access to decision information for business enterprise. They are vulnerable to the general business cycle and the product cycles of their standardized, highly tradable commodities. The low-cost producer dominates competition in commodity markets. This translates into extreme dependence on low wages or, alternatively, high productivity in resource uses.

What Is a Rural Community?

Like the urban community, the rural community fits many different descriptions of its people and their places of work, residence, shopping, entertainment, and recreation. Its size and distance from a metropolitan area has much to do with the daily activity patterns of its residents. A large rural community has more diversity of jobs and daily activities than a small community. Proximity to a metropolitan area affects the size and character of its business population as well as the jobs available locally to the resident workforce.

Place, Purpose, and Proximity

We start with an economic definition of *rural area* and *community.* Again, as earlier, we define both in terms of place, purpose, and proximity. A rural economic community is the commuting area of a smaller central place—still the local labor market area (LMA) and, again, roughly the functional economic area for rural people and places.

Cornelia Butler Flora views rural communities of place as being important for a substantial number of people in the United States (Goreham, 1997, p. 113). Flora refers to the capacity of rural communities to acquire capital resources for investing in their people and places to keep from following the overall trends of continuing decline in population and economic activity. These capital resources would include financial and manufactured capital, human capital, environmental capital, community social capital, horizontal social capital (that is, egalitarian forms of reciprocity in both giving and receiving), and hierarchical social capital. Flora suggests that each form of capital can enhance the productivity of the other forms of capital and that increasing social capital greatly cuts transaction costs, making other resource use more efficient. This depends on entrepreneurial social infrastructure (ESI), that is, "the capacity of a community to initiate and adapt" (p. 115). Flora identifies the key components of ESI as legitimization of alternatives, diversity of community networks, and widespread resource mobilization. These essential attributes of a healthy and viable community occur in the metropolitan areas in larger measure and with fewer risks involved in their achievement than in many rural areas—a reality that only the exceptional rural communities may have the resources to face successfully.

Nonetheless, a growing number of rural communities within an easy overnight ride from a large, growing metropolitan statistical area (MSA) are achieving a turnaround in economic activity. This occurs in part from lower site costs for manufacturing and related activities. These lower site costs, together with lower labor costs for highly productive workers, allow the owners of these activities to compete successfully in the nearby metropolitan area markets.

Quantitative Measures

We delineate rural areas and communities, like urban areas and communities, in many ways—by natural boundaries as well as artificially built barriers. Most often, we resort to the simple definition of nonmetropolitan areas, that is, areas outside the MSA boundaries. In Chapter 3, rural areas are defined as outside the labor market areas of MSAs. This results in a total area somewhat smaller than the strictly nonmetropolitan area since MSA boundaries lag behind the commuting, or labor market, area boundaries of growing metropolitan places.

Of the more than 23,000 places within the U.S. boundaries in 1990, nearly 18,000 (77 percent) were outside the metropolitan areas. More than 14,000 (60 percent) had a population of less than 2,500. Nearly 90.5 million persons (37 percent of the total U.S. population) lived outside metropolitan areas in 1990, and nearly 25 percent lived in places with populations of less than 2,500.

Delineating Rural Areas and Communities

A first step in deciding how a rural community might restructure itself is to delineate its economic base—the main sources of its prosperity and economic well-being. Without the economic base the community would not exist. It provides the means of trade with the rest of world and the income payments for maintaining the industry production processes. Rural areas and communities are extremely vulnerable to changing market conditions and government policies, especially when they affect the community economic base. Revitalizing a natural resource-based rural community under existing global economic conditions is a challenge that entails much uncertainty about its successful resolution.

Differentiating Rural Areas and Communities

The 130 rural LMAs, according to their definition in Chapter 1, lack an MSA, that is, a contiguous urbanized area with a resident population of 50,000 or more. The transitional LMAs closest to the economic region's core experience rapid population and job growth. They have an expanding manufacturing base as a result of low site costs—rent, labor, and environmental protection—coupled with excellent access to metropolitan area markets. For many counties in the transitional areas, the percentage rates of growth exceed those in the metropolitan core area. The peripheral rural areas are supply sources for the standardized products that compete on a price basis. Low unit cost of production translates into a competitive market price,\ which, in turn, depends on high labor productivity or low wages. The periphery lacks cities that trade with other cities. These include areas in varying stages of development and decline. A few of the LMAs are the resident locations of branch plants from the core areas seeking low site and/or labor costs. A few others are recreation and retirement destinations. Most, however, have declining local economies, in part because of labor-reducing technological advances in local production.

One interpretation is that a particular region's location in the national and global regional settlement and trading systems imposes severe constraints on regional development options. A rural LMA located outside the commuting area of any metropolitan LMA lacks prospects for long-term economic viability beyond the lifetimes of its principal product cycles. It lacks the resources for innovation and improvisation—essential attributes for import replacement and export diversification. Rather than belabor the obvious, our discussion of rural areas and communities shifts to the activities in various rural counties. Such a discussion may provide some new insights regarding the sorts of opportunities distant rural areas may face in the future and the necessary means for taking advantage of these opportunities.

Regional Economic Organization

We address the challenge of global competition from two national perspectives—Norway and the United States—on regional economic organization. We focus on the wood products (millwork) manufacturing industry in each of the two countries. Establishments in this industry are primarily engaged in manufacturing fabricated wood millwork, including wood millwork covered with materials such as metal and plastic. Doors, windows, and related products account for much of the industry output, which is purchased mostly by the domestic construction industry. This industry exists in both the core metropolitan area and the rural periphery.

The millwork industry is highly import-dependent. Several large single-establishment firms dominate segments of the industry in locally diverse economic environments. We compare this industry and its local economic setting in three localities—two in the United States and one in Norway. Each locality is part of an extended labor market area, which, in turn, is part of a larger economic region. Two of the localities are rural, and one is urban—a single county in the Minneapolis-St. Paul core metropolitan area.

A dominant firm accounts for most of the specific local industry output in the case studies. Each dominant firm purchases most of its intermediate inputs from sources outside the region and exports most of its product beyond its region. Its income payments to primary resource owners in the form of value added—employee compensation, indirect business taxes, and property income—account for most of its local impact.

Our objective is to assess (1) the special contributions of an area's core competencies—a skilled workforce and management with cost-effective access to product markets and input supply sources, including the special conditions for the occurrence of these core competencies in both the principal or inner metropolitan area of an economic region and its rural periphery, (2) additional conditions, if any, that account for the success of the millwork industry in two different business environments.

Interregional trade theory and economic base theory provide the broad analytical framework for this initial phase of the study (Porter, 1990; Tiebout, 1962). Exports of the local millwork businesses in each of the two sets of case studies reach their markets outside the local area because of their comparative advantage in price, quality, and/or marketing effort involved in introducing and selling the product to current and new customers. Exports, in turn, define the area's base economy. Innovation theory (von Hippel, 1988) provides the models and paradigms for understanding and predicting the role of innovation in establishing the competitiveness of a business enterprise. We apply these findings to new forms of regional economic organization in affecting a region's competitive position among its trading partners.

The millwork industry in each of the two sets of case studies is an important player in its local community. Even in the county with the largest population and workforce, which is also a part of the Minneapolis-St. Paul LMA, it accounts for one-third of the local exports. Its rural counterpart in the same economic region accounts for more than two-thirds of the local base economy. We conceive of regional economic organization at three levels, as described below.

A long-established millwork manufacturing company with more than $250 million in sales dominates the millwork industry in the Minneapolis-St. Paul local labor market area. Its proximity to a large metropolitan market gave it an early market advantage. This advantage was sustained by above-average growth of the local housing market and residential construction industry. Product output and value added per worker for the millwork industry rose to levels well above the average in other industries as the dominant millwork company cluster improved its own competitive position in regional markets. In the metropolitan county, value added per worker in 1990 was $38,000 for all industry compared to $62,000 in millwork manufacturing.

Each of the local areas containing the three millwork firms is marked by a high level of local product specialization and correspondingly high levels of local exports and imports. Total exports, as a proportion of total commodity output, range from 30 percent in the economic region to 33 percent in the inner core area to 74 percent in the rural area. Local labor and capital resources, rather than natural resources, thus sustain the high levels of exports per worker that are possible in each case because of cost-effective access to product markets and supply sources outside the local area.

The local workforce and the capital and entrepreneurial resources of local industry provide one set of critical core competencies for successful business enterprise. They include the skilled labor and exceptional know-how needed for reaching distant markets with competitively priced products. In addition, larger state and federal transfer payments sustain high levels of public expenditures in the Duluth-Superior LMA, which, in turn, support high levels of imports.

Table 10-1 shows the contribution of the millwork manufacturing industry to the local, area, and regional economies. This is the value added by the

TABLE 10-1 Millwork industry income payments for specified services, Minneapolis-St. Paul Economic Region and Norway, 1990

		Economic Region		Labor Market Area			Locality		
							U.S.		
Economic Indicator	Units	Mpls-St. Paul	Norway	Mpls-St. Paul	Duluth-Superior	Trøndelag	Metro	Rural	Norway Rural
Employee compensation	mil. $	520.0	476	256.9	83.0	46.3	230.3	79.3	1.9
Indirect business taxes	mil. $	10.7	*	4.3	2.9	*	3.6	2.8	*
Proprietary income	mil. $	36.3	*	10.0	6.4	*	8.0	6.1	*
Other property income	mil. $	105.2	*	52.0	16.8	*	45.0	15.5	*
Total value added	mil. $	672.2	646	323.2	109.1	62.3	286.9	103.7	2.5
Total industry output	mil. $	1,451.6	2,124	587.8	271.7	215.3	510.8	261.4	12.6
Valued added per $1 million output	mil. $	0.5	0.3	0.6	0.4	0.3	0.6	0.4	0.2
Total output per $1 million employment	mil. $	2.6	4.5	2.2	3.0	4.7	2.1	3.1	6.6

Source: University of Minnesota IMPLAN System Report No. 114; Manufacturing Statistics, Central Bureau of Statistics, Norway.
*Information not available.

workforce and the capital and entrepreneurial resources of the millwork industry. Other millwork firms join the single, dominant firm and the cluster of small firms at the county level in the United States and at the municipal level in Norway in the larger geographic delineation. Even then, the two dominant millwork firm clusters at the county level in the United States account for 89 to 96 percent of the millwork production in their multicounty labor market areas and for 50 percent in the economic region.

Another long-established millwork manufacturing company with more than $150 million sales is in the extreme northwest corner of the Duluth-Superior LMA, which borders on Canada. Its imports of intermediate inputs also are much greater than its local purchases. It has a major impact on local economic activity because of the recycling of its large payroll by local residents. Its county has a total population of 15,000, a gross local product of $308 million, and 10,600 jobs. A large discrepancy occurs for this county between the resident employed labor force and the job count for local industry because of commuting from nearby counties. In addition, the job count is larger because some individuals hold more than one job. The millwork industry in the rural county comes close to the overall industry average in the county in value added per worker because of high labor earnings paid by the second largest employer in the county—a snowmobile manufacturer.

A key difference in the income payments of the millwork manufacturers in the rural county and the metropolitan county is in the balance of payments between primary inputs, like labor. This is also true for intermediate inputs, particularly the purchases of imported goods and services used in production or in acquiring the production inputs.

The Norway case study focuses on the Trøndelag labor market area. This area has a total population of 377,000 and a total employed population of 182,000. The labor market centers on Trondheim, a city in central Norway with a population of 137,000. The rural periphery is a dominantly natural resource-dependent labor market.

The immediate locality for the millwork case study has a total population of 3,800 and 1,900 total jobs. The municipality (Overhalla) has primarily agricultural and natural resource-based production and some service center functions. The local millwork industry has a total employment of 53, and total employee compensation is 11.2 million Norwegian kroners ($1.6 million in 1992 U.S. dollars). The local municipality is the destination of roughly 20 percent of the local millwork industry.

The method of approach to the study of market linkages in the millwork manufacturing industry is to first estimate the total millwork market and production in each labor market area. We compare the excess production (that is, gross shipments to markets outside the local area) with the total demand for this product in each of the economic regions, starting with the Minneapolis-St. Paul Economic Region. The total demand for intermediate inputs is partitioned between local sources and imports from sources outside the local production area. The supply sources of other inputs are similarly identified, given information provided by the University of Minnesota IMPLAN (*Im*pact Analysis for *Plan*ning) System and the millwork companies. Finally, the primary inputs—labor, capital and entrepreneurship—are entirely local, including the workers commuting to their respective production sites, largely from adjacent counties in the same labor market area.

Inter-industry Linkages—Rural and Urban

Table 10-2 compares the commodity disbursements of the millwork industry in each of the three areas. Less than 0.1 percent of the local millwork industry output is purchased locally; the new residential construction is the largest market. The remaining 99-plus percent of the total goes mostly to the private construction industry outside the immediate locality. The $96 million of millwork industry disbursements to the construction industry in the inner core and the $13.3 million in the outer core area are shared about equally between regional production and imports. At the regional level, $101 million of the $243 million in total purchases are from the millwork industry in the economic region.

The total demand for the millwork industry output in the Minneapolis-St. Paul Economic Region was $1.5 billion, or slightly more than twice the combined millwork production of the two localities. Only 42 percent of the total regional requirements were purchased locally in 1990. Individual firms in this industry, of course, are heavily export-dependent. They have extensive distribution systems that link the firms to customers well beyond their immediate locality. Successful product innovation depends on close ties with their customers in the construction industries (Table 10-3).

Table 10-4 compares the millwork industry purchases for the economic region, the labor market area, and the locality. Intermediate purchases of the millwork industry from all sources total $779 million in the Minneapolis-St. Paul Economic Region. Purchases from suppliers in this region total $488 million, or 63 percent of the total intermediate input requirements. These purchases are evenly balanced between own production and imports at the labor market level. At the local level, however, local suppliers account for only 15 percent of total input purchases in the rural county and 30 percent in the metropolitan county in this economic region. Input mix differences between firms in the two localities are matched by corresponding differences in both product mix and market strategies.

TABLE 10-2 Millwork industry sales to specified intermediate demand, Minneapolis-St. Paul Economic Region and Norway, 1990

		Economic Region		Labor Market Area			Locality		
							U.S.		
Intermediate Demand Sector	Units	Mpls-St. Paul	Norway*	Mpls-St. Paul	Duluth-Superior	Trøndelag†	Metro	Rural	Norway Rural
Local, area or region:									
New residential structures	mil. $	71.1		31.6	4.2	60.5	1.4	0.1	1.3
Industrial and commercial buildings	mil. $	9.1		3.7	0.5	‡	0.2	0.0	‡
Farm structures	mil. $	3.7		0.9	0.1	‡	0.0	0.0	‡
Government facilities	mil. $	1.6		0.4	0.1	‡	0.0	0.0	‡
Residential repair	mil. $	3.3		1.6	0.2	‡	0.1	0.0	‡
Maintenance	mil. $	7.2		2.9	0.4	‡	0.1	0.0	‡
Millwork manufacturing	mil. $	2.6		0.9	0.6	31.2	0.8	0.6	0.0
Others	mil. $	2.7		0.4	0.0	17.4	0.0	0.0	0.0
Total	mil. $	101.3		42.4	6.0	109.2	2.6	0.7	1.3

Sources: University of Minnesota IMPLAN System Table No. 115; Manufacturing Statistics 1990, Central Bureau of Statistics, Norway.
*Information not available.
†All Norwegian figures are exclusive of imports.
‡The value of the item on the line is included in new residential structures.

TABLE 10-3 Millwork industry sales to specified final demand, Minneapolis-St. Paul Economic Region and Norway, 1990

		Economic Region		Labor Market Area			Locality		
							U.S.		
Indicator	Units	Mpls-St. Paul	Norway*	Mpls-St. Paul	Duluth-Superior	Trøndelag	Metro	Rural	Norway Rural
Intermediate demand, total†	mil. $	243.3		96.0	13.4	109.2	5.5	1.5	1.3
Capital formation	mil. $	0.0		0.0	0.0	16.8	0.0	0.0	0.7
Inventory addition	mil. $	9.1		3.8	1.8	‡	3.3	1.7	‡
State and local government	mil. $	0.0		0.0	0.0	0.0	0.0	0.0	0.0
Federal government	mil. $	0.0		0.0	0.0	0.0	0.0	0.0	0.0
Consumption	mil. $	0.0		0.0	0.0	10.1	0.0	0.0	0.6
Foreign exports	mil. $	3.8		1.5	0.7	12.3	1.2	0.6	2.5
Domestic exports	mil. $	1,263.2		508.3	246.8	67.4	471.7	242.1	7.5
Final demand, total	mil. $	1,276.1		513.6	249.3	106.2	476.2	244.4	10.0
Total commodity output†	mil. $	1,519.4		609.6	262.6	215.3	481.7	245.9	12.5

Sources: University of Minnesota IMPLAN System Table No. 115; Manufacturing Statistics, Central Bureau of Statistics, Norway.
*Information not available.
†Import is not included in the Norway figures.
‡Inventory addition is included in Capital formation.

TABLE 10-4 Millwork industry intermediate input purchases, Minneapolis-St. Paul Economic Region, 1990

		Economic Region	Labor Market Area		Locality	
					U.S.	
Intermediate Input	Units	Mnpls-St. Paul	Mnpls-St. Paul	Duluth-Superior	Metro	Rural
Total, including imports	mil. $	779.4	264.7	162.6	223.9	157.8
Imports	mil. $	291.7	128.0	79.2	156.9	133.9
Locality, area or region	mil. $	487.7	136.6	83.4	67.0	23.9
Distribution of locality, area or region purchases:						
Maintenance	pct.	1.6	2.0	1.7	3.5	3.4
Sawmills	pct.	22.4	8.2	28.6	7.4	20.5
Hardwood dimension/flooring	pct.	0.8	0.1	0.5	1.4	2.3
Millwork	pct.	0.5	0.7	0.7	1.3	2.5
Veneer and plywood	pct.	3.6	2.0	4.9	2.0	2.1
Reconstituted wood products	pct.	3.6	0.3	4.9	3.4	3.4
Wood products not elsewhere classified	pct.	2.2	2.1	3.5	5.4	0.2
Paper board containers	pct.	1.6	2.0	0.0	0.0	0.0
Paints and allied products	pct.	0.0	0.0	0.0	0.0	0.0
Adhesives and sealants	pct.	1.3	1.8	1.4	0.8	0.0
Misc: plastic products	pct.	0.0	0.0	0.0	0.0	0.0
Glass and glass products	pct.	2.8	2.8	0.9	0.8	4.4
Hand and edge tools	pct.	0.3	0.3	0.5	0.0	0.0
Hardware not elsewhere classified	pct.	1.1	1.1	0.5	3.4	0.1
Screw machine products	pct.	2.2	3.9	0.0	0.0	0.0
Metal stampings	pct.	1.8	3.0	0.9	0.1	0.0
Fabricated metal products	pct.	0.3	0.3	0.2	0.3	0.0
Power driven hand tools	pct.	0.5	0.4	0.2	0.2	0.0
Woodworking machinery	pct.	0.0	0.0	0.0	0.0	0.0
Motor freight and warehousing	pct.	6.8	8.3	5.7	11.1	10.9
Communications	pct.	1.3	1.9	1.3	1.5	0.7
Electric services	pct.	4.8	4.9	6.0	9.1	5.9
Wholesale trade	pct.	15.9	20.5	11.3	13.9	14.0
Real estate	pct.	1.8	3.7	1.6	5.0	1.5
Advertising	pct.	9.8	11.9	12.0	14.4	10.1
Consulting services	pct.	0.7	1.3	0.7	1.0	0.0
Other	pct.	12.2	16.4	12.0	13.9	18.0
Total	pct.	100.0	100.0	100.0	100.0	100.0

Source: University of Minnesota IMPLAN System Table No. 114.

Sawmill products, wholesale trade, and transportation are the largest purchases of the millwork industry in the rural county. We add screw machine products, electric services, advertising, and real estate to this list for the metropolitan county. The sawmill output from Oregon and the Pacific Northwest is the principal source of the sawmill lumber used in both the metropolitan and the rural millwork industry clusters in the United States. It is the largest single purchase of the millwork industry.

An important complement to labor and entrepreneurial skills of globally competitive local enterprise is the education and training offered local residents to improve their work skills. Large differences in education and training requirements occur among individual economic sectors. For the two localities in the Minneapolis-St. Paul Economic Region, however, the labor skill requirements for millwork manufacturing are essentially the same. In both localities, the millwork industry succeeds because its dominant firm is well established, with a long-standing record of producing a dependable product with a strong brand name that is preferred by construction contractors. Each firm is increasing its market share by both improving its existing product line and introducing new products. An extensive distribution system provides low-cost access to input suppliers and product markets. The Trøndelag labor market area has also a well-developed education system on all levels that are relevant for the millwork industry.

The dominant firms in the U.S. case studies now face the prospect of new restrictions on facility expansion because of local concerns expressed both by landowners adjoining the proposed expansion sites and by local environmental organizations. New forms of environmentally motivated industry regulation offset the existing comparative advantage of the dominant firms, thus imposing a threat to these firms' workforces of relocation to less restrictive local environments. The reshaping of local institutions that simultaneously support both successful export-producing enterprise and legitimate concerns about environmental degradation is minimized or avoided by local and state governments. They are, however, unsure of popular support for their efforts.

Coping with Open Markets

Rural regions with a long agricultural history usually have many small businesses. In the United States, small businesses account for most of the job growth.[1] Large companies often buy the expanding small businesses as a way of increasing market shares. This happens quite often because of the advantages large companies have in financing expansion, coupled with the difficulties faced by small businesses in acquiring capital financing. Small businesses must work closely with their markets and customers to produce salable products that provide a reasonable return to the producer. They must find ways to make the most of their linkage not only with local customers, but also with more distant customers that have many sources of supply for inputs that go into the products these firms produce. Small businesses must also seek ways to work closely with regional schools and colleges to keep abreast of changing skill requirements in the workplace. Given today's economy, building market access bridges for small businesses in rural regions is no easy task.

Central Norway is not alone in coping with open markets. Rural Minnesota has its problems, also. Clay County, bordering North Dakota in western Minnesota and part of the Fargo-Moorhead LMA, has much in common with central Norway and other rural areas in Scandinavia threatened by entry into the European Union and its Common Market. This section is about building bridges to new markets, which the revitalizing of rural regions involves.

Free trade agreements and global competition trigger new forms of economic interdependence that in varying degree link rural and urban areas. The underlying challenge of this process is the reshaping of existing local institutions to provide the essential regional infrastructure for successful export-producing business enterprise. Such enterprise defines the scope and character of every trading region and determines its viability in global competition. An essential element of this viability is a supportive economic environment for innovative business and government enterprise.

Rural Responses to the North American Free Trade Agreement

The Clay County experience in coping with the North American Free Trade Agreement (NAFTA)[2] provides a case study of the varied local responses to a perceived threat. The potential stakes are high. They deal with the economic well-being and viability of a dominant sector of a highly specialized and prosperous local economy based on agriculture.

Clay County Economy

In 1990, Clay County businesses and governments produced $1 billion of goods and services. Total population was more than 50,000. The workforce consisted of more than 22,000, with more than $420 million earned in wages, salaries, and self-employed income. Exports totaled more than $445 million. The county economy added nearly $600 million to Minnesota's gross state product.

We compare Clay County with North Trøndelag County in total population, employment, and exports in Table 10-5. In 1990, North Trøndelag's population was 127,000, more than 2.5 times the Clay County population. Its employment was nearly 2.5 times as large. The Clay County employment represents jobs rather than persons employed—the jobs per 1,000 population being larger, in

TABLE 10-5 Total employment and export share, North Trøndelag County, Norway, and Clay County, Minnesota

Title	SIC Code	North Trøndelag Total Employment 1980 (no.)	North Trøndelag Total Employment 1990 (no.)	Clay County 1990 (no.)	Proportion of Total: North Trøndelag (pct.)	Proportion of Total: Clay County (pct.)	Export Share, 1990: North Trøndelag (pct.)	Export Share, 1990: Clay County (pct.)
Total employment		48,175	50,389	22,263	100.0	100.0	100.0	100.0
Agriculture	01–07	7,874	6,275	1,983	12.4	8.9	3.4	14.5
Mining, forestry, fisheries	08–14	575	1,668	14	3.3	0.1	0.8	1.6
Construction	15–17	4,566	4,107	1,348	8.2	6.1	3.3	0.0
Manufacturing	20–39	9,713	8,681	1,274	17.2	5.7	56.4	56.5
Services-producing, total	40–89	25,447	29,658	17,644	58.9	79.3	32.8	27.2

Sources: University of Minnesota IMPLAN System; Maki and Westeren, 1992.

part, because of multiple job-holdings, many available jobs for both genders, and some commuting to Clay County jobs from outside the county.

Comparing the two areas, we find that the distribution of employment differs, with North Trøndelag having more than from Clay County in farming and manufacturing and less in services-producing industries—transportation, communications, and public utilities; retail and wholesale trade; finance, insurance, and real estate; and private services. Dependence on exports, however, is much the same in the two areas, except for the slightly smaller percentage of farm products exported from North Trøndelag than from Clay County (because local residents and processing plants utilize more of the total farm production locally in North Trøndelag than in Clay County). The services-producing sector generates exports, also, because of the marketing services provided by local transportation and other marketing businesses that accrue to the commodity exports as they move to their final destinations.

The export-producing sector of Clay County is almost entirely agriculture and agribusiness. In 1990, this sector accounted for nearly 30 percent of the county's total industry output and 70 percent of its total exports. Sugar beet production and beet sugar manufacturing are the largest contributors to these exports that represent the county's economic base—the activities that bring the "first dollars" into the county. Any threat to a sector this important to the local economy becomes a vital concern of its workers and families.

NAFTA—A Clay County Perspective

Several important businesses and many individuals in Clay County view NAFTA as a double-edged sword. There is promise of expanded markets and increased economic opportunity for some U.S. industries and workers. But economics, like politics, is local when it really counts. There was a strongly held belief in Clay County that NAFTA would eventually destroy a mainstay of the local economy—the beet sugar industry. Our analysis shows, however, that these fears were ill-founded if you looked at only the short run, say up to five years. The analysis concluded that in the short run an increase in Mexican sugar exports to the United States that would force a sharp drop in sugar prices was unlikely. It provided us with a series of arguments, as follows:

1. Sugar consumption is increasing faster than its production in Mexico.
2. Sugar production costs in Mexico are nearly twice those in the United States.
3. NAFTA identified sugar as a "sensitive" product, with provisions to phase out tariff rate quotas against Mexico sugar over a fifteen-year period. Furthermore, tariff rate quotas on sugar will be maintained, meaning that the amount of sugar Mexico exports to the United States would automatically decrease the amount of sugar other countries are allowed to ship to the United States.

In the short run, the more likely possibility is an increase in sugar exports from Clay County. This would result when Mexico reduces its 55 percent tariff on imported sugar while its sugar consumption continues to increase.

Long-run effects—those beyond NAFTA year 15—are the ones that worry Clay County sugar producers. For Mexico to become a net surplus producer, however, the Mexican sugar-producing and sugar-using industries must extensively restructure in the following ways by year 15.

1. Mexico's beverage industry must switch from sugar to high-fructose corn syrup—a decision that Pepsi and Coca-Cola would probably not make because of its high direct and indirect costs.
2. Mexico's sugar-producing area must increase by 20 percent, from 530,000 to 620,000 hectares.
3. Mexico's government must continue efforts to privatize its previously nationalized sugar industry—a very difficult step that would force closure of some mills but allow the remaining mills to become more competitive.
4. Transportation and infrastructure problems, including electricity for pumping water through irrigation systems, must be solved, along with land tenure and long-term capital financing problems that would allow Mexico to produce 800,000 metric tons of surplus raw or semirefined sugar.

Even if Mexico could produce 800,000 metric tons of surplus sugar by 2008, built-in provisions in NAFTA would soften the blow on the Clay County beet sugar industry. For example, if all restrictions on Mexico's exports of surplus sugar were eliminated by year 15, tariff rate quotas for sugar producers outside the United States would still be in place. This would mean that the imports of sugar from other countries could be adjusted down to make up for the expanded Mexican sugar production. Of course, countries that experience a reduction in their quota from trade restriction practices such as this may offer challenges under the General Agreement on Tariffs and Trade (GATT).

The analysis for this portion of the text concluded that the Clay County beet sugar industry had the least to fear of any segment of the sugar industry. Hawaii and Louisiana producers would be among the first to fail because of their high production costs. The Clay County beet sugar industry, with its historically low production costs and superior access to the large midwestern markets, would stand the best chance of survival beyond year 15 of NAFTA.

The study findings show that NAFTA makes Clay County wheat producers slightly better off than they are now. Over the first ten years, Mexico barriers to U.S. wheat through state trading, import licensing, and tariffs will be eliminated. Projected U.S. wheat exports to Mexico thus show an increase from

740,000 to 1.5 million metric tons, with 20 percent of the increase being attributed to NAFTA.

Wheat consumption per person per year in Mexico is small compared to U.S. consumption—52 kg versus 78 kg, respectively. As personal income and population increase, however, and the price of wheat falls, consumption will increase. This still would account for only a small portion of total U.S. wheat exports. Moreover, because of a transportation advantage, the hard red winter wheat producers in the southern plains region of the United States would be the greatest benefactors. Wheat shipped from the southern plains region to Mexico would reduce shipments to other markets, thus increasing their demand for Clay County wheat. This increase would be small compared with total production.

To present these findings, the study team prepared three scenarios of alternative industry futures, in which three sets of conditions are postulated that lead to three different results for the Clay County beet sugar industry. In the *worst-case scenario,* Mexican sugar producers would reach a net surplus of 800,000 metric tons by NAFTA year 15. Mexico sugar producers would ship this surplus to the United States. The United States could then adjust its tariff rate quotas for other countries to make up for the 800,000 metric tons of new sugar. This is, however, in violation of GATT rules. The United States would be forced to amend the sugar import tariff rate quotas and take the additional foreign sugar to the amount of 1.25 million metric tons. The additional sugar would not be absorbed by the combined United States, Canada, and Mexico sugar markets. Therefore, the U.S. Commodity Credit Corporation would be required to buy U.S. sugar to prevent prices from falling below the market stabilizing price. This, of course, would interfere with the "no cost" provision of the sugar legislation. The U.S. sugar industry would lack its former government protection and support, thus resulting in an eventual decline in the price of sugar that the Clay County beet sugar industry could no longer absorb. One spokesperson for the sugar industry claimed that the results portrayed in this scenario "would expose producers and consumers to large and unpredictable swings in price. One very possible outcome would be the end of the sugar industry in the Red River Valley."

The study team also presented a *medium scenario* and a *best-case scenario.* They similarly listed a set of conditions accompanying a corresponding set of results for the industry. The medium scenario, for example, shows Mexican sugar producers reaching a net surplus of 10,000 metric tons of raw and semi-refined sugar by year 15. U.S. sugar mills continue to supply Mexico with refined sugar. This small amount of additional sugar imports into the United States would have no measurable net effect. Such a conclusion assumes that tariff rate quotas of other foreign imports would be adjusted downward, domestic sugar consumption would continue expanding, and high cost production areas in the United States would continue dropping out of sugar production. The best-case scenario presents an even more optimistic set of market prospects for the Clay County beet sugar industry. The scenario approach thus gave the study team a way of probing more deeply into the concerns of sugar producers on one hand, and on the other, the possibilities of negotiated solutions within the context of NAFTA provisions.

Revitalizing a Rural Region's Economic Base

The final chapter in the Clay County-NAFTA saga is yet to be written. We can say, however, that the NAFTA agreement on sugar put the beet sugar industry in the neutral camp, which, in turn, gave President Clinton additional votes for the U.S. Congress to pass the North American Free Trade Agreement by a very close margin. A majority of Clinton's own party opposed the agreement. The negotiated agreement is certainly not everything that the industry had wanted. It nonetheless gives the Clay County economy a much appreciated respite from having the sugar markets immediately adjust to the new uncertainties. These undoubtedly would have had adverse long-term effects on Clay County sugar beet growers and processors.

Negotiating Adjustments to Prospective Market Changes

Before the fateful day arrives when the last sugar beet grower hauls the last load of sugar beets to a nearby processing plant, every conceivable effort will have been explored to keep the industry competitive. The Red River Valley beet sugar growers and processors are still in a stronger financial position than any other segment of the industry. The competition they face from cane sugar and corn fructose, however, is fierce, and it will grow as NAFTA gradually reduces import quotas on sugar. Local communities are well advised, therefore, to look for new sources of income-generating exports to replace the losses that would occur with the demise of the beet sugar industry.

Clay County has been mentioned as a place to locate one or more value-added agricultural processing facilities. A corn fructose manufacturing facility was almost built, but high taxes and employee compensation costs, coupled with Clay County's peripheral location to the major corn-producing areas, finally scared away promoters to another location. A second possibility is an ethanol plant. Again, the peripheral location and high production costs, coupled this time with the uncertainty of state and government subsidies, make this possibility also an unlikely one. What remains is full-time or part-time employment in manufacturing, marketing, and the private services industries. Because of the recent rapid growth of these industries in western Minnesota, this area actually faces a housing shortage. If housing were adequate, it would most likely face a skilled-worker shortage.

The remarkable news about rural Minnesota is the growth of its manufacturing sector. This is at a time when manufacturing is declining in total employment elsewhere in the country. What contributes to this turnaround for rural Minnesota is the high cost of doing business in metropolitan areas. This includes the additional costs of labor, rents, taxes, pollution abatement, and industry regulations—costs that are lower in the rural areas. These are negatives for existing manufacturing plants in metropolitan areas that become positives for the rural areas in Minnesota. In addition, rural Minnesota has some positives—a skilled and productive workforce; a strong work ethic; abundant recreational amenities; a strong commitment to active participation in the emerging knowledge-driven, learning society; and access to regional markets by interstate highway, rail, and air. These advantages are especially strong for rural areas within 75 to 100 miles of Minneapolis-St. Paul—one of the 29 U.S. metropolitan centers that serves also as an air transportation node in the global communication and transportation networks.

Finding a Competitive Advantage

The Minnesota Extension Service and its extension educators throughout the state, with the support of researchers in applied economics, dairy and poultry sciences, food technology, and other fields at the University of Minnesota, assist rural communities in finding their competitive advantage as they cope with changing market conditions. This, again, is no easy task, and there is no one way of doing it. Efforts in accomplishing this task point to the importance of understanding clearly the significance of the following:

1. Economic and demographic trends—identifying present and prospective export-producing businesses.
2. Community economic base—delineating, differentiating, and redefining the functional economic community[3]; assessing the competitive position of local businesses in export markets; business linkages with other businesses, industries, and areas; role of tourism.
3. Community-based economic development strategy—capacity-building and networking; organizing community-based economic development corporations or cooperatives; establishing accountability for the management of the development corporations and/or cooperatives.

Clay County is part of the Fargo-Moorhead LMA—the *functional economic community*—a daily commuting area that straddles the Red River, with counties in both North Dakota and Minnesota. The two-county Fargo-Moorhead

Metropolitan Statistical Area listed by the U.S. Bureau of the Census has a total population of more than 100,000. It is the most densely populated part of the LMA. Markets for the export-producing businesses in the Fargo-Moorhead LMA lie outside the LMA boundaries, which may change in size, of course, from one year to the next. Thus, the destination for some of the exports from a single county is another county in the same LMA. Total exports for the LMA are, therefore, less than the sum of the individual county exports. This difference represents the intra-area, county-to-county trade. This sort of local trading provides opportunities for small businesses to acquire the production and marketing know-how for succeeding in larger and more distant markets. We call the process *import replacement* or *import substitution,* depending upon the local market destination, whether intermediate (that is, semifinished products) or final sales. It eventually may become a form of *export expansion.*

Studying Clay County economic and demographic trends without assessing its changing status and role in the larger LMA is likely to result in seriously underestimating or overestimating these trends. What they show for the Fargo-Moorhead LMA are the changing patterns of business investment and growth from one sector to the next, patterns that also vary among individual counties. Identifying future growth sectors with profitable investment prospects is difficult, even with an abundance of decision information for the individual investor. Not taking into account the overall LMA growth prospects means, of course, that the chances of making profitable investments are extremely low.

Study of Clay County only as part of the Fargo-Moorhead LMA will not show the important business linkages, present and potential, with the larger metropolitan core area—the Minneapolis-St. Paul LMA—200 miles to the southeast. Clay County manufacturing businesses may extend their export markets beyond the local LMA, currently and in the future. They may serve as production input suppliers, for example, to manufacturing businesses in the Minneapolis-St. Paul LMA. They may even have several intermediate product outlets in the Minneapolis-St. Paul market. Alternatively, they may ship finished products to consumer markets that have expanded beyond the local LMA. Individual business employees exchange ideas and suggestions with one another for improving product delivery and even the product itself. This is a common practice of process and product innovation that helps both sides of the market to remain competitive. Thus, in many ways, not only in manufacturing, the Fargo-Moorhead economy is closely linked to the Minneapolis-St. Paul economy.

One LMA's economic base differs from another's for many reasons. One LMA, like Fargo-Moorhead, may specialize in agriculture and related industry. Another, like Minneapolis-St. Paul, may specialize in technology-intensive manufacturing and high-order services. A multiplicity of criteria would apply to these two types of areas in an assessment of their economic base. Among such criteria are the *risks and costs* of production and the *productivity and flexibility* of resource use in production (Maki and Westeren, 1992).

1. *Risks* are high in an area with extreme specialization of economic activity. These risks correlate with both early and later stages of a product cycle. Introduction of a new product incurs costs of investing in its development and marketing. Cost recovery depends on the market success of the new product. An old product, on the other hand, is in danger of losing its market to competing products on a quality or cost basis. When these are export-producing businesses, adverse multiplier effects hurt the entire community. In either case, high-risk businesses face reduced prospects for business loans and favorable credit terms from suppliers. Rural regions, removed from access to large consumer and producer markets, are singularly dependent on farming, fishing, and mining. They lack prospects for diversifying their local economic base. In such cases, local credit sources soon disappear while distant sources heavily discount their high-risk loans.
2. *Costs* increase with risks that are heavily discounted. On the other hand, high wages and extreme regulatory and environmental requirements in metropolitan areas add to the costs of doing business. Many rural areas offer cost advantages because of low site costs—land and buildings, labor, environmental regulations, and congestion. Other costs often are less in rural than in metropolitan areas. These would include business and personal services and housing.
3. *Productivity* of resource use usually correlates with investment. It is high where investment per worker is high, which is generally high in metropolitan areas, low in rural areas. Capital substitutes for labor, which increases overall resource productivity. Despite an initial disadvantage in low investment per worker, residents in rural areas have opportunities for reducing these disadvantages by wisely directed investment in education and training that closely tracks a changing job market. A strong work ethic and supportive attitudes and values, when coupled with new and superior skill development, make some rural areas superior places for new business formation, branch plant location, and expansion of existing businesses.
4. *Flexibility* in work practices and organization is important for successful business enterprises facing increasingly stiff competition in both producer and consumer markets. Flexibility of access to input supply and procurement practices is also important. In both cases, flexibility is a function of the learning processes in an organization and the ability to gauge the likely outcomes of improved learning capacities. Successful business organizations

have the robustness and capacity for adaptation to changing market conditions and new production technologies. They are the successful learning organizations.

These four criteria—risks, costs, productivity, and flexibility—help in differentiating vulnerable from less vulnerable rural areas. Unanticipated changes in export markets and income flows are endemic with vulnerable rural areas and natural resource-based economies. Strong linkages with core area businesses help reduce a rural area's vulnerability to unanticipated economic change by reducing the risks and costs faced by individual businesses. The decentralized markets of economic theory function well because of the variety of contractual arrangements between buyers and sellers. Recurring negotiation sustains and improves these contracts. Core area manufacturing businesses, for example, initiate numerous contractual arrangements with businesses in rural areas to produce their product in peak periods or in periods of tight labor supply. They may initiate other contractual arrangements for integrating rural input supply sources into their production scheduling through professional and technical training programs in supply management, inventory control, and production scheduling.

Vulnerable rural areas learn from viable rural areas when an assessment of the differences between the two focuses on productivity and flexibility. Businesses forming, locating, or expanding in transitional rural areas adjacent to the metropolitan core area have an important advantage over those in the more distant labor market areas. They have proximity to world-class information systems and producer services for adapting rural area businesses to changing production technologies and global markets. In these areas, business volatility—changes in total employment due to births and deaths, expansions and contractions of business enterprises—correlates with economic growth.

Finally, capacity-building, as a purpose of community-based economic development strategy, refers to the successful application and integration of the means of control and foresight—good management coupled with realistic anticipation about the future—in both private and public management. From an economic perspective, a strategy for capacity-building must eventually focus on the community economic base and the economic environment for successful export-producing enterprise. Skill-based education and employment counseling activities, for example, enhance the availability of various labor skills that, in turn, improve the competitive position of a community's export-producing businesses. Readily accessible and well-managed private and public delivery systems that provide adequate health care at competitive prices and costs (relative to the competition in export markets) also contribute to a favorable local economic environment.

A series of development strategies were identified in a recent symposium on rural development in the United States.[4] None of the proponents of these strategies, however, cited linkage or proximity to metropolitan core areas as relevant to their success. The underlying rationale for rural development in this case assumed a primacy of federal or central government leadership and support, rather than the empowerment of state, local, and area organizations and their spokespersons as participants in the formulation of a rural development strategy. The conference organizers had overlooked the fact that rural economies vary greatly from area to area and that local capacity-building is an important precondition in successfully adapting the tools of community-based economic development to recognized differences in each community's economic base.

Tracking Rural Area Performance

As mentioned in Chapter 4, the Puerto Rico Study Team, with the cooperation of Richard Weisskoff, University of Coral Gables, and Angel Ruis, University of Puerto Rico, together with professional staff of the Puerto Rico Planning Board, constructed a new economic modeling system for extending the Weisskoff model of the Puerto Rico economy to more recent years and providing greater industry disaggregation and detail. The new modeling system can be upgraded from its 1982 database to recent-year databases when these become available. (Puerto Rico Study Team, 1994; Puerto Rico Planning Board, 1990).

Labor Markets in Developing Economies

Potential future uses of the new Puerto Rico model include the assessment of alternative economic development strategies and their consequences for specific industries, occupations, socioeconomic groups, and labor market areas in Puerto Rico. The areas differ sharply in their resident populations, both household and business, and they have correspondingly different requirements for basic infrastructure and essential community services. If Weisskoff's economic modeling efforts of the mid-1980s were replicated for the mid-1990s and beyond, they could again address the earlier concerns about export specialization and import dependency—the hallmarks of an economy that lacks the local industry for competing successfully in local markets. In addition, these modeling efforts can start with the possibilities of locally initiated cooperation and linkages among small- and medium-size enterprises for building a new industrial structure of locally owned businesses. These modeling efforts identify goods and services for niche markets in Puerto Rico and explore the effects of these developments on jobs and incomes among its labor market areas.

The local labor market approach contrasts with the earlier ways of thinking about economic development by its attention to local institutions in building a civic community delineated by a certain geographic space, namely, the daily commuting area of the local workforce. This approach provides the conceptual environment for building both a theory and a model for economic development. The theory addresses the "local" of local economic development, but within the framework of the larger region of which it is an integral part. The theory is rooted in the notion of a functional economic community, circumscribed by the invisible commuting area boundaries of the local labor market. This functional community serves also as a building block in the formation of an economic region composed of several LMAs, of which one or more form its core area, while the more distant ones form its periphery.

Lessons from Central Norway and Puerto Rico

We now return to the earlier question: What are the lessons from central Norway and Puerto Rico that apply to regions like Chiapas, Mexico, and its economic development? Clearly, none of the alternative ways of thinking about economic development apply directly to the two geographic areas, although each offers insights about historic processes of growth and change affecting local economies. The alternatives range from the general to the specific and the simple to the complex, each portraying the role of government and the private sector and how one relates to the other in economic development. If government intervenes in economic development, however, the questions about the criteria and rules of engagement directing this intervention and determining its magnitude, focus, and duration still remain largely unanswered. We seek evidence of the importance of these elements in the outcomes of the central government's interventions in local and regional economic development in Norway and Puerto Rico and the lessons they convey for Chiapas in its current and future economic development efforts.

Economic development efforts in Norway, these past 25 years, can be divided into three periods. The decade of the 1970s represents a shift in the regional development pattern from the preceding postwar years. During this time, a concomitant stabilization and consolidation in the regions accompanied the movement of population from rural areas. The stabilization of the 1970s came about in no small part because of expansion of the government sector, particularly the commune (municipality), which had rapid growth in employment. The first half of the decade evidenced growth in manufacturing, until the oil crisis of 1973–74. An increase in government support of the primary sectors, especially agriculture, occurred during the 1970s.

During the 1980s, regional imbalances increased once again, according to three important indicators—population growth, employment, and internal migration. The peripheral areas underwent a reduction in employment during the 1980s, as well as negative population growth due to increased out-migration, especially of persons in the 20-to-40 age range. The reduction in employment in primary sectors continues, but the changes are not as dramatic as they were in the earlier period. The reductions in manufacturing employment are greater in rural areas than in the rest of the country. At the same time, the growth in the private service sector is well under the average for urban areas for the nation. The growth in the public sector has a continued effect on the rural areas, but it alone cannot compensate for the reductions in other sectors. The trends from the 1980s continue into the 1990s. Additional factors now come into play as rural Norway enters a new era of competition.

Regional development policy observed in rural Norway can also be seen in many western nations. Within the European Union and the United States many governmental programs exist for rural areas under the slogan "Rural Revitalization and Rural Rebirth." Types of recommended activities are quite different, however. Generally speaking, the initiatives in the European Union and United States can be divided into three groups:

1. *Traditional initiatives.* Here the effort is for increase in production and employment in traditional agriculture, with subsidized loans to all types of production and infrastructure improvements, such as roads, sewer, and water.
2. *Small-scale rural development initiatives.* In this model one finds, for example, professional consultation to establish new activities, increase entrepreneurship, and develop production based on high technology.
3. *Natural and human resource development.* Here we often see, for example, in-service training and other skill-development efforts. It is also important to try to use the wilderness as a resource for job creation. We can harvest and utilize different raw materials as a basis for industrial production. The wilderness can also serve as an attraction for tourists.

In both the European Union and the United States, we have seen many examples in which the three types of programs have been combined in unique ways, but there is no common trend toward the strategy eventually selected. In the European Union, the recognition surrounding regional development and the willingness to provide funds for these purposes have grown in recent years. Pure rural development programs have never had a real breakthrough in the European Union, where one finds a clear preference for central development of regional politics. In the United States we have seen a flowering of efforts for rural development both in number and scope, at the federal level and among states that have their own special programs for rural development. Such programs are often in collaboration with state universities. These programs include efforts to develop small- to medium-size communities in peripheral areas. We have used a "shotgun" approach to rural policy formulation, trying whatever we can when a problem shows up with no long-term perspective among the public policy approaches.

Traditionally, Norway has allocated substantial resources for development in rural areas. This support has come from both general economic policy and special efforts for rural development.

The 1990s marked a shift to a new competitive age for rural areas. One important change is having more open markets. This creates difficulties for rural areas with an economic base dominated by primary production—agriculture, forestry, fisheries, and mining. These products are readily produced in other countries. Of special importance is agriculture, which now faces a change from a system based on import restrictions to a system based on duties. The GATT agreement does not seem to have substantial effects on agriculture in the short term, but long-term effects can dramatically change the competitive situation for agriculture in rural areas in Norway as well as other developed countries.

Changes in productivity in agriculture and manufacturing have led to a significant reduction in employment in rural areas. Agriculture in North Trøndelag had a yearly increase in labor productivity of 3 percent during the 1970–90 period. In comparison, manufacturing, with its sheltered wood products and food-processing sectors, had productivity increases of only 1 to 2 percent. Export industries, like pulp and paper, metals, and chemicals, have on average increased labor productivity by 5 to 7 percent per year. Within the agricultural sector, we anticipate that the labor productivity will continue on a high level with a yearly increase of 3 to 5 percent. To keep employment reductions small, one solution is to enhance product value. The trend has already started but has severe technological and marketing limitations.

In the 1980s, employment growth in the private sector was much greater in urban than in rural areas. This led to substantial shifts in regional population and industry—a trend continuing in the 1990s. Further growth will depend on the continuing financing of municipal activities by the central government. Reorganization of important public sector activities, such as postal service, telecommunications, and railroads, has occurred in the 1990s. The cost savings from breaking up these government monopolies are real enough, but the effects often impose problems of rising levels of unemployment in some, if not all, rural areas.

We relate to the economic development of Puerto Rico since 1950 in three periods: the first from 1950 to approximately 1975, the second continuing to 1990, and the third beyond 1990. The beginning of this new era of Puerto Rico's economic development marked the end of its earlier plantation culture. The export-producing coffee plantations of the 1800s were followed in the late 1890s by a new export-producing sugar plantation-based agro-industrial economy. Sugar production and milling remained a dominant export-producing industry through the sugar crisis of the 1920s until the Great Depression, which shook this industry and left many of its former workers unemployed. As an industry, it was too weakened from World War II and postwar trade restrictions and international competition to fully recover. In 1950, agriculture as a whole accounted for slightly more of the gross product ($400 per capita in 1954 dollars) than all of manufacturing, including sugar production—19.4 versus 17.6 percent. By 1970 the percentages were reversed and then some—3.4 versus 24.8 percent. Manufacturing had decisively overtaken agriculture as the dominant player in Puerto Rico's basic economy. Both the trade balance and unemployment increased sharply, leaving the entire economy dangerously dependent on foreign imports. This meant diminishing prospects for job creation in local labor markets, especially on the part of small and newly founded businesses. Because of large transfer payments from the United States, per capita gross product in constant dollars nearly tripled during this 25-year period.

The search for federal funds became "big business" in the 1970s. Net federal transfers increased from $80 million in 1970 to approximately $600 million in 1975 and more than $1.7 billion in 1980. Introduction of the Food Stamp Program in 1975 accelerated the growth of transfer payments. Other federal transfer payments increased even more rapidly than food stamps in the late 1970s. The subsequent increases in disposable income led to corresponding increases in household purchases of imported, rather than locally produced goods and services. Rapidly expanding exports and imports, without corresponding increases in employment, characterized this second period of economic growth.

The failures of externally induced growth in Puerto Rico gradually precipitated a rethinking of its economic development strategy. Tourism is now targeted as the new growth industry for special attention by the central government. At the same time, labor-intensive industry is being favored through various tax incentive programs. Thus, the 1990s marked the beginnings of a new approach to economic development that emphasizes Puerto Rico's comparative advantage as a subtropical tourist destination, and its favorable environment for the production processes of technology-intensive industry, such as pharmaceuticals, medical devices, and other scientific and controlling instruments.

Lessons for Chiapas

The Mexican state of Chiapas, which lies 200 miles south of Mexico City, is probably best known for the Zapatistas armed movement and uprising. Much has been written about the social, economic, and political conditions in the highlands of Chiapas, especially the extreme poverty and the lack of health care and housing, along with near starvation of the Chiapan peasants, who are mostly Indians. Public attention has focused mainly on the reasons for the armed movement, why it occurred when it did, and how the central government must now cope with its immediate national consequences. Very little is being written about the economic potential or the most effective ways of tapping this potential to help redress the grievances of the past and the present that precipitated the armed revolt.

The main difference between the Zapatistas movement and the guerillas of the past decade in Latin America is that they do not want to replace the existing central government and usurp its powers. Their participatory democracy would address immediately and directly the economic development issues of their impoverished highland communities as well as the effective utilization, on a sustainable basis, of the presently untapped and underutilized natural resources of the tropical lowlands, especially of the Lacandon jungle.

We start with some ideas about the future economic growth of Chiapas following the signing of a peace agreement.[5] We take into account present conditions for regional economic development in Mexico and learn from the economic development experiences of central Norway and Puerto Rico. We compare, first, the two more similar economies—Chiapas in 1990 with Puerto Rico in 1970 and 1980, (Table 10-6). The two compare closely in population, but diverge in other economic indicators. Chiapas mirrors several of the Puerto Rico indicators for 1970, more so than for 1980, while other indicators compare more closely with those of Puerto Rico of an earlier vintage. For example, Chiapas is more rural than Puerto Rico, with a much larger proportion of its people dependent on agriculture. Chiapas has fewer transfer payments than Puerto Rico, along with lower earnings per worker, which results in lower disposable personal income and imports. However, net out-migration is greater for Chiapas than for Puerto Rico. Emigration remains a critical "safety valve" for the unemployed labor workforce in both economies.

Table 10-7 summarizes selected measures of economic success in the form of quality-of-life indicators for Puerto Rico and Chiapas. These measures again reveal the results of lagging economic growth. Few of the benefits of a modernizing commercial economy are available to large numbers of Chiapans, especially in health care and education.

TABLE 10-6 Selected economic indicators, Puerto Rico and Chiapas, Mexico

		Puerto Rico		Chiapas
Indicator	Units	1970	1980	1990
Population:				
Total	thousands	2,710	3,184	3,210
Rural	thousands	NA	NA	1,914
Urban	thousands	NA	NA	1,296
Labor Force:				
Total	thousands	686	783	875
Male	thousands	NA	NA	755
Female	thousands	NA	NA	120
Participation rate	percent	45	43	27
Unemployment rate	percent	10	17	25
Employment:				
Total	thousands	617	650	655
Agriculture	thousands	62	32	386
Manufacturing	thousands	182	162	72
Services	thousands	370	456	297

Source: Puerto Rico Study Team, 1994.

Education is one of the most important aspects in the development of any region. In this case Chiapas, as an all-Indian region, has a serious lack in education. In 1990, 70 percent of the people older than 15 years could read and write, while 30 percent could not. Urgently required are programs to improve the literacy of the adult people, mainly women. Most young people not attending school live in rural and remote areas. In addition, programs are needed for training adults to acquire job skills in all productive sectors. While both public and private schools can provide education and training, the government has a central role in providing the needed funding and leadership.

Crop agriculture is the most important productivity sector in the region. One very important problem in this sector is its system of land tenancy. Chiapas has been characterized as one state where the agrarian reform did not take place. Improving the quality of crop plantings and better using available production services can lead to an increase in the value of locally produced agricultural products. Only 26 percent of cultivated land was sown with genetically superior seeds, while 66 percent was cultivated without use of the technical extension services provided by government institutions and private organizations. Only 16 percent of the cultivated area was attended by plant sanitation services. Fertilizers were used on 53 percent of the cultivated land. Only 23.4 percent of the cultivated area was worked with mechanization.

TABLE 10-7 Measures of economic success, Puerto Rico and Chiapas, Mexico

		Puerto Rico		Chiapas
Measure	Units	1970	1980	1990
Literacy of population over 10 years	percent	89	91	70
Teachers	thousand	22	31	40
Registered motor vehicles	thousand	614	1,035	115
Maintained roads	thou. km	5.9	6.4	14.6
Telephones	thousand	319	693	82
Sugar production	thou. tons	0.5	0.2	93.6
Tourist visitors registered	thousand	735	823	641
Electricity consumed	thou. kwh	6,495	11,121	1,027
Electricity authority customers	thousand	691	978	544
Water authority customers	thousand	530	534	316
Persons per physician	number	758	534	1,500
Hospital beds	thousand	11.9	12.1	2.7

Source: Puerto Rico Study Team, 1994.

Manufacturing is the second-most important export-producing sector in Chiapas, although its characteristic businesses are small- to medium-size and short on capital. The petroleum and metal products industries are economically more successful, followed by food, beverages, and tobacco, which are the agro-industries. Retail and wholesale trade generate more revenue than any other productive sector. Table 10-8 summarizes the principal economic indicators of industries from the 1994 Economic Census.

The manufacturing sector is typically composed of small labor-intensive businesses with limited capital investments. The biggest manufacturing industries are petroleum refining and electric power production. These are owned and operated by parastate enterprises.

Producers in every sector lack adequate transportation. Currently, deficiencies exist in all forms of transport—roads, rail, airports, and seaports. Sea transportation is among the most economic for the shipments of the primary commodities to distant markets. Chiapas has a long coastline, where additional shipping centers can be constructed.

The overall economy remains highly dispersed. Tuxtla Gutiérrez, the capital and largest city in the state of Chiapas, has fewer than 300,000 inhabitants. Most people in Chiapas live in small towns and villages. Any strategy of industrial and manufacturing development must contemplate this reality and take into account the necessity of adding certain basic resources and services now lacking, but needed for any successful industrial production process.

TABLE 10-8 Total establishments, employment, employee compensation, purchases, and revenues, by industry, Chiapas, Mexico, 1994

Industries	Establishments (no.)	Employment (no.)	Employee Comp. (mil. $ U.S.)	Total Purchases (mil. $ U.S.)	Revenues (mil. $ U.S.)
Mining and petroleum	84	3,232	43	834	1,459
Petroleum and natural gas	1	2,578	41.2	821	1,452
Metal and nonmetal	83	654	1.8	13	7
Manufacturing	8,239	27,017	93.6	1,432	1,825
Food, beverages, and tobacco	2,888	12,068	42.5	359	421
Textiles, clothing, and feathers	1,852	2,618	1.5	6	10
Wood products	1,156	2,697	3.8	14	21
Paper and printing	244	1,411	5.1	25	36
Petrochemicals	62	3,332	34.2	1,001	1,298
Minerals, nonmetal	1,006	2,527	2.7	14	19
Machinery and equipment	941	2,188	3.6	13	19
Other manufacturing	90	176	0.1	1	1
Electricity and water	1	1,893	nd	nd	nd
Trade	36,599	70,691	128.6	2,265	2,683
Wholesale	1,064	12,291	56.1	781	911
Retail	35,535	58,400	72.4	1,484	1,772
Business services	20,904	41,481	134.5	310	535
Transport and communications	261	1,898	26.8	25	92
Financial rents	648	1,174	2.4	10	49
Building and equipment rents	269	541	1.7	4	41
Furniture rents	379	633	0.7	6	8
Social and communal services	24,220	124,521	nd	nd	nd
Public administration, army	1,458	33,756	nd	nd	nd
Health, education services	5,114	49,588	44.7	37	71
Restaurants and hotels	7,865	17,715	16.8	99	154
Cultural, sports services	1,208	5,315	5	10	19
Professional services	2,969	8,015	24.5	28	70
Maintaining	5,304	9,240	8.4	29	53
Related services	302	892	5.8	72	27
Total	91,604	284,081	537	5,161	7,086

Source: Puerto Rico Study Team, 1994.
nd=not disclosed

The local economies of central Norway, Puerto Rico, and Chiapas face a future of considerable uncertainty because of geographic location in the periphery of their respective trading regions and core metropolitan area economies. All share a common experience in their dependence upon a central government that is no longer available, as it was in the past, to enrich their lives with transfer payments and other support measures. Even if available, the experiences of developing and developed countries during the past 50 years suggest that these measures would likely fail. All local economies thus share a common fate of looking inward to their own ingenuity and resources, starting with a few critical measures for improving the delivery of essential education and health care to their resident populations.

The most important lessons from central Norway and Puerto Rico are those demonstrating the increasingly important role of endogenous growth factors: local entrepreneurship, cooperation among small- and medium-size businesses, and improving and intensifying essential linkages—downstream to input suppliers and upstream to product markets and market niches. These lessons carry an additional message—the downsizing of central governments and a diminishing role for these governments in local economic development. The new civic community and its labor market area roughly define the territorial limits of the building blocks of the economic development strategy.

Predicting the impact of a certain change in a rural economy differs from the accounting of its impact, that is, identifying and describing the factors contributing to the change and the local environment in which this change and any subsequent changes take place. The different ways of thinking about economic growth discussed in Chapter 4 provide a number of different interpretations of the causes and sources of economic change. Table 10-9 lists a variety of theories and models of economic growth and development and some plausible interpretations of each with reference to the local economic conditions in central Norway, Puerto Rico, and Chiapas. As one might guess, the various theories and models have varying degrees of relevance in accounting for economic change in the three areas.

Earlier, we presented industry- and region-specific assessments of the factors and environments affecting firm births, deaths, and volatility, based on a variety of economic models for testing a wide range of hypotheses about economic growth and change attributed to the business sector of a locality or region. From this review, we conclude that flexible employment practices, industry volatility, career opportunities, personal wealth, and economic diversity contribute to above-average rates of firm births and hence economic growth. These factors occur preponderantly in the larger metropolitan areas, especially those with strong ties to urban and rural areas outside their particular local labor market.

TABLE 10-9 Relation of theory and model to development strategy in Central Norway, Puerto Rico, and Chiapas, Mexico

Theory/Model	Central Norway	Puerto Rico	Chiapas
Labor surplus economy	Not relevant; stable employment situation in agriculture. Medium-level wage in agriculture.	Some relevance with continuing high rates of unemployment, but with an emigration safety valve.	Very relevant; at present, about 25 percent unemployment. Very low wage rate in agriculture.
Unbalanced growth process	Relevant; emphasizes the importance of investments as a factor to explain differences in growth.	Some relevance; Puerto Rico is at a development stage in which other theories also are relevant.	Little relevance; generally little investment activity. Imports of some size in only a few sectors.
Endogenous growth process	Relevant, interesting, and important to analyze why endogenous growth occurs. Central Norway has an environment for business development based on high education, flexible production, and high-wage labor.	Especially relevant, given increasingly higher proportions of local populations participating in their institutions of local governance.	Relevant for future developments. Until now, relatively little support from federal government for measures to create a favorable business environment.
Competitive advantage strategy	Relevant and much used; puts emphasis on the entrepreneur and industrial clusters.	Relevant at the firm level for mature firms.	Relevant for a few developed firms in agro-industry and energy.
Disassociated growth economy	Some relevance; puts emphasis on imports.	Relevant; describes an important change in the economy.	Little relevance because the region has a limited export base. Few imported household goods because of low income levels.
Civic community and local labor market	Relevant in defining backward and forward linkages of small- and medium-size enterprises.	Relevant in defining local units of governance and their economic base.	Relevant in defining local units of governance and their economic base.

Resource Management Strategies

We illustrate next an application of regional economic analysis in an assessment of alternative forest management strategies in the rural areas of the Pacific Northwest (Waters, Holland, and Hayes, 1997).

The main purpose of the study was to measure the economic impact of Forest Service policies on the northeast Oregon economy. The five counties making up this area are largely rural, forested, and enriched by exceptional scenic assets, like the Snake River, which forms its eastern boundary with Idaho; the Columbia River on its northwest boundary with Washington; and the Blue and Wallowa Mountains in between. Total population ranged from 112,000 to 116,000 during the 1980s and early 1990s. The three largest cities—Pendleton, LaGrande, and Baker City—with total populations ranging from roughly 9,000 to 15,000, are about 40 miles apart on Interstate Highway 84. This well-traveled highway connects Salt Lake City—via Ogden, Utah, and Boise, Idaho—with Portland and Seattle.

Agriculture and natural resource-based jobs accounted for 17,500 (30 percent) of the 59,800 area jobs in 1992. Total jobs in agriculture and food processing were nearly three times as many as in logging and wood processing—11,900 compared with 4,100. Total personal income increased from $416 million in 1979 to $1.8 billion in 1992. Per capita income increased from slightly less than $10,000 to $16,000 during this period.

The study findings show the economic impacts of a series of "resource shocks" using a northeast Oregon Computable General Equilibrium model to represent the area economic structure (Waters, Holland, and Hayes, 1997, pp. 15–19). The main scenarios are two moderate and two severe resource shocks. The two moderate shock scenarios represent a 25 percent reduction in federal grazing and a 50 percent reduction in logging on federal lands. The two severe shock scenarios represent a 99 percent reduction in federal grazing and an 80 percent reduction in logging on federal lands. The findings show results based on both fixed-price input-output and flexible-price CGE methods. In addition, the CGE method includes a short run of one to two years, an intermediate run of two to five years, and a long run of five to ten years. The CGE models also have different assumptions about unemployed labor leaving the area and proprietors not leaving the area.

We summarize the northeast Oregon model simulation results in Table 10-10. The moderate-range shock scenario, for example, shows a $279,000 loss in household income, $128,000 in government revenue, and twelve jobs. The estimates increase almost threefold for the long-run scenario. The values also are higher for the fixed-price input-output models, even higher than any of the long-

TABLE 10-10 Comparisons of total employment and household income impacts, northeast Oregon

Shock	Units	Short-run	CGE Mode Intermed-run	Long-run	I-O Model Fixed Price
Range shocks, moderate:					
Household income	thousand dollars	–279	–220	–989	–6,850
Government revenue	thousand dollars	–128	–102	187	*
Equipment	number	–12	5	–41	–308
Multiplier	units	1.33	–0.16	1.41	1.31
Range shocks, severe:					
Household income	thousand dollars	–10,558	–6,292	914	–27,510
Government revenue	thousand dollars	–1,471	–849	–3,601	*
Employment	number	–594	53	–1,371	–1,237
Multiplier	units	1.15	–0.04	1.32	1.31
Logging shocks, moderate:					
Household income	thousand dollars	–13,989	–35,002	–40,452	–52,230
Government revenue	thousand dollars	–11,156	–12,185	–13,327	*
Employment	number	–1,228	–1,500	–1,766	–2,160
Multiplier	units	1.32	1.31	1.55	1.73
Logging shocks, severe:					
Household income	thousand dollars	–27,189	–65,178	–75,071	–83,500
Government revenue	thousand dollars	–18,627	–21,407	–22,599	*
Employment	number	–2,366	–2,799	–3,329	–3,453
Multiplier	units	1.32	1.32	1.57	1.73

Source: Water, Holland, and Haynes, 1997, Tables 8 and 9.
*Not available

run scenarios for the CGE models. The only exceptions are the employment multipliers for the CGE long-run scenarios.

When evaluating the findings, two issues call for careful examination of the propriety and validity of the findings. The first issue deals with its definition of the economic base. The authors define basic economic activity in terms of money inflows, including transfer payments. They count only transfers into an area, not transfers out in the form of federal tax payments, including Social Security. Location quotients, in contrast, are surrogate estimates of net, not gross, exports. Net money inflows, for some areas, are actually negative values. By the same logic, location quotient values, based on surrogate estimates of gross exports, would be much larger than those currently estimated.

Use of both a flexible-price computable general equilibrium model and a fixed-price, demand-driven input-output model make comparisons of results for the two approaches possible. As noted earlier, the results from two models differ markedly, as they do from one CGE model scenario to the next. Detailed documentation of the coefficients and assumptions for the CGE models are not available, however, to help account for these differences.

Wetlands Project Development

This example of a detailed analysis and discussion is centered on the economic impact of a proposed wetlands project on two counties in southern Indiana. Local economic effects of project development originate from changes in business activity associated with wetlands land acquisition and development. For the most part, the initial changes are concentrated largely in two activities: crop, timber, and coal production and project visitations. Losses and gains in these activities trigger changes in federal income payments to local residents and property tax payments to local governments.

The project is in the Evansville Economic Area, within an hour's drive for much of the economic area population. The two-county area is an "open" economy with extreme dependence on imports, as shown by the sources of its production inputs. It has a correspondingly high dependence on exports—the central measure of its economic base. The economic base of the two-county area consists of major industry groups ranked in importance as follows: (1) coal mining; (2) electrical and electronic equipment manufacturing, food and kindred products manufacturing, and rubber and miscellaneous plastics manufacturing; (3) public utilities; and (4) several miscellaneous sectors.

Durable goods manufacturing accounts for slightly less than half of the economic base of one county, while mining accounts for more than half of the economic base of the other county. The transportation, communications, and utilities sector is important in both counties, accounting for slightly more than a third of the economic base of each county. Access to nearby coal fields underlies the above-average importance of public utilities in the two counties.

Table 10-11 gives a statistical cross-section of the two-county economy in 1990. Total industry output exceeded $1.5 billion. Much of the output value includes the production inputs of the manufacturing and public utilities sectors purchased from local mining and manufacturing businesses and out-of-county suppliers of these inputs. From this output, the two-county economy contributed nearly $738 million to Indiana's gross state product. This includes $444 million of employment earnings—the income payments to wage and salary workers and their social insurance programs and self-employed income. Other property income—capital consumption allowance, dividend and corporate income tax liabilities, and retained earnings—totaled slightly more than $209 million.

TABLE 10-11 Employment and final payments of specified industries and commodity trade, Gibson and Pike counties, Indiana, 1990

			Imput Purchases		Value Added		Commodity Trade	
Industry Group	Total Employ. (no.)	Total Industry Outlays (mil. $)	Local Suppliers (mil. $)	Total Imports (mil. $)	Employ. Earnings (mil. $)	Other Val.Add. (mil. $)	Total Export (mil. $)	Total Import (mil. $)
Total	18,991	1,555.9	194.4	619.5	444.3	213.5	968.6	1,066.1
Agriculture	1,454	102.6	16.2	50.9	20.9	11.7	57.4	169.8
Mining	1,125	121.9	6.1	23.1	58.5	16.1	93.3	52.4
Construction	1,060	73.9	9.9	35.3	26.8	1.7	0.6	25.2
Manufacturing	3,435	514.6	81.5	311.6	99.9	19.1	413.8	345.9
Transportation, communications, utilities	1,874	384.2	51.2	120.7	74.3	112.1	311.8	78.7
Trade, services, government	10,043	358.8	29.5	77.9	163.8	52.9	91.7	394.1

Source: University of Minnesota IMPLAN regional modeling system.

Total exports for the two-county area were nearly $1.1 million in 1990—roughly twice the total local sales—the mark of a highly trade-dependent local economy. Manufactured products and public utilities topped the list of export sales. Manufactured products also topped the list of imports, followed by agricultural products. Exports exceeded imports for mineral products, largely coal, manufactured products, and the electricity produced by the public utilities in the transportation, communications, and public utilities sector of the two-county economy. In addition, manufacturers were the largest importers, followed, again, by the public utilities—in this case, as purchasers of production inputs. Without shipments of coal and agricultural products and of manufactured products back and forth across county lines, this economy would collapse. Inter-area trade is central to the viability of the local economy, provided that the trade sustains a diversity of industry that can survive the adverse repercussions of the product and business cycles characteristic of highly specialized producing areas.

The direct effects of project development appear in sharp contrast with the levels of local economic activity represented earlier. Estimates of the direct effects (in 1990 dollars) of project development for each of the ten alternatives for years 10 and 20 range between a negative $36.2 million for coal production and a positive $3.8 million for the local purchases of project visitors. The largest negative changes occur for the coal mining sector under project alternative 6 because of prohibition on coal mining on project lands once the service has purchased the land. This results in the subsequent loss of local jobs and related economic activity. Agricultural and timber production decline also, but much less than coal mining in alternative 6. All other sectors show a net increase in revenues, including local governments. Oil and gas extraction are excluded from this list because they are unaffected by project development under all alternatives.

The ten project alternatives relate to the economic activities in the two-county area that are the focus of the economic impact assessments. Eight of the ten project alternatives involve changes in local economic activity, including changes in the number of visitors to the project area and the production of agricultural and timber products and coal mining on land acquired for project purposes. For example, alternative 1 is no action—simply the continuation of existing patterns of economic activity. The U.S. Department of Commerce Office of Business Economics Regional Series 1988 projections for the Evansville Economic Area provide estimates of prospective employment, earnings, population, and personal income under this assumption. Similarly, the baseline estimates serve alternative 3, which relies on an unlikely expansion of state and local land use and zoning regulations. Thus, no estimates of direct changes due to project activity are available for either alternative 1 or alternative 3. Alternative 2, on the other hand, relies on the U.S. Department of Agriculture Wetland Reserve and Water Bank programs. The two programs are limited to existing or converted wetland areas. They generally do not provide for protection of complementary upland areas for wildlife protection; neither presently exists in Indiana. If the Wetland Reserve Program were to be available in Indiana in the future, it would be limited to 10 percent of agricultural land in the county. Similarly, if the Water Bank Program were to be available in Indiana in the future, it would be limited to existing, privately owned wetland areas and lands adjacent to certain designated types of wetlands for a ten-year period. Thus, alternative 2 would result in small, compensating gains and losses in economic activity in the two-county area.

Area economic effects from Wetlands Project Area (WPA) visitor expenditures start with the initial purchases of food, lodging, transportation services, recreation equipment, and various permits and licenses. These final transactions trigger a long series of local business purchases from suppliers both in the project area and outside—in large part within the 25-county Evansville Tri-State Region. Finally, most items are brought into the area from manufacturers and wholesalers outside the state. Large imports into the area result in "leakage" of basic dollars coming into the area from export sales, which erodes the multiplier effect of exports.

WPA visitor expenditures fall into two activity categories—consumptive and nonconsumptive use of natural resources. Hunting and fishing fall into the first category. Canoeing, bird watching, and hiking fall into the second category. Alternatives 1, 2, 3, and 5 have no visitors and, hence, no recreation impacts on the local economy. Visitor expenditures are for the initial period of ten years and the completed project period. Total visitor expenditures are derived from the 1990 database, given the expected expenditures per visitor day for the two types of visitor activities. Visitors engaged in hunting and fishing spend $26.23 (adjusted to 1990 dollars) per visitor day based on estimates by Joseph O'Leary, forestry professor at Purdue University. Nonconsumptive visitors would spend less in the two-county area—on the average only $13.71 per day.

The average daily visit to a recreation destination area, like the two-county project area, is one day, as reported from recreation visitor surveys conducted by the U.S. Forest Service. To the extent that visitors originate from outside the two-county project area, the total impact will be greater because of the new dollars entering the local economy. Estimates based on time-distance distribution of expected recreation visitation show 50 percent of the visitors driving less than one hour to reach their travel destination. Nonlocal residents include all those traveling more than one hour to reach their destination. Less than 15 percent of all visitors would drive more than three hours to their recreation destination.

Long-term direct effects of recreation expenditures are generated by the 50 percent of visitors driving more than one hour.

Notes

1. Birch found that small businesses were the real generators of new jobs—a theme picked up by the U.S. Small Business Administration in its efforts to support the growth of private entrepreneurship and assist small- and medium-size businesses (Birch, 1981; Bingham and Mier, 1993).

2. NAFTA provides for the gradual elimination of all restraints on trade between the United States and its two neighbors—Canada and Mexico. This includes all farm and export subsidies, import quotas, and other forms of government assistance that result in an unfair trade advantage of one country over another.

3. Bendavid (1974) and Maki and Westeren (1992) provide guidelines for identifying and assessing the community base and its contribution to local and regional economic viability and well-being.

4. Seven strategies presented: business recruitment, business loans, tax increment financing, university-based research parks, enterprise zones, self-development, and rural telecommunications. Sears and Reid (1992), the symposium organizers, devised a set of thirteen criteria for scoring the individual strategies: breadth of studies, relevance of existing research to the strategy, number of success dimensions, appropriateness of success measures, quality of experimental design, data quality, timeliness of research, objectivity of research, settlement, representativeness, sample size, rural applicability, recognition of rural diversity, and policy relevance (pp. 304–307). In an earlier article, Kenneth Deavers (1992, pp. 184–189) comments on what is rural—namely, small scale, low density, and distance from urban centers, with the rural areas closest to the urban centers being most affected by the urban overspill of industry and population.

5. All the data used here and in the tables were taken from the Instituto Nacional de Estadística Geografía e Informática (INEGI). Anuario Estadístico del Estado de Chiapas. Ed 1994. Censo Agrícola y Ejidal Resultados Definitivos del Estado de Chiapas, and Censos de población y vivienda del Estado de Chiapas, Mexico. 1995.

11

Economic Regions

Preoccupation with the Cold War left some economic regions with large, outmoded military bases, others with a vast industrial-military complex of technology-intensive manufacturing plants, and still others with neither. At the same time, central cities were clamoring for more resources to fight poverty, crime, and economic decline. These are the central cities that house the new engines of economic growth in their respective regions. Their downtown districts are the nerve centers of trade and commerce, but their neighborhoods are hurting for lack of jobs and services. Meanwhile, commuters from outlying suburbs take the highest-paying jobs in the downtown district, driving to these jobs on state-subsidized expressways to city-subsidized parking ramps. The economic region thus faces a diversity of issues from the changing linkages and relationships among its residents—business, household, and government.

What Is an Economic Region?

Jane Jacobs writes about cities and how they generate economic regions—"city regions," as she calls them. This requires something more than exports or drawing visitors or serving as a cultural, religious, or political capital. "This something more is the capacity of the city to replace wide ranges of its imports exuberantly and repeatedly. Cities that generate city regions of any significance possess that capacity, or have possessed it in the past. The very mechanics of city import-replacing automatically decree the formation of city regions" (Jacobs, 1984, p. 47). We focus on the role of imports, as well as exports, in the delineation of an economic region and its structure and activities.

Place, Purpose, and Proximity

City import-replacing of any significance, as Jane Jacobs sees it, unleashes "five great forces of expansion: city markets for new and different imports; abruptly increased city jobs; technology for increasing rural production and productivity; transplanted city work; city-generated capital" (Jacobs, 1984, p. 47). Some regions lacking large metropolitan areas as within-region markets for their rural areas become specialized supply regions, like the Northern Great Plains, exporting their surplus products to distant markets. Some regions are abandoned by their workers, like the "cutover region" extending over the northern parts of Minnesota, Wisconsin, and Michigan. Still others become clearance regions, like the Scottish clearances described by Jacobs that once pastured the small native sheep with wiry hair-like wool unsuitable for export. The native sheep were replaced by a new breed of sheep—the Cheviot—that required very little labor for its care, which, in turn, prompted the sheep owners to clear the lands of their tenant farmers. The transplant regions, like Puerto Rico with its many transplanted manufacturers from the United States, suffer from a heavy dependence on the transplants that can fail and leave the region without the means to support itself. These regions are characterized by a certain commonality of experiences that left their inhabitants much worse off, and sometimes better off, than they were formerly.

Jacobs offers a helpful clue in characterizing and delineating economic regions in her discussion of imports and import replacement. Economic regions with continuing and growing linkages between their metropolitan core area and rural area activities, and the same for other metropolitan areas in the region, have strong possibilities for continuing growth and development. The core area serves as a "growth pole" for the entire region, with its important spillover effects on surrounding areas within the region and trade linkages with the more distant ones. However, the early growth pole literature focused on firms, groups of firms, and industries as growth poles rather than urban centers (Hansen, 1972, pp. 102–122).

Amos views the growth pole process in three stages: concentration at a single center, diffused concentration at multiple centers, and diffusion to the periphery (Amos, 1990, p. 38). The first stage is triggered by technological innovations that eventually lead to the creation of new industries—much like the early literature describes. In the second stage the growth pole becomes an industrial complex with many distinct growth centers organized into an urban hierarchy. This stage induces improvements in local transportation systems, reducing costs and improving transportation efficiency. In the third stage, the growth centers diffuse their innovations to surrounding areas, opening new markets, and developing new supply sources. Transportation system improvements now extend to the periphery.

Quantitative Measures

Clearly, the economic development regions formulated by federal agencies in the 1960s, like Appalachia, administered by the Appalachian Regional Commission—the first joint federal-state agency—followed a rationale that differed from the idea of stimulating a growth pole process. Appalachia covered the rural, poverty-stricken parts of twelve states (Miernyk, 1982, pp. 48–54). Existing metropolitan core areas that could serve as regional growth centers were outside Appalachia. Measures of unemployment, income, and population decline were the critical measures of inclusion or exclusion of counties from Appalachia and the economic development regions that followed.

In Chapter 3, we presented a system of urban-centered economic regions based on the proximity of each of the 382 U.S. labor market areas to the 29 air node core areas. The 29 core areas experienced widely different rates of growth from 1980 to 1990—ranging from –0.7 percent per year for Detroit to 4.3 percent per year for Orlando. These rates varied sharply by region and from one time period to the next. For the most part, a lagging core area was accompanied by a lagging region.

Delineating the Economic Region

Addressing the urban crisis from a city-region perspective is a new challenge. For the city-region, economic dislocation is a central issue. The driving force of economic dislocation is global competition, compounded by lagging resource productivity and a growing burden of industry regulation. When left to market forces, economic dislocation imposes a heavy burden on local communities as manufacturing plants shut down, leaving jobless workers behind with well-paying employment alternatives out of reach. The urban crisis further reduces access to education and financing for improving labor and management skills.

Without remedial efforts that successfully address underlying problems, the urban crisis spills to the downtown. The oldest suburbs bordering the central city soon follow in economic decline. Eventually, residents of the surrounding labor market areas lack access to the previously available jobs, markets, and producer services. The urban crisis consumes the city-region.

Mobilizing local resources to cope with the adverse effects of global competition is increasingly a cooperative effort between the central city and its region. Businesses in the metropolitan area already collaborate with input suppliers in the surrounding areas through contractual arrangements that improve just-in-time product delivery and reduce production costs. State and local governments also cooperate in financing and maintaining urban and rural infrastructure—roads, highways, airports and other transportation terminals, water and waste disposal systems, educational institutions, and hospitals.

Local Relationships among Firms

New research findings confirm earlier insights that the most strategic relationships among firms in network systems serving global markets "are often local because of the importance of timeliness and face-to-face communication for rapid product development" (Saxenian, 1994, p. 5). The government sector similarly puts a high priority on maintaining face-to-face relationships through the annual sessions of legislative bodies and the even more frequently recurring meetings of legislative and administrative groups in the state's capital city. These findings also have important implications for the role of downtown districts and the core metropolitan labor market areas as the new "engines" of regional economic growth.

Michael Porter argues that to explain "competitiveness" we must understand "the determinants of productivity and the rate of productivity growth. To find the answers, we must focus not on the economy as a whole but on specific industries and industry segments" (Porter, 1990, p. 9). Following this suggestion, we focused on the medical devices industry in Chapter 9. We show the economic environments that exist for the globally competitive industries in a core metropolitan labor market area (LMA) and other labor market areas in the larger region of which both are an integral part. Again, we make use of our reference region as the territorial entity that involves many separate local governments and several state governments in the building of local economic environments for globally competitive business enterprise.

Table 11.1 shows the four areas of the reference region—the first three being all or part of the sixteen-county Minneapolis-St. Paul LMA and the fourth the remainder of the 281-county Minneapolis-St. Paul Economic Region , as shown in Chapters 1 and 3. Thus, the total value added—the region's contribution to the gross domestic product—was more than $151 billion in 1990, of which the sixteen-county area accounted for nearly 48 percent. Total employment was more than four million, of which one-third worked in the sixteen-county area.

The contribution of individual sectors in the two areas differed sharply, as shown in Table 11.2. Private services—including hotels and motels, personal services, business services, health care, education, and other services—was the top-ranked sector of the sixteen-county area in value added. The same sector ranked fifth from the bottom in the rest of the economic region. The two manufacturing sectors ranked high in both areas, along with finance, insurance, and

TABLE 11-1 Value added and employment in specified areas, 1990

Area	VA/Employ. ($)	ValueAdded (pct.)	Employment (pct.)
Central counties, Hennepin and Ramsey	59,556	37.7	23.7
Rest of Twin Cities Metropolitan Region	43,265	7.6	6.5
Rest of Mpls.-St. Paul Labor Market Area	32,668	2.5	2.9
Rest of Mpls.-St. Paul Economic Region	29,190	52.2	66.9
Economic Region	37,405	100.0	100.0

Source: University of Minnesota IMPLAN regional modeling system.

TABLE 11-2 Rank order of industry value added, Minneapolis-St. Paul LMA Economic Region, 1990

Sector Title	Mpls.-St. Paul LMA (mil.$)	Sector Title	Other Econ.Reg. LMAs (mil.$)
Private services	16,836	Finance, insurance, real estate*	10,544
Finance, insurance, real estate*	10,889	Government	10,454
Manufacturing, total durables	9,480	Agriculture and mining	8,761
Manufacturing, total nondurables	8,012	Manufacturing, total durables	7,929
Government	6,548	Retail trade	7,155
Wholesale trade	5,594	Manufacturing, total nondurables	6,758
Retail trade	5,466	Wholesale trade	4,571
Construction	3,289	Construction	4,153
Other transportation	1,588	Private services	4,153
Communications	1,420	Other transportation	2,733
Air transportation	1,229	Sanitary services, public utilities	2,600
Sanitary services, public utilities	1,183	Communications	1,445
Agriculture and mining	783	Air transportation	100

Source: University of Minnesota IMPLAN regional modeling system.
*Includes imputed rental value of home ownership.

real estate, which includes the imputed rental value of resident-owned housing. Manufacturing in total was the top-ranking sector in total value added in both areas.

A region's industrial system has three dimensions: local institutions and culture, industrial structure, and corporate organization. Saxenian notes, however, that, "spatial clustering alone does not create mutually beneficial interdependencies" (1994, p. 7).[1] An industrial system may be geographically agglomerated and yet have limited capacity for adaptation. This is overwhelmingly a function of organizational structure, not of technology or firm size. Moreover, the clustering is self-reinforcing through its vast network of industry and individual business and professional linkages in production, marketing, procurement, and personnel staffing. In this paradigm of a geography-bounded industry linkage network, customers and suppliers, as well as researchers, drive the product and process innovations. These, in turn, help sustain the industry clusters that account for an above-average share of total U.S. value added in each of the industries in the industry cluster.

Rosegrant and Lampe (1992) summarized the lessons learned from Boston's high-tech community and included the following topical headings (pp. 182–190):

1. The Boston area's extensive educational and research infrastructure is the key to the region's ability to innovate.
2. The Route 128 high-tech community was not planned, but evolved for specific reasons.
3. State government has not played a major role in the development of the research infrastructure or the high-tech community.
4. Innovations have occurred when the "traditional" roles of these sectors and the boundaries between them have been challenged.
5. Innovations, at all levels, have often been driven by individuals, not organizations.

The business and professional linkages in the local labor market extend well beyond the local area as alliances between companies. They are a "fact of life in business today," comments Rosabeth Moss Kanter, whether these alliances are from "different parts of the world or different ends of the supply chain" (1994, p. 96). The most common difficulties faced in business alliances deal with money—capital infusions, transfer pricing, licensing fees, compensation levels, and management fees, as well as the complexity of roles that one partner has with another. "Active collaboration takes place when companies develop mechanisms—structures, processes, and skills—for bridging organizational and interpersonal differences and achieving real values from the partnership," according to Kanter (1994, p. 105). These relationships range from mutual service consortia (whereby similar companies in similar industries may pool their resources to gain access to new and expensive technology) to joint ventures and, finally, to value-chain partnerships, such as supplier-customer relationships—the strongest and closest collaboration.

These resources provide the means for rapidly changing the productivity and product mix of the city-region's export-producing and import-replacing industries. They include the strategic management functions of corporate enterprise. The many small businesses in the downtown district provide most of these services. Their lack spells the difference between growing and declining city-regions.

Highly productive levels of collaboration achieve integration across the board—strategic, tactical, operational, interpersonal, and cultural—when each party is willing to let the other party inside the organization. This requires serious communication across functions. Even better are companies with a high order of cross-functional teamwork and exchange of ideas, according to Kanter. Many benefits derive from establishing "multiple, independent centers of competence and innovation" to companies that can be flexible and open to new ideas. For many companies, however, internal barriers to communication limit learning to those directly involved in the relationship. Again, the metropolitan core area with a well-functioning labor market, a diversity of industry and skilled workers, and one or more advanced technology research centers, with informal, as well as formal, interaction between the market-driven businesses and the concept-driven advanced technology researchers, has a clear advantage over other areas.

Producer services (including commercial and personal transportation, communications, banking, insurance and real estate, business services, legal, educational, management and consulting, and other high-order professional services) facilitate the collaboration between businesses. They are an important complement of a globally competitive manufacturing sector that is most abundant and accessible in the core metropolitan LMA. They are available to the export-producing businesses in the rural locality, but from a greater physical distance. The rural business enterprise provides one or more of the producer services from within its own premises. Both businesses maintain strong linkages backward in the production process to suppliers as well as forward to customers.

Delineating areas and regions for purposes of identifying critical market linkages of core and peripheral areas depends heavily on data concerning journey to work and place of work. Such data, collected by the U.S. Bureau of the Census every ten years, provide the underlying information for delineating the commuting areas of local labor markets and thus the boundaries of an economic region. We use the LMAs, based on commuting-to-work data from the 1980 U.S. Census of Population (Tolbert and Killian, 1987). Commuting areas often overlap state boundaries. In 1980, the 3,124 U.S. counties formed 382 LMAs that serve as the relevant geographic units for evaluating area economic performance.

Emphasis on jobs rather than the performance of the job-creating activity—the exports that sustain a region's economy—can easily misdirect attention from the original sources of job-creating and wealth-creating activity to the mere counting of jobs. Granted, this is a much-used measure of economic well-being, especially in political contests.[2] However, sustainable economic development depends on outcomes, that is, the competitive standing of a region's production of goods and services as measured by changes in a region's market shares—its export base rather than its employment shares.

Economic Base Linkages

We have said often that the regional economic base sustains a region's overall economy. It is formed by the export-producing activities of local resident populations—the goods and services produced by businesses, governments, and households. These include, also, the activities that cater to visitors purchasing locally available goods and services that bring "new dollars" into the community. The components of the economic base vary from one part of an LMA to another and from one time period to the next.

The differentiation of activity within an economic region and the corresponding differences in the economic base of individual labor market areas correlate with the differentiation of regional infrastructure—transportation, communications, finance, and various producer services. Both the center and periphery have the basic infrastructure for trade and commerce. The center, however, has the high-order infrastructure of communications, transportation, education, health care, and producer services. The overspill of manufacturing and producer services industries into the surrounding countryside attracts a growing labor force that, in turn, supports a gradually increasing number and variety of consumer-oriented businesses, especially retail trade and personal services. The peripheral labor market areas adapt to their more remote locations by exploiting the gifts of nature that become the principal means, when coupled with price-competitive production technologies, of acquiring income from outside sources for the purchase of their many imported goods and services.

Table 11.3 presents several measures of the economic activity occurring in the Minneapolis-St. Paul Economic Region. Starting from the left side of the table, each column under one of the two major headings, other than the one for the two central counties, represents the remaining counties in the LMA or the economic region. The two measures of industry activity—employment and the equivalent value added of exports from each of the five areas—provide a summary portrait of the differences that exist among groups of counties in an economic region. The employment profile presents a portrait of the region's job-creating industry. The export-defined value added profile presents a totally different portrait of the region's economy, but one that potentially has more predictive value about the region's future than a portrait based on total employment. Employment numbers alone fail to show the economic base and the competitive position and prospects of the region's economy for future growth and change.

TABLE 11-3 Rank order for exports value added of specified sectors, Minnneapolis-St. Paul Economic Region, 1990

Sector	Total Exports Value Added				Total Employment			
	Mpls-St. Paul LMA			Other EconReg LMAs (pct.)	Mpls-St. Paul LMA			Other EconReg LMAs (pct.)
	Central Counties (pct.)	Metro Region (pct.)	All Counties (pct.)		Central Counties (pct.)	Metro Region (pct.)	All Counties (pct.)	
Manufacturing, durables	22.8	28.0	30.0	22.6	11.9	11.9	10.3	6.9
Manufacturing, nondurables	24.2	26.2	28.0	27.0	9.3	10.1	9.8	6.0
Private services	17.0	12.7	10.2	3.5	10.6	7.3	9.5	23.2
Wholesale trade	11.2	10.2	9.6	2.9	8.3	5.8	3.5	4.2
Finance, insurance, real estate*	11.1	9.4	6.0	1.8	11.2	7.2	5.7	5.0
Retail trade	5.4	5.0	4.8	4.3	20.5	25.3	19.9	17.5
Other transportation	4.5	3.7	3.5	2.0	2.7	4.3	3.2	2.7
Air transportation	1.6	2.1	1.2	0.0	2.2	0.1	0.0	0.1
Agriculture and mining	0.8	1.4	6.0	35.1	0.5	3.2	12.3	11.7
Government	1.0	0.7	0.1	0.2	14.9	14.4	15.9	15.3
Construction	0.1	0.5	0.5	0.2	6.3	9.6	8.8	5.9
Communications	0.1	0.1	0.1	0.2	1.1	0.5	0.4	0.7
Sanitary services, public utilities	0.2	0.0	0.0	0.3	0.5	0.4	0.6	0.7
Total	100.0	100.0	100.0	100.0	100.0	100.0	100.0	100.0

Source: University of Minnesota IMPLAN regional modeling system.
*Includes imputed rental value of home ownership.

Table 11.3 shows the relative importance of the so-called "one-digit" industry groups in each of the given areas in the Minneapolis-St. Paul Economic Region and LMA. The retail trade and service industries, including government, provide much the same consumer sector infrastructure in both metropolitan and rural areas. The important difference occurs in the producer infrastructure, particularly high-order producer services and technology-intensive manufacturing. Even the aggregate data for transportation, wholesale trade, finance, insurance, and real estate show these differences. As the distance from the central counties and their business districts increases, the relative importance of high-order producer services and manufacturing declines.

More striking is the existence of high levels of intraregional, inter-area trade, with the metropolitan core area providing high-order services and the rural areas providing the primary resource-based exports, both intermediate inputs and manufactured products. The largest differences between the central counties and those in the periphery of the core LMA and the economic region are in the upstream and midstream sectors of each local economy. Much of the primary production occurs in the rural areas. Efficient and timely transportation for the movement of these commodities to the manufacturing facilities in metropolitan areas, both in the United States and abroad, is important to these primary producers.

Total exports are smaller for the larger area simply because of large intra-area purchases of goods and services that come under the export classification. Industry imports are much less in total value than total exports. They include only purchases by the producing sector of the local economy, not the consuming and investing sectors. A comparison of the total exports of an industry with its total imports provides a measure of the net new dollars the given industry generates. An import-dependent industry adds value to its imports and thus generates new dollars for the local economy. An import-dependent consuming sector, on the other hand, spends the new dollars to cover the cost of these imports.

Table 11.4 provides measures of labor and capital productivity in the four groups of counties—output per worker and value added per worker. Proximity to the two central counties correlates with greater output and higher value added per worker. Exceptions occur, however, in manufacturing, communications, finance, insurance, and real estate. These industry groups have greater output per worker in the five outlying metropolitan counties than in the two central counties—generally the case, also, for value added per worker. The newer facility investments in these industry groups are in the newer suburbs rather than the central cities and the first-ring suburbs. Again, location within a metropolitan regional system correlates with certain industry attributes and activities that may account for a particular industry's concentration in one part

TABLE 11-4 Value added and gross output per worker in specified sectors, by area, 1990

	Value Added per Worker				Gross Output per Worker			
	Mpls-St. Paul LMA				Mpls-St. Paul LMA			
Sector	Central Counties (thou.$)	Metro Region (thou.$)	All Counties (thou.$)	Other Econ.Reg. LMAs (thou.$)	Central Counties (thou.$)	Metro Region (thou.$)	All Counties (thou.$)	Other Econ.Reg. LMAs (thou.$)
Sanitary services, public utilities	177.8	172.3	172.8	146.1	446.4	423.7	425.5	445.9
Private services	157.8	137.4	127.6	18.6	177.7	157.5	147.6	65.0
Communications	121.1	120.9	120.5	79.0	145.1	146.1	145.9	218.9
Finance, insurance, real estate*	77.5	81.9	82.5	77.7	118.8	123.7	123.9	231.4
Manufacturing, nondurables	64.7	65.2	63.2	41.4	134.4	146.4	144.9	307.4
Manufacturing, durables	60.5	61.6	60.5	42.3	127.1	127.9	127.0	210.0
Air transportation	59.4	59.2	59.2	34.0	100.6	100.5	100.6	70.2
Wholesale trade	58.8	57.9	57.1	39.8	61.2	60.2	59.4	92.1
Average	59.6	56.0	54.0	29.2	89.3	86.9	85.1	59.2
Other transportation	39.4	39.3	38.6	37.1	67.4	65.7	64.5	96.9
Construction	36.4	35.2	34.5	25.8	87.0	85.7	84.8	123.3
Government	34.5	33.7	32.9	25.3	33.4	32.8	31.9	43.6
Agriculture and mining	39.2	33.5	28.2	27.6	81.5	75.9	73.5	75.0
Retail trade	19.9	19.5	19.2	15.2	29.7	29.1	28.8	39.4

Source: University of Minnesota IMPLAN regional modeling system.
*Includes imputed rental value of home ownership.

of the economic region and labor market area rather than another. The core metropolitan area and its highly active and diverse local labor and capital markets assure both employers and employees of comparatively low private costs of possible business failure.

For most businesses, whether their shipments or purchases are domestic or foreign is a moot question since this information is not readily available to the individual exporting business. The important issue is the contribution of each export-producing industry to the local economic base that is measured by the value of its exports, including local purchases by visitors to the area. In 1990, the manufacturing sector accounted for more than one-half to more than two-thirds of the total exports in the three areas. Private services accounted for slightly more than 14 percent to slightly more than 8 percent of total exports. The two central counties were more dependent on producer services than either of the other two areas, while the outlying counties in the LMA were more dependent on manufacturing than either of the other two areas.

Factors Affecting Market Competitiveness of Local Industries

Economic trends among U.S. trading regions affecting the international competitiveness of local industries originate from both demand-side and supply-side determinants of regional growth and change. They include the changing product composition in final sales with more rapid growth in the demand for services than commodities, like the final products manufactured from the raw materials of farms, forests, and mines. They also include the product and process technologies that increase the productivity of human effort and provide new products that make existing products obsolete. The rates of discovery and diffusion of the productivity-increasing and market-expanding new technologies in each of the U.S. trading regions are in large part a function of the core competencies that sustain a favorable economic environment for successful business enterprise.

Jobs and Income

The findings reported in the early 1990s on business volatility, regional economic growth, and related matters (Reynolds and Maki, 1990 a,b, 1991, 1994 a,b) addressed issues still current in studies and deliberations of the European Union. These findings on U.S. regional growth, for example, compare closely with more recent findings on the sources of growth of the European Union (Mack and Jacobson, 1996). Included in these studies is a shift-share analysis based on annual time series from the U.S. Department of Commerce Regional Economic Information System, which yields a "regional-share effect" for each of 81 labor market areas. These areas are located in the thirteen-state east-west transportation corridor region extending from Michigan to Oregon and Washington. Historically, the study region has experienced much economic volatility due to the many natural resource-based local economies in the interior states and manufacturing of cyclically sensitive durable goods elsewhere. The remaining twenty-three LMAs in the current study (of 100 U.S. LMAs) include both rapidly growing and generally declining base economies that vary in income volatility and overall growth from the lowest to among the highest. A totally different data series (Dun's Marketing Indicators) provides estimates of employment change, by size, type, and location of firms, associated with the estimates of total earnings change, total earnings, business volatility, and spatial structure.

Sources of income volatility—that is, period-to-period shifts in labor earnings—are illustrated by the shift-share analysis that includes the two long periods of economic recovery—1970–80 and 1982–86—separated by two recessions occurring in the 1980–82 period. Data for the recession from 1973 to 1975 are not included in the analysis. Both the income volatility index and the income growth index are based on income change over the entire 1970–86 period. Excess earnings from each two-digit Standard Industrial Classification (SIC) code industry group in the county level labor earnings series, compiled and reported by the U.S. Department of Commerce, were calculated for each county and aggregated by LMA.

The results of the shift-share analysis, summarized here from the reports cited earlier, show vastly different growth patterns for the four regional groupings. Over the sixteen-year period, total labor earnings—the principal source of personal income—increased by more than $782 billion (in 1982 dollars), from $1.426 trillion in 1970 to $2.208 trillion in 1986. The overall increases ranged from $40.5 billion in midcontinent West to $50 billion in midcontinent East, $132 billion in comparison LMAs, and $560 billion in the remaining LMAs in the United States. The comparison LMAs increased in importance from 10.6 percent of total U.S. labor earnings in 1970 to 12.8 percent in 1986. Midcontinent East dropped from 9.5 percent of the total to 8.4 percent.

The statistical series illustrate important differences among U.S. regions in industry performance attributable to corresponding differences in each region's core competencies. We view these differences from the perspective of economic base theory, strengthened in part by a generous use of interregional trade theory and regional income and product accounts.

Table 11.5 presents the summary results of this effort. All values represent relative (to the United States) differences. An "excess earnings" value for each industry group within a labor market area measures its importance in the area's economic base. This value represents labor earnings in excess of an

TABLE 11-5 Proportion of total excess earnings and annual rates of change for specified industry group, U.S., 1974–86

Industry Groups	Metro LMAs		Nonmetro		All Regions		Relative Annual Change		
	1974 (pct.)	1986 (pct.)	1974 (pct.)	1986 (pct.)	1974 (pct.)	1986 (pct.)	Metro (pct.)	Nonmet. (pct.)	Total (pct.)
Agricultural products and agricultural services	0.2	0	28.9	15.2	20.9	9.5	−10.3	−6.4	−6.4
Mining and extractive industries (SIC)	2.1	4.4	10.8	14.3	8.4	10.6	9.3	1.2	2
Construction (SIC 15–17)	1.7	1.2	1.7	3.1	1.7	2.4	−0.4	3.9	2.9
Manufacturing nondurables (SIC 20–2)	9.7	5.4	16.6	17.7	14.7	13.1	−2.4	−0.7	−1
Manufacturing durables (SIC 24–25, 32)	47.9	40.2	32.9	34.4	37.1	36.6	1.1	−0.8	−0.1
Transportation, communications, utilities	4.1	3.7	2	2.9	2.6	3.2	1.5	2	1.7
Wholesale trade (SIC 50–51)	14.9	12.4	1.3	1.2	5.1	5.4	1	−2	0.5
Retail trade (SIC 52–59)	1.4	0.7	1.5	2	1.5	1.5	−3.6	1.2	0.1
Business services (SIC 60–67, 73, 81, 86)	12.6	29.8	1.4	3.6	4.5	13.4	10.2	7	9.6
Consumer services (SIC 70, 72, 75, 76)	0.8	1.2	1.4	4.1	1.2	3	6.2	8	7.7
Human services (SIC 80, 82, 83)	4.6	0.9	1.5	1.6	2.3	1.3	−10.4	−0.6	−4.6
Total	100	100	100	100	100	100	2.5	−1.2	0

Source: Reynolds and Maki, 1990a.

amount calculated for each area industry from the U.S. distribution of labor earnings for the same industries. Manufacturing, for example, declined in relative importance from 57.6 to 45.6 percent of excess earnings for all industries in the 29 metropolitan core labor market areas from 1974 to 1986. It increased in relative importance in the remaining LMAs, from 49.5 to 52.1 percent during the same period. For all LMAs it declined in relative importance from 51.8 to 49.7 percent. The relative annual change columns show manufacturing declining in relative importance, largely due to lagging or lack of growth in nondurable goods manufacturing.

Two of the largest shifts in the economic base of individual LMAs are indicated by the percentage changes in agriculture and business services. Agriculture declined in relative excess earnings in the noncore LMAs from 28.9 percent in 1974 to 15.2 percent in 1986. Business services increased in the core LMAs from 12.6 percent in 1974 to 29.8 percent in 1986. The core LMAs, like Minneapolis-St. Paul, are characterized by the increasing importance of business and related services in the area's economic base. The business services sector is replacing many of the job losses in manufacturing and retail trade. The new jobs include well-paying jobs in management and administration, as well as various professional and technical occupations.

Because of the overall rapid growth of labor earnings in human services (health care, education, and social services), relative positive differences were less in 1986 than 1974. Manufacturing, on the other hand, declined in relative importance during this period. It is an industry that is highly differentiated by product and place specialization and its high propensity for importing and exporting among trading regions. The large increase in business services is in part attributed to the shift of business services within the manufacturing sector to an independent business services sector. A similar phenomenon took place many years earlier in agriculture. The shift to services is thus positive for business services but negative in human services in the core LMAs.

The principal reasons for the contrasting growth patterns rest with the base economies of the two regions. Not only are the local base economies of mid-continent East dominated by below-average growth industries, but they also are marked by a continuing decline in the competitive position of their principal exports. An overall above-average industry-mix effect and an overall above-average regional-share effect distinguish the base economies of the comparison regions.

A distinguishing difference between LMAs with high and low income volatility is the direction of relative change. It is strongly negative for high-volatility areas and strongly positive for low-volatility areas. For most LMAs with high income volatility, a positive regional-share effect for the 1970s turned negative in the 1980s, contributing to the significant negative relative changes in the 1980s.

The ranking of total change in labor earnings in the 1970–86 period confirms the unique role of the local base economy in accounting for regional job and income growth. For the 30 fastest-growing LMAs, total labor earnings increased from $182.6 billion in 1970 to $345 billion in 1986—an increase of 89 percent. During the same period, total labor earnings increased by only 22 percent for the 30 slowest-growing LMAs—from $96.1 billion in 1970 to $116.9 billion in 1986.

High labor income growth is as frequently associated with high as with low labor income volatility—nine LMAs in both cases. In comparison, low-income LMAs include thirteen with the highest and five with the lowest volatility. Thus the midrange LMAs include twelve with high and twelve with low income volatility. The findings show a lack of strong correlation between income growth and income volatility when further differentiation of local base economies is lacking.

The base economies of the high-volatility LMAs are marked by high levels of industry specialization in farming, mining, or manufacturing. In these areas, the high-income volatility is associated with a high degree of vulnerability to the vicissitudes of cyclically sensitive export markets. Moreover, the extreme specialization of industry in the base economies of the LMAs with high income volatility persisted through the 1970s and much of the 1980s. Where high-income volatility was accompanied by slow income growth, the local base economies also faced shrinking export markets.

High-income growth areas differ from high-income volatility areas and low-income growth areas in the diversity of their base economy. Even specialized base economies support high-income growth when the export-producing sectors remain competitive in their export markets and maintain their market shares. Generally, however, the specialized fast-growing economies had lost their earlier momentum by the mid-1980s and faced, instead, much reduced income growth.

Each measure of regional growth varies in relative value from one period to the next. For some areas, volatility in rates of regional growth is due to the cyclical sensitivity of the local economy. For others, period-to-period changes in jobs and earnings are related to long-term changes in industry product cycles. Changes in industry mix reveal both short-term and long-term changes in the importance of individual industries in the U.S. economy. Changes in regional share reveal changes in the competitive position, or economic performance and importance, of a given industry relative to the corresponding industry in the United States. A distinguishing characteristic of declining and growing areas is the rapidity and direction of change in jobs and labor earnings. Once the volatility in jobs and income is removed, the residual "regional-share effect" becomes a measure of regional growth and decline.

Regional Policy Initiatives

Regional policy refers to the activities and initiatives of state and local legislative and administrative bodies that address regional concerns. Much of the literature on regional policy pertains to "top-down" approaches for enhancing overall national well-being (Miernyk, 1982; Sears and Reid, 1992). Such an approach views regions as constructs for the devolution of national policies in agriculture, environmental concerns, and public works. Benjamin Chinitz, Assistant Secretary for Regional Development in the U.S. Department of Commerce during the Johnson administration, defined regional policy to include pieces of legislation, administrative mechanisms, and programs and projects aimed at affecting regional indicators. The regional indicators cited by Chinitz were overall economic growth, the rate of unemployment, and per capita income (Chinitz, 1974, p. 23).

Regional policy generally favors top-down approaches to the amelioration of regional disparities, particularly unemployment, as in Great Britain. The much publicized Technopolis project adopted by Japan's Ministry of International Trade and Industry (MITI) in 1983 depended on its central government to accelerate the development of technology-intensive industry outside the current areas of industry concentration (Okamura, Sakauchi, and Nonaka, 1993, p. 134). The Twin Cities Metropolitan Council, established in 1967 by the Minnesota Legislature, formulated a strategy with two objectives: "1) delivery on a regional basis of services that can no longer be provided effectively by existing units of local government; and 2) establishment of policies, plans, programs and controls to guide the physical, social and economic development of the Twin Cities Metropolitan Region" (Naftalin and Brandl, 1980, p. 1). More than a quarter-century later, measures of the council's primary issues and performance are still largely top-down indicators of regional disparities, like poverty rates, and earnings from employment and occupational distributions, but without specific information about the vitality and viability of its economic base (Metropolitan Council, 1994).

Four of the nine contributors to *Regional Conflict and National Policy,* edited by Kent Price (1992), obviously were uncomfortable with the concept of the region. Regions exist, but their size and shape depend on the criteria for delineating them, according to one writer.

In contrast to the expressed concerns of resource economics about regional conflict, the seven contributors to the symposium on rural development strategies and public policy were untroubled by concerns about the notion of a region. None of the criteria for evaluating the research focusing on the seven strategies (one from each of the seven authors) cited linkage or proximity to metropolitan core areas, for example, as relevant in evaluating research and methodology. The underlying rationale for rural development in this symposium was an assumed primacy of federal leadership and support, rather than the empowerment of state, local, and regional organizations and their spokespersons as participants in the formulation of a rural development strategy.[3] Except for occasional efforts in delineating rural development areas, the U.S. Department of Agriculture defines regions largely for the purposes of federal agricultural policy. This agency has an overriding political commitment to treat all states and localities alike, except for differences in agricultural production and closely related activities.

Thus, several belief systems exist contemporaneously with different definitions of a region and hence different frameworks—geographic, economic, and political—for the study and implementation of regional policy. One belief system may emphasize legal, rather than economic, structures because of its institutional biases, the other a metropolitan–nonmetropolitan dichotomy because of its administrative and organizational biases. Still another may seek public subsidies to reduce the loss of local jobs and income from business closures or relocation. These readily recognized belief systems have one common feature: they define a region for the study of top-down regional policy.

Tracking a Region's Performance

Economic trends among U.S. trading regions affecting the international competitiveness of local industries originate from both demand-side and supply-side determinants of regional growth and change. They include the changing product composition of final sales, with more rapid growth in the demand for services than commodities, like the final products manufactured from the raw materials of farms, forests, and mines. They also include the product and process technologies that increase the productivity of human effort and provide new products that make existing products obsolete. The rates of discovery and diffusion of productivity-increasing and market-expanding new technologies in each of the U.S. trading regions are in large part a function of the core competencies that sustain a favorable economic environment for successful business enterprise.

International competitiveness means producing price- and quality-competitive goods and services for domestic and international markets and having a world-class distribution system to both quickly and efficiently reach these markets. A region's industries that are internationally competitive typically acquire the relevant know-how by succeeding, first, in domestic markets. Thus, the study of domestic interregional trade is likely to offer critical insights for a deeper and broader understanding of the industry attributes and the market conditions that lead to superior business performance in international markets.

Overview of Interregional Air Transportation

Because of its critical importance to a region's growth and development in the new global economy, we focus on a region's air transportation. Air transportation performance measures—like the immediately recognized measures of passenger safety, travel costs, and on-time reliability, and the less immediately recognized measures of workforce productivity—apply to other modes of travel and transportation. The immediately recognized measures relate to demand-induced changes in air travel and transportation. Here the quantity demanded is a function of price, given measures of safety and reliability and the income of traveler or value of product.

These measures also relate to supply-induced changes affecting the availability and cost of transportation. Whether cost reductions (such as close-to-capacity payloads) translate into reduced prices of air travel depends on the competitive environment for air carriers. Rising levels of personal and business incomes, for example, shift the downward sloping demand curves to the right, thus intersecting a given supply curve at higher and higher levels, that is, prices.

Productivity improvements in air transportation are the long-term changes in organization, technology of aircraft and airport use, and maintenance. New aircraft technologies, for example, make feasible the building and use of larger aircraft that reduce per-passenger prices of air travel, while the organization of airport occupancy with effectively competing airlines reduces the cost of air travel to the individual air traveler. In short, the airport and the competing airlines operate on different "production functions" with increasing product or service output from a given level of overall operating inputs.

Individual enterprise-induced changes via the introduction of changes in air transportation organization and technologies translate into market-induced changes in the availability, safety, and cost of air transportation services. The amount of change depends on various local conditions for the service providers and their clientele. Among the most important of these conditions is the kind and amount of management oversight established by the local governments charged with the responsibility for fair and efficient use of critical infrastructure.

Local Markets for Interregional Air Transportation

Forecasting the local market for air transportation services starts with the forecasting of air transportation purchases by local businesses, households, and governments. This task has several parts, including forecasts of the number of purchasers of air transportation and the amount and value of purchases by each purchaser. The local market forecasts also include the air transportation purchases originating in the extended regional service area.

Destination market purchases of air transportation services are the counterpart of the local market purchases. In the forecast models these purchases originating from an interregional air transportation services center, like one of the 29 air nodes in the contiguous 48 states, are attributed to destination residents—business, household, and government. The forecast problem is very much like the one encountered in forecasting local markets for interregional air transportation, except that here the focus is on the passengers and freight originating from a given air transportation center and its service area.

Improving the accuracy of interregional air travel and transportation forecasts of local markets and their destination areas for public purposes is a task that probably exceeds the capabilities of our existing forecast models and data. The public purpose, of course, is to provide local agencies and organizations, like the Metropolitan Airports Commission and its airport management organization, with accurate information about current and prospective demands for air transportation services and the deficiencies and excesses, if any, in the provision of these services.

For major airports in densely populated areas, retaining the airport close to the downtown district both enhances the productivity of business travel by minimizing travel distance to the airport and adds to the profitability of the resident airlines by retaining the airport at its existing capacity-limited site. A dominant hub airline, in effect, builds on the proclivity of local businesses for the short view of minimizing current travel time to the airport. Much of the traveling public shares this short view.

The long view for managing the total air node facility would favor some alternative approaches to (1) introducing competition in the pricing of originating air fares and (2) monitoring the likely impact of each alternative on the region's export-producing businesses and traveling public. We focus on two related issues associated with the long view—nodal airport management and interregional business travel. It owes much of its inspiration and content to Charlie Karlsson and Peter Hugoson from the Jönköping International Business School, Jönköping University, and Ake Andersson and Folke Snickers from the Royal Institute of Technology, who are researching the restructuring of the Swedish air transportation system.

The findings of the Hugoson-Karlsson (1996) study support a series of hypotheses about business travel, of which two are particularly important to this study: that business travel is primarily for meetings at higher levels within organizations and that growing regions will, all other factors being equal, exhibit a higher order of business-trip intensity than nongrowing regions. The study found that long-distance business trips are generated particularly by geographically dispersed firms engaged primarily in the distribution of knowledge.

Airport Management and Organization

The airport-managing organization is an important part of the picture and should be included in any economic impact assessment (Andersson and Snickers, 1996). The organization generates revenues from the airlines as well as the resident commercial establishments, and it makes important investments in the airport facilities. One suggestion is the use of auction markets for buying and selling slots and landing and takeoff rights every five years or so. The airlines can do this now. On the other hand, use of 99-year leases is a sure way of limiting local competition. Slots on short-term leases at an airline node are eagerly sought by dominant airlines to restrict entry. They should be monitored accordingly to ensure that new airport tenants come to the airport and offer added competition to existing carriers.

Airport management provides a place of business both for airlines operating from the airport and for other commercial enterprises occupying space in the airport. For some airports, the rental revenues from the commercial enterprises, including parking, may exceed the rental and facility-use revenues from the airline occupants. Marketing space in the airport for further revenue enhancement as well as increasing the revenues derived from the airlines becomes an important management activity in the effort to secure additional funding for airport relocation and/or expansion.[4]

A related issue cited earlier is the structure and organization of the airport oversight authority and its biases in airport management and operations, whether engineering or economic. An engineering bias would favor limited accountability for airport performance—an accountability confined to activities within airport perimeters. An economic bias would favor having the oversight agency explore alternative ways of opening additional gates and offering landing and takeoff rights for additional competing airlines, and would ensure a level playing field to accomplish the intended purpose of providing lower fares to as many or even more destinations than are already available.

Whether the airport-managing organization asserts an engineering or economic bias in managing its assets makes a big difference in the results from a public-interest perspective. The organization with the engineering bias does well to save a few dollars with improved efficiencies; the savings accrue to the airlines. Shortcomings of the engineering bias are evident when the charge given to the airport management's economic-impact study consultants addresses only the direct effects of the airport payroll and nearby real estate activity on the local economy and not the transportation-dependent businesses and industries that are part of the locality's real economic base. The savings achieved with superior airport management are small compared with the profit premiums captured by the dominant airlines facing minimal competition at their originating airports. Besides, the management savings accrue to the airlines, while the economic losses in higher-than-average fares for similar flights accrue to the traveling public. On the other hand, the organization with an economic bias may limit the number of slots to the dominant airline and thus introduce added competition that could reduce the average fare, with the savings accruing directly to the ticket buyer.

The cost of air transportation, as well as its availability and the timely market access it provides, are decision variables affecting business location and travel. Monopoly or near-monopoly markets appear to have higher fares. Market concentration also influences level of air service. Airline profitability thus becomes a function of market power rather than superior service and resource management. The few largest airlines exercise their market power at each of the major air nodes. Introducing new sources of competition is extremely difficult, however, because of the complexity of the issues. In addition, the Metropolitan Airport Commission may opt for a passive role in its oversight responsibility by performing essentially engineering functions. The regional economic impact of a major airport thus depends, in part, on its modus operandi—the particular biases it asserts in airport management and operations that affect access to air transportation and the rational expectations of businesses when making new investment decisions, whether on-site expansion or relocation to and/or expansion at another metropolitan statistical area (MSA).

Competing Air Node Regions

The hub concept permits an air carrier to provide service to a multicity network with minimal investment in aircraft and crews (Braslau, 1998). For example, a hub with four spokes can be serviced with four aircraft, each flying between the hub and one spoke city (naturally, the system can be considerably more complex). For service to be provided between all five of these communities, ten aircraft would be needed to provide nonstop service between each of the communities. For this simple case, the airline would avoid buying six additional aircraft while still providing a reasonable level of service. Originating passengers of the hub have the benefit of increased nonstop service to most of the cities on the airline network.

Studies have shown that the market share of an airline is more than proportional to the number of flights it provides (Borenstein, 1992, pp. 45–73).[5] For example, an airline with 60 percent of the flights from an airport will likely attract 65 to 70 percent of the demand. Because of this pattern (which can also be seen in the category "killer" stores today), the "loser" gets farther and farther behind and eventually capitulates the market to the dominant carrier. In this way, the dominance of hubs around the country by particular airlines has developed. Where the airline network is sufficiently large, more than one hub may be needed to service the market since both distance and airport capacity play a role. The selection of secondary hubs depends upon a number of factors, but generally aims to provide a reasonably central service center for an important part of the airline network.

Table 11.6 lists (as in Chapter 3, Table 3.1) the 29 primary air nodes in the U.S. hub system for the contiguous 48 states. It ranks the 29 core MSAs by their core MSA 1990–96 population change. Also listed (in column 4 next to the economic region identification number) is the ranking based on the 1980–90 population change for each MSA. The fifteen core areas with above-average growth account for a smaller proportion of the total economic region population and total core area population than the fourteen areas with below-average growth.

TABLE 11-6 Total population and land area in specified air node regions in 1990, ranked by 1990–96 core area population change, U.S.

					Total Population					Land Area (sq. mi.)			Pop Density	
						Metropolitan (MSAs)								
90–96 Chg.	Air Node Region	Core State	80–90 Chg.	No.	Region Total (thou.)	Total (thou.)	Core* (thou.)	Other (thou.)	Non-metro (thou.)	Total (thou.)	MSAs (thou.)	Nonmet. (thou.)	MSAs (sq.mi.)	Nonmet. (sq.mi.)
Above-average growth, 1990–96:														
1	Atlanta	GA	2	24	11,077	7,211	2,960	4,251	3,866	455	120	325	60	12
2	Raleigh	NC	3	28	7,299	4,934	858	4,076	2,365	222	77	145	64	16
3	Orlando	FL	1	26	6,835	6,205	1,225	4,980	630	140	73	67	85	9
4	Denver	CO	10	2	8,287	5,724	1,623	4,101	2,563	2,023	178	1,845	32	1
5	Houston	TX	5	8	12,741	10,393	3,322	7,071	2,348	537	163	375	64	6
6	Dallas	TX	4	7	11,071	8,003	2,676	5,327	3,068	957	140	816	57	4
7	Charlotte	NC-SC	8	27	5,401	3,911	1,162	2,749	1,490	156	67	89	58	17
8	Salt Lake City	UT	7	5	3,252	1,709	1,072	637	1,543	1,295	32	1,263	53	1
9	Nashville	TN	11	29	4,756	2,908	985	1,923	1,847	199	60	139	48	13
10	Seattle	WA	6	1	7,775	5,875	2,033	3,842	1,900	708	121	587	49	3
11	Minneapolis-St. Paul	MN-WI	14	10	7,792	4,122	2,539	1,583	3,670	1,041	134	907	31	4
12	Washington, D.C.	MD-VA-WVA	9	18	6,019	5,370	4,223	1,147	650	94	48	46	112	14
13	Miami	FL	12	25	5,425	5,193	1,937	3,256	231	58	41	17	127	14
14	Memphis	TN-AK-MS	17	11	5,996	2,392	1,007	1,385	3,604	434	47	387	51	9
15	Kansas City	MO-KS	15	6	7,566	4,974	1,583	3,391	2,592	515	102	413	49	6
	Total or average				111,292	78,924	29,205	49,719	32,367	8,824	1,403	7,421	56	4

TABLE 11-6 Continued

90–96 Chg.	Air Node Region	Core State	80–90 Chg.	No. (thou.)	Total Population: Region Total (thou.)	Metropolitan (MSAs) Total (thou.)	Core* (thou.)	Other (thou.)	Non-metro (thou.)	Land Area (sq. mi.) Total (thou.)	MSAs (thou.)	Nonmet. (thou.)	Pop Density MSAs (sq.mi.)	Nonmet. (sq.mi.)
Below-average growth, 1990–96:														
16	Cincinnati	OH-KY-IN	19	21	6,747	4,813	1,526	3,287	1,935	184	59	125	81	16
17	Chicago	IL	20	20	16,185	13,581	7,411	6,170	2,604	296	104	191	130	14
18	Baltimore	MD	16	14	5,211	4,478	2,382	2,096	733	65	33	32	134	23
19	San Francisco	CA	18	4	12,018	11,040	1,604	9,436	978	436	200	237	55	4
20	Los Angeles	CA	13	3	22,727	22,025	8,863	13,162	702	905	490	416	45	2
21	St. Louis	MO-IL	21	9	4,751	2,794	2,492	302	1,957	232	33	199	84	10
22	Boston	MA-NH	25	16	9,920	8,678	5,686	2,992	1,242	235	78	157	111	8
23	Newark	NJ	27	15	6,411	6,002	1,916	4,086	409	50	38	12	160	34
24	Detroit	MI	28	23	10.017	8,422	4,267	4,155	1,595	179	70	109	121	15
25	New York	NY-NJ	23	12	15,020	14,855	8,547	6,308	165	33	29	5	517	36
26	Philadelphia	PA-NJ-DE-MD	24	17	5,380	5,380	4,922	458	0	21	21	0	258	†
27	Syracuse	NY	22	13	6,348	5.016	742	4,274	1,332	182	78	104	64	13
28	Dayton	OH	26	22	4,810	3,689	951	2,738	1,121	88	41	47	90	24
29	Pittsburgh	PA	29	19	10,216	7,815	2,395	5,420	2,400	247	84	163	93	15
	Total or average				135,761	118,588	53,704	64,884	17,173	3,153	1,358	1,797	87	30
	Total or average, 29 nodes				247,052	197,512	80,310	117,202	49,541	11,977	2,761	9,215	72	5

Sources: U.S. Bureau of the Census, *Statistical Abstract of the United States*, 1995.

*MSAs, PMSAs, and NECMA (Boston).

†Not available.

These are the more rural regions with lower population densities and more of the metropolitan area population outside rather than inside their respective core areas. They also were the faster growing core areas over the ten-year period from 1980 to 1990.

Because of large differences in other MSA and nonmetropolitan area populations, the total region populations depart sharply from the rank order of their core areas. Both land area and population density measures also show much variability from one region to the next, thus suggesting large differences in the length and duration of daily trips to work, school, and cultural activities. Thus, the 29 proximity-based air node regions reveal a diversity of demographic and geographic attributes that affect the accuracy of forecasting interregional travel and transportation to and from each of these regions.

Table 11.7 lists the population of the core labor market areas for the 29 air node regions as in Chapter 3, Table 3.2, but ranked by 1990–96 population change. The Bureau of Economic Analysis, of the U.S. Department of Commerce, prepared year 2000 and 2010 forecasts in 1994, with 1993 the last year of available historical data. For the fifteen metropolitan core areas with above-average growth, the estimated 1990–96 annual rate of change exceeded the projected 1990–2000 annual rate for twelve of the areas. For the fourteen areas with below-average growth, the estimated rate exceeded the projected rate for only one area. Hence, the average for each of the two groups differs. Forecast values generally are biased toward their overall means. The estimated and projected overall growth rates are essentially identical.

We compare, next, two metropolitan core areas—Detroit and Minneapolis-St. Paul—that serve as air nodes for the same dominant airline. These are air nodes with contrasting growth histories and export-producing activities that constitute their local economic base. Yet, from an airline ownership perspective, the investment strategy also contrasts sharply with the growth histories of the two airports, given an airline priority to establish and retain its dominant position in the slower-growing metropolitan area. Thus, the comparisons of economic measures for the two metropolitan areas have different implications for the different decision centers—the dominant airline, the metropolitan airport management, the local corporate community, and the extended civic community—and their clients, constituents, and customers.

Table 11.8 summarizes the employment profiles for the two core MSAs cited earlier—Minneapolis-St. Paul and Detroit. In 1982, total employment was slightly more than 1.3 million in the Minneapolis-St. Paul MSA and 1.8 million in the Detroit MSA. It increased by 12.9 percent and 12.4 percent, respectively, for the two MSAs over the 1982–90 period. Total jobs grew slightly during the 1990–91 recession in the Minneapolis-St. Paul MSA, but they declined by nearly

TABLE 11-7 Total population change in core metropolitan areas and ranking (1990–96 change), U.S., 1980–96 and projected 2000–10

Area Title	State(s)	80–90 Chg.	No.	Estimated 1980 (April) (thou.)	1990 (April) (thou.)	1993 (July) (thou.)	1996 (July) (thou.)	Projected 2000 (July) (thou.)	2010 (July) (thou.)	Annual Change 1980–90 (pct.)	1990–93 (pct.)	1990–96 (pct.)	1990–00 (pct.)	2000–10 (pct.)
Above-average growth, 1990–96:														
Atlanta, MSA	GA	2	24	2,233	2,960	3,229	3,541	3,682	4,231	2.9	2.7	2.9	2.2	1.4
Raleigh, MSA	NC	3	28	665	858	938	1,025	1,092	1,269	2.6	2.8	2.9	2.4	1.5
Orlando, MSA	FL	1	26	805	1,225	1,334	1,417	1,605	1,963	4.3	2.7	2.4	2.7	2.0
Denver, PMSA	CO	10	2	1,429	1,623	1,762	1,867	1,970	2,235	1.3	2.6	2.3	1.9	1.3
Houston, PMSA	TX	5	8	2,753	3,322	3,589	3,792	4,020	4,492	1.9	2.4	2.1	1.9	1.1
Dallas, PMSA	TX	4	7	2,055	2,676	2,844	3,048	3,161	3,538	2.7	1.9	2.1	1.6	1.1
Charlotte, MSA	NC-SC	8	27	971	1,162	1,234	1,321	1,361	1,520	1.8	1.9	2.1	1.6	1.1
Salt Lake City, MSA	UT	7	5	910	1,072	1,155	1,218	1,327	1,547	1.7	2.3	2.1	2.1	1.5
Nashville, MSA	TN	11	29	851	985	1,045	1,117	1,155	1,288	1.5	1.8	2.0	1.6	1.1
Seattle, PMSA	WA	6	1	1,652	2,033	2,159	2,235	2,416	2,800	2.1	1.9	1.5	1.7	1.5
Minneapolis-St. Paul, MSA	MN-WI	14	10	2,198	2,539	2,655	2,765	2,877	3,148	1.5	1.4	1.4	1.2	0.9
Washington, D.C., PMSA	DC-MD-VA-WVA	9	18	3,478	4,223	4,416	4,563	4,860	5,455	2.0	1.4	1.2	1.4	1.2
Miami, PMSA	FL	12	25	1,626	1,937	2,003	2,076	2,171	2,370	1.8	1.0	1.1	1.1	0.9
Memphis, MSA	TN-AK-MS	17	11	939	1,007	1,042	1.078	1,122	1,217	0.7	1.1	1.1	1.1	0.8
Kansas City, MSA	MO-KS	15	6	1,449	1,583	1,631	1,690	1,752	1,908	0.9	0.9	1.1	1.0	0.9
Total or average				24,014	29,205	31,032	32,754	34,571	38,979	2.0	1.9	1.9	1.7	1.2

Below-average growth, 1990–96.:														
Cincinnati, PMSA	OH-KY-IN	19	21	1,468	1,526	1,572	1,597	1,643	1,745	0.4	0.9	0.7	0.7	0.6
Chicago, PMSA	IL	20	20	7,246	7,411	7,612	7,734	8,030	8,826	0.2	0.8	0.7	0.8	1.0
Baltimore, PMSA	MD	16	14	2,199	2,382	2,444	2,474	2,597	2,791	0.8	0.8	0.6	0.8	0.7
San Francisco, PMSA	CA	18	4	1,489	1,604	1,638	1,655	1,701	1,808	0.7	0.6	0.5	0.6	0.6
Los Angeles, PMSA	CA	13	3	7,477	8,863	9,134	9,128	9,669	10,455	1.7	0.9	0.5	0.9	0.8
St. Louis, MSA	MO-IL	21	9	2,414	2,492	2,528	2,548	2,637	2,798	0.3	0.4	0.4	0.6	0.6
Boston, NECMA	MA-NH	25	16	5,336	5,686	5,700	5,796	5,993	6,456	0.6	0.1	0.3	0.5	0.7
Newark, PMSA	NJ	27	15	1,964	1,916	1,929	1,940	1,979	2,050	−0.2	0.2	0.2	0.3	0.3
Detroit, PMSA	MI	28	23	4,388	4,267	4,304	4,318	4,337	4,408	−0.3	0.3	0.2	0.2	0.2
New York, PMSA	NY-NJ	23	12	8,275	8,547	8,573	8,643	8,604	8,636	0.3	0.1	0.2	0.1	0.0
Philadelphia, PMSA	PA-NJ-DE-MD	24	17	4,781	4,922	4,940	4,953	5,136	5,396	0.3	0.1	0.1	0.4	0.5
Syracuse, MSA	NY	22	13	723	742	755	746	777	803	0.3	0.5	0.1	0.5	0.3
Dayton, MSA	OH	26	22	942	951	959	951	990	1,037	0.1	0.2	0.0	0.4	0.5
Pittsburgh, MSA	PA	29	19	2,571	2,395	2,407	2,379	2,442	2,537	−0.7	0.2	−0.1	0.2	0.4
Total or average				51,273	53,704	54,493	54,864	56,534	59,745	0.5	0.4	0.3	0.5	0.6
Total or average, 29 nodes				75,287	82,909	85,526	87,618	91,105	98,724	1.0	1.0	0.9	0.9	0.8

Source: U.S. Bureau of the Census, State and Metropolitan Area Data Book, 1997–98 (5th ed.), Washington, DC, 1998.

TABLE 11-8 Total employment in specified industry, by MSA and year, 1982 through projected 2000

	MSA 5120, Mpls.-St. Paul					MSA 2160 Detroit				
Industry	1982 (thou.)	1990 (thou.)	1991 (thou.)	1992 (thou.)	2000 (thou.)	1982 (thou.)	1990 (thou.)	1991 (thou.)	1992 (thou.)	2000 (thou.)
Farm	20.2	17.5	17.2	16.8	16.0	10.0	8.3	8.2	7.9	7.3
Agricultural services, forestry	5.9	9.8	11.0	10.7	13.5	7.7	12.4	12.8	13.0	16.0
Mining	1.8	2.0	2.0	2.0	2.1	2.9	2.7	2.6	2.3	2.3
Construction	50.9	71.8	68.9	70.7	81.0	56.3	85.6	81.2	82.5	87.8
Manufacturing, durables	146.6	158.3	154.7	153.3	159.9	350.0	362.1	338.4	339.3	325.7
Manufacturing, nondurables	98.5	114.9	114.9	115.9	123.5	72.1	84.0	81.1	78.8	77.0
Transportation, communications, utilities	69.5	88.4	89.8	88.6	96.8	84.8	99.9	93.3	90.7	93.8
Wholesale trade	83.4	99.5	100.5	102.5	112.6	87.4	119.1	118.5	120.7	129.5
Retail trade	224.8	285.0	281.4	285.1	327.0	312.7	395.9	385.0	383.1	403.5
Finance, insurance, real estate	118.1	149.1	151.2	151.5	172.6	132.0	174.0	174.3	171.0	183.6
Services	330.3	495.4	501.9	522.6	649.1	455.5	640.2	636.2	652.9	773.2
Federal civilian	19.8	23.4	22.7	22.8	21.9	31.4	32.7	30.5	30.0	27.9
Military	11.3	15.1	15.0	14.7	13.1	14.1	15.8	15.3	14.8	12.2
State and local government	134.2	167.1	170.4	173.0	195.6	190.2	205.0	204.0	203.2	201.8
Total	1,315.4	1,697.3	1,701.6	1,730.2	1,984.7	1,807.1	2,237.7	2,182.9	2,190.2	2,341.6

Source: U.S. Department of Commerce, Regional Economic Information System, 1969–94.

150,000 in the Detroit MSA. Total jobs in the Detroit MSA failed to reach their 1990 level in 1992, while projected job growth from the 1990 peak to 2000 is only one-third of the projected job growth over the same period in the Minneapolis-St. Paul MSA. The projected job growth is larger, or the job losses are smaller, in the Minneapolis-St. Paul MSA in every major industry group. The disproportionate increases in total employment strongly favor the Minneapolis-St. Paul MSA, which results in an increase from 75 percent of the Detroit MSA total employment in 1982 to a projected 85 percent in 2000.

Table 11.9 summarizes the total employment change from one period to the next, with large differences for the two MSAs. Total employment change over the 1980–2000 period for the Minneapolis-St. Paul MSA exceeds that of the Detroit MSA by 135,000, mainly as a result of a large negative change in the 1990–91 period and a small positive change for the projected 1992–2000 period. If the total employment change followed the U.S. rates of change, as represented by the national growth effects, the total employment change would be 154,000 greater than projected. The differences are attributed to negative industry-mix effects and even larger negative regional-share effects.

For most areas, the U.S. growth effect is the largest change source, which is negative only in recession periods, if at all. However, a negative industry-mix or regional-share effect will adversely affect overall area growth. We attribute the industry-mix effect, which is a proportional change paralleling the changing composition of industry in an area, to the particular distribution of industry in the area. The Detroit MSA shows a much larger proportion of durable goods manufacturing with negative growth rates, and a smaller proportion of services with positive growth rates than the Minneapolis-St. Paul MSA. The overall regional-share effects for the Detroit MSA are largely negative for the historical period and also negative for the projected period.

Tables 11.10 and 11.11 show the individual industry contributions to the overall industry-mix and regional-share effects that explain, at least in part, the period-to-period differences in the total employment growth in the two MSAs. Of course, the two obvious deficiencies, namely, lack of industry detail and an essentially accounting approach (that is, without measures of the variability of period-to-period estimates) limit the usefulness of the findings. The shift-share model simply accounts for the change sources under the headings of national growth, industry mix, and regional share, as defined earlier.

Table 11.10 tackles the first of the two change sources—industry mix. It summarizes the industry composition of the two sets of industry-mix effects for each of the four periods listed earlier. The critical difference in the two sets of forecasts is the relative importance of each industry for each of the two MSAs. In the calculation of the industry-mix effect, we assume an industry group composition

TABLE 11-9 Employment change sources in specified industry, Minneapolis-St. Paul and Detroit, 1982–2000

	MSA 5120, Mpls.-St. Paul					MSA 2160 Detroit				
Change Source	1982–90 (thou.)	1990–91 (thou.)	1991–92 (thou.)	1992–00 (thou.)	1982–00 (thou.)	1982–90 (thou.)	1990–91 (thou.)	1991–92 (thou.)	1992–00 (thou.)	1982–00 (thou.)
Total change	382.0	4.3	28.6	254.5	669.4	430.6	–56.3	8.8	151.4	534.5
National growth	286.1	–15.2	7.1	242.0	520.0	393.1	–20.1	9.2	306.3	688.5
Difference, total	95.9	19.5	21.5	12.5	149.4	37.5	–36.2	–0.4	–154.9	–154.0
Industry-mix effect	19.8	1.1	0.4	9.0	30.3	14.9	–3.9	–2.5	–0.6	7.9
Regional share effect	76.2	18.4	21.1	3.5	119.2	22.6	–32.4	2.2	–154.3	–161.9

Source: U.S. Department of Commerce, Regional Economic Information System, 1969–94.

TABLE 11-10 Employment change due to industry-mix effect in specified industry, by MSA, 1982–2000

Change Source	MSA 5120, Mpls.-St. Paul					MSA 2160 Detroit				
	1982–90 (thou.)	1990–91 (thou.)	1991–92 (thou.)	1992–00 (thou.)	1982–00 (thou.)	1982–90 (thou.)	1990–91 (thou.)	1991–92 (thou.)	1992–00 (thou.)	1982–00 (thou.)
Farm	–7.1	–0.3	–0.3	–2.5	–10.2	–3.5	–0.1	–0.1	–1.2	–4.9
Agricultural services, forestry	1.6	0.5	–0.2	1.6	3.5	2.0	0.6	–0.2	1.9	4.3
Mining	–0.9	0.0	–0.2	–0.5	–1.6	–1.5	0.0	–0.2	–0.6	–2.3
Construction	6.6	–4.0	–0.2	1.7	4.1	7.3	–4.7	–0.3	2.0	4.3
Manufacturing, durables	–29.7	–6.4	–4.9	–23.0	–64.0	–71.0	–14.6	–10.7	–50.9	–147.2
Manufacturing, nondurables	–17.4	–1.3	–0.7	–12.0	–31.4	–12.7	–0.9	–0.5	–8.1	–22.2
Transportation, communications, utilities	–3.4	0.5	–1.5	–0.6	–5.0	–4.1	0.6	–1.6	–0.7	–5.8
Wholesale trade	–3.6	–0.4	0.4	–3.3	–6.9	–3.8	–0.4	0.5	–3.9	–7.6
Retail trade	11.1	–0.3	1.7	1.9	14.4	15.4	–0.4	2.4	2.6	20.0
Finance, insurance, real estate	–2.9	–0.4	–4.0	–4.0	–11.3	–3.2	–0.5	–4.6	–4.5	–12.8
Services	75.7	10.2	8.6	59.6	154.1	104.4	13.2	10.9	74.4	202.9
Federal civilian	–2.0	–0.6	–0.1	–4.0	–6.7	–3.1	–0.9	–0.1	–5.3	–9.4
Military	–2.2	–0.3	0.1	–3.3	–5.7	–2.7	–0.3	0.1	–3.4	–6.3
State and local	–5.9	3.8	1.5	–2.4	–3.0	–8.4	4.7	1.8	–2.9	–4.8
Industry mix, total	19.8	1.1	0.4	9.0	30.3	14.9	–3.9	–2.5	–0.6	7.9

Source: U.S. Department of Commerce, Regional Economic Information System, 1969–94.

TABLE 11-11 Employment change due to regional share effect in specified industry, by MSA, 1982–2000

	MSA 5120, Mpls.-St. Paul					MSA 2160 Detroit				
Change Source	1982–90 (thou.)	1990–91 (thou.)	1991–92 (thou.)	1992–00 (thou.)	1982–00 (thou.)	1982–90 (thou.)	1990–91 (thou.)	1991–92 (thou.)	1992–00 (thou.)	1982–00 (thou.)
Farm	0	0.2	–0.2	–0.7	–0.7	–0.4	0.1	–0.2	–0.5	–1
Agricultural services, forestry	1	0.8	–0.2	–0.2	1.4	1	–0.1	0.3	–0.7	0.5
Mining	0.8	0	0.2	0.3	1.3	0.7	0	–0.1	0.2	0.8
Construction	3.3	1.7	1.7	–1.3	5.4	9.8	1.1	1.2	–8.2	3.9
Manufacturing, durables	9.6	4.2	2.8	8.2	24.8	7	–5.9	10.2	–10.2	1.1
Manufacturing, nondurables	12.3	2.3	1.2	3.4	19.2	9	–1.2	–2.1	–4.7	1
Transportation, communications, utilities	7.2	1.7	–0.1	–3.5	5.3	0.8	–6.3	–1.4	–9	–15.9
Wholesale trade	1.6	2.2	1.2	–0.9	4.1	16.5	0.9	1.2	–4.2	14.4
Retail trade	0.2	–0.8	0.8	0.1	0.3	–0.2	–7	–5.9	–35.7	–48.8
Finance, insurance, real estate	8.2	3.8	3.6	3.9	19.5	16.5	2.3	0.5	–6.8	12.5
Services	17.6	0.7	10	–6.1	22.2	–18.7	–11.4	3.1	–45.4	–72.4
Federal civilian	1.2	0.2	0.1	0	1.5	–2.4	–1	–0.5	–1	–4.9
Military	3.5	0.3	–0.4	–0.3	3.1	1.4	0	–0.6	–1.3	–0.5
State and local	9.6	1	0.4	0.8	11.8	–18.2	–3.8	–3.5	–26.9	–52.4
Total	76.2	18.4	21.1	3.5	119.2	22.6	–32.4	2.2	–154.3	–161.9

Source: U.S. Department of Commerce, Regional Economic Information System, 1969–94.

that duplicates the national industry distribution for each MSA. For the United States, the growth in durable goods manufacturing is consistently less than national growth, as it is in nondurable goods manufacturing; finance, insurance, and real estate; and federal civilian employment. These industry groups, if present in disproportionately large numbers, generate below-average overall growth for an area, as the results show for the Detroit MSA.

Table 11.11 completes the comparisons of industry employment growth in the Minneapolis-St. Paul and Detroit MSAs, with its focus on the industry-specific regional-share effects for each of the four periods. Lack of industry detail, however, limits the usefulness of this comparison since it will not differentiate industry change attributed to differences in industry composition from that attributed to above-average or below-average performance of a specific industry, like motor-vehicle manufacturing. Because of the combination of these differences, the Minneapolis-St. Paul MSA has consistently positive regional-share effects for the two manufacturing groups, while the Detroit MSA has consistently negative regional-share effects, except for the 1982–90 period.

Even during the 1990–91 recession nationally, both the manufacturing and the overall regional-share effects are positive for the Minneapolis-St. Paul MSA. In comparison, the regional-share effect is negative for the Detroit MSA, except in the two periods of recovery (1982–90 and 1991–92). A majority of the industry groups show negative regional-share effects in every period except 1982–90. Similarly, the industry-mix effects are mostly negative for the Detroit MSA. Thus, the negative effects of below-average growth industries in the Minneapolis-St. Paul MSA are counterbalanced by the above-average growth of the private services sector, which accounts for a disproportionately large share of the local economy. But even with services, the regional-share effect is negative in two of the four periods for the Detroit MSA.

The shift-share results thus reveal a more robust and rapidly growing economy in the Minneapolis-St. Paul MSA than the Detroit MSA. These findings complement the earlier indications of lagging growth in the Detroit MSA and the likelihood of only partial economic recovery from the restructuring of the automobile industry. In the Minneapolis-St. Paul MSA the strong performance of the two manufacturing sectors accounted for correspondingly strong growth in other sectors, which contributed to positive regional-share effects in services. With more detailed industry data, we could spot the specific industries contributing to the superior performance. The same is the case with services. The positive industry-mix effect in services relates, in part, to the positive effect in manufacturing. Because of period-to-period changes in the composition of these industries, the overall forecasts based on the aggregate industry groups are less accurate and useful than the forecasts based on the industry-specific breakdown.

Table 11.12 introduces personal income and labor earnings (in constant 1987 dollars) for the recession year (1982), the peak year (1990), the most recent recession year (1991), and projected 2000, as additional measures of regional economic growth for the core areas of two competing air node regions—Minneapolis-St. Paul and Detroit. They compare closely in their time series performance. Further anomalies emerge, however, in the comparisons of personal income and labor earnings with population growth. While per capita income in the Minneapolis-St. Paul core MSA is greater than in the Detroit core MSA, for example, earnings per worker are reversed. High wages and salaries fail to generate high per capita incomes because of low labor force participation rates. Lagging population growth in the Detroit MSA may be linked to limited employment opportunities.

Growth in real disposable income per capita is recognized as an important determinant of the growth in passenger traffic (Ellison, 1982; Ghobrial and Kanafani, 1995). Disagreement occurs, however, in the rate of growth, which relates to the measure of growth. Business enplanements—takeoffs and landings—will grow at a declining rate in the future, according to a recent study (Rapheal and Starry, 1995). This measure, of course, obscures the increase in the number of passengers and tons of freight shipments attributed to increasing size of aircraft. Still other studies find measures showing interregional business travel and face-to-face communication increasing with the localization of information-intensive businesses at the major airport cities and their downtown districts engaged in transfer of nonstandardized information (Hugoson and Karlsson, 1996). Standardized information readily transfers without face-to-face contact.

Table 11.13 lists the industry outlays for air transportation, other transportation (except local), and communications (except radio and television) in the Minneapolis-St. Paul core labor market area in 1990. The top twenty input-purchasing industries, ranked according to their total purchases of air transportation, are among the 528 industries and sectors listed in the University of Minnesota IMPLAN (*Im*pact Analysis for *Plan*ning) regional economic modeling system. Also listed are the outlay shares, that is, the percent of total outlays accounted for by the specified input purchases, and the market shares of the three input purchases. A SIC code identifies each industry group. Finally, the two measures—outlay share and market share—show, respectively, the relative importance of air transportation purchases to and by this industry. Air transportation accounts for only 1.2 percent of total spending for the top twenty industries. Other transportation, exclusive of rail and passengers, is four times larger in outlay share than air transportation. For much of the local resident market, superior access to air transportation may be more important than purchase price.

TABLE 11-12 Total and per capita personal income and earnings per worker and labor force participation, by MSA, 1982–2000

	MSA 5120, Mpls.-St. Paul					MSA 2160 Detroit				
	1982	1990	1991	1992	2000	1982	1990	1991	1992	2000
Total personal income (mil. 1987$)	36,092	47,378	47,165	50,034	57,955	62,795	75,910	73,733	75,404	84,223
Per capita income (mil. 1987$)	15,990	18,593	18,268	18,849	20,147	14,682	17,779	17,208	17,538	18,422
Earnings per worker (mil. 1987$)	21,493	22,218	22,085	22,701	23,459	25,439	25,094	24,733	25,787	26,966
Labor force participation (pct.)	58.3	66.6	65.9	66.1	69.0	42.3	52.4	50.9	50.9	54.0

Source: U.S. Department of Commerce, Regional Economic Information System, 1969–94.

TABLE 11-13 Total outlays of specified industries for intermediate inputs, Minneapolis-St. Paul metropolitan core area, 1990

Rank	Database Sector Name	SIC Code	Industry Outlays		Air Transportation			Other Transport, Exc. Local			Comm Exc. TV, Radio		
			Total (mil.$)	Market Share (pct.)	Total (mil.$)	Outlay Share (pct.)	Market Share (pct.)	Total (mil.$)	Outlay Share (pct.)	Market Share (pct.)	Total (mil.$)	Outlay Share (pct.)	Market Share (pct.)
	Total		111,998	100.0	176.3	0.2	100.0	1,483.3	1.3	100.0	1,003.0	0.9	100.0
1	Air transportation	45	2,042	1.8	24.4	1.2	13.8	194.3	9.5	13.1	32.6	1.6	3.3
2	Business associations	861, 862	269	0.2	9.0	3.3	5.1	11.0	4.1	0.7	3.6	1.4	0.4
3	Security and commodities brokers	62	1,037	0.9	8.4	0.8	4.8	10.6	1.0	0.7	22.5	2.2	2.2
4	Electronic computers	3571	2,347	2.1	7.9	0.3	4.5	11.3	0.5	0.8	13.9	0.6	1.4
5	Paper coated and laminated	2672	3,208	2.9	7.2	0.2	4.1	64.5	2.0	4.3	15.5	0.5	1.6
6	Management and consulting	874	665	0.6	6.7	1.0	3.8	8.6	1.3	0.6	15.4	2.3	1.5
7	Newspapers	271	586	0.5	6.0	1.0	3.4	11.2	1.9	0.8	24.9	4.2	2.5
8	U.S. Postal Service	4311	591	0.5	4.9	0.8	2.8	15.2	2.6	1.0	1.1	0.2	0.1
9	Research, development, testing	873	405	0.4	4.2	1.0	2.4	5.4	1.3	0.4	9.7	2.4	1.0
10	Colleges, universities, schools	822	387	0.3	3.6	0.9	2.0	4.9	1.3	0.3	6.3	1.6	0.6
11	Other nonprofit organizations	84, 865, 869, 6732	259	0.2	3.0	1.2	1.7	4.7	1.8	0.3	7.0	2.7	0.7
12	Legal services	811	1,317	1.2	2.8	0.2	1.6	3.9	0.3	0.3	26.0	2.0	2.6
13	Wholesale trade	50, 51	5,760	5.1	2.6	0.0	1.5	5.8	0.1	0.4	17.2	0.3	1.7
14	Commercial printing	275	1,396	1.2	2.4	0.2	1.4	20.8	1.5	1.4	6.2	0.4	0.6
15	Doctors and dentists	801, 802, 803, 804	2,113	1.9	2.4	0.1	1.4	4.9	0.2	0.3	40.2	1.9	4.0
16	Motor vehicles	384	909	0.8	2.1	0.2	1.2	20.0	2.2	1.3	1.8	0.2	0.2
17	Electric services	443	1,831	1.6	2.1	0.1	1.2	12.1	0.7	0.8	4.1	0.2	0.4
18	Periodicals	175	231	0.2	2.0	0.9	1.2	3.4	1.5	0.2	1.5	0.6	0.1
19	Computer storage devices	340	500	0.4	1.9	0.4	1.1	2.7	0.5	0.2	3.3	0.7	0.3
20	Top 20, total or average		25,852	23.1	103.6	58.8	58.8	415.4	28.0	28.0	252.9	25.2	25.2

Source: University of Minnesota IMPLAN Regional Modeling System.
Note: exc. = except.

The Minneapolis-St. Paul core area is an *attractor* as well as a *generator* of air transportation passengers and cargo. This dual role of any international air node is easily overlooked in efforts to concentrate simply on destination markets for air transportation (Charlie Karlsson, Jönköping International Business School, personal communication, May 25, 1998). From a passenger transportation perspective, we can describe the area in terms of its potential both as an *attractor* for business as well as leisure travelers, and as a *generator* of business and leisure traveling. Incoming and outgoing travelers actually represent two different market segments. Incoming travelers generally originate from other regions than those of interest to outgoing travelers, and their service needs are different.

Determinants of business travel differ markedly from those affecting leisure travel (William Gartner, Tourism Center and Applied Economics Department, University of Minnesota, personal communication, May 14, 1998). The two types of travelers are interdependent in their use of common facilities, although their linkages differ from place to place and from one period to the next. Like pleasure travel, business travel is largely for a single purpose. For business travel, the purpose is proprietary, nonroutine information transfer (as noted by Charlie Karlsson in a number of his books and papers) . These are high-valued information transfers involving high-valued personnel of a company; hence, the direct travel costs appear rather small relative to travel purpose. Unlike pleasure travel, however, business travel is highly inelastic. This difference goes back to the difference in purpose, which, in turn, provides some benefits of effective price discrimination to the leisure traveler. Unlike pleasure travel, business travel occurs year-round (although weather accounts for some seasonal rescheduling of travel plans for some destinations, such as Minneapolis-St. Paul and Seattle-Tacoma). Like pleasure travel patterns, business travel patterns differ between the United States and Europe; that is, interregional distances are shorter and rail service is faster and more convenient (with downtown access) in Europe than in the United States. Like pleasure travel, business travel has its own destination images, such as Boeing and Microsoft for Seattle, computers for Silicon Valley, Ford and the auto industry headquarters for Detroit, Medical Alley and food processing for Minneapolis.

Table 11.14 lists the purchases of all transportation and other inputs by purchasing sector in the Minneapolis-St. Paul core labor market area (roughly the Minneapolis-St. Paul service area) in 1990. The totals show the much greater importance of the final demand sectors (that is, household, government, and export) over the intermediate demand sector (that is, industry input purchases) in the purchases of air transportation. Of the $2,088 million of air transportation

TABLE 11-14 Purchases of transportation and other inputs, by purchasing sector, Minneapolis-St. Paul core area, 1990

Title	SIC Code	Output (mil.$)	Total Intermed. Demand (mil.$)	Total (mil.$)	Final Demand Hshlds. (mil.$)	Gov't (mil.$)	Exp., Tot. (mil.$)	For'gn (mil.$)	Output Share Inter (pct.)	Hshlds. (pct.)	Govt (pct.)	Exp., Tot. (pct.)	For'gn (pct.)
Air transportation	45	2,088	180	1,908	627	30	1,233	351	9	30	1	59	17
Arrangement of passengers	472	245	169	77	26	0	50	47	69	11	0	20	19
Transportation services	47 export 472	41	37	4	0	0	4	4	90	0	0	10	10
Other transportation	40–44, 46	2,404	1,367	1,037	513	116	378	185	57	21	5	16	8
Transportation, total	40–47	4,737	1,715	3,022	1,166	147	1,660	583	36	25	3	35	12
Communications, export radio and TV	48, export 483	1,697	1,014	684	545	66.5	33	33	60	32	4	2	2
Other input purchases	–	108,888	25,759	83,129	31,196	7,470	35,796	5,767	24	29	7	33	5
Total input purchases	–	115,322	28,487	86,835	32,907	7,683	37,489	6,383	25	29	7	33	6
Air transport as pct. of transportation	–	44.1	10.5	63.1	53.8	20.8	74.3	60.2	na	na	na	na	na
Air transport as pct. of total	–	1.8	0.6	2.2	1.9	0.4	3.3	5.5	na	na	na	na	na

Source: University of Minnesota IMPLAN regional modeling system.

purchases, local industries account for only $180 million, or 9 percent. Households account for 30 percent of the total and exports for 59 percent. Exports include visitors to the area and airfare purchases from destination market areas.

We turn, finally, to the widely held belief that access to airports and air transportation is restrained at several of the 29 major airports, as manifested by high ticket prices and lack of competing carriers. According to the U.S. General Accounting Office (GAO), access to airports is impeded by (1) federal limits on takeoff and landing slots at the major airports in Chicago, New York, and Washington; (2) long-term, exclusive gate leases; (3) 'perimeter rules' prohibiting flights at New York's La Guardia and Washington's National airports that exceed a certain distance. The GAO study supports the concept of an auction market for the buying and selling of takeoff and landing slots at airports—a right the airline carriers have exercised in their own private interest since 1986 (Gale, 1994).

The various access barriers may affect any airline, but particularly the newer airlines entering the key markets in the East and upper Midwest, where the established carriers hold nearly all the slots through long-term gate leases.[6] According to the General Accounting Office, airlines extract a large monopoly price premium at the most congested large airports.[7] The GAO notes that travelers pay 31 percent more for the average fare at the ten most congested airports than at 33 other large airports. Comparing some of the 29 airports—for example, Atlanta, Denver, Detroit, and Minneapolis-St. Paul—we find each with a dominant airline, but with varying levels of competition among the airline tenants. In 1993, Delta had an 83.4 percent share of enplanements at Atlanta, while United had a 51.8 percent market share at Denver. Northwest Airlines had a 74.8 percent enplanement share at Detroit and an 80.6 percent enplanement share at Minneapolis-St. Paul. The enplanement share increased over the 1978–93 period for each of the three airlines at each of the four airports (Morrison and Winston, 1993). Economic activity also increased at each of the airport MSAs, according to the selected aggregate economic indicators cited earlier.

Table 11.15 brings together estimates of price differences from the Borenstein study (1996) and enplanement ratios from the Morrison-Winston study (1993). These are ranked according to the 1993 values of the airport price differences, but only for those airports included in both studies. They again show an apparent high correlation, in this example between airport and airline price differences and dominant airline enplanement ratios. The differences are consistently higher for the largest airline. They also show that airline enplanement ratios continue increasing over the 1979–93 period for each of the ten airports.

TABLE 11-15 Prices at major U.S. airports and prices of largest airline at U.S. hub airports compared to national average prices for same-distance trips, 1984, 1993, and 1995, and enplanement share, 1979 and 1993

		Airport Price Differences*			Largest Airline Price Differences*			Enplanement Share†	
Major Airport	Dominant Airline	1984 (pct.)	1993 (pct.)	1995 (pct.)	1984 (pct.)	1993 (pct.)	1995 (pct.)	1979 (pct.)	1993 (pct.)
Cincinnati	Delta	26	38	54	24	37	64	35.1	89.8
Charlotte	Eastern, US Air	27	36	56	26	36	59	74.8	94.6
Minneapolis-St. Paul	Northwest	8	20	32	10	21	34	31.7	80.6
Atlanta	Delta	31	18	12	35	33	41	49.7	83.5
Detroit	American, Northwest	5	13	14	8	22	25	21.7	74.8
Denver	United	–15	7	7	–9	9	12	32	51.8

*Testimony of Severin Borenstein before the Minnesota State Legislature, June 14, 1996.
†Steven A. Morrison and Clifford Winston, *The Evolution of the Airline Industry*, The Brookings Institution, Washington, D.C., 1993.

Borenstein (1996) has somewhat larger estimates of hub premiums than those reported in the Morrison-Winston study (1993) cited earlier. These estimates, which are ranked according to the 1995 values for the major airports, represent price differences from the national average price for each of the three years.[8] They show an apparent high correlation between airport price differences and largest airline price differences, although the differences are consistently higher for the largest airline. They show, also, that airport price differences continued increasing over the 1984–95 period in some airports, while in others they vacillated from one year to the next or actually declined. The price increases, of course, are an essential part of the improved viability and profitability of the major airlines. They had suffered severely from the boom-and-bust character of their industry during the 1980s.

Economic Models of Air Transportation

An analytical framework for assessing the role and importance of air transportation to a locality or region must account for more than the passengers and freight originating from and arriving at a given airport. It must include, also, the air traffic to and from competing airports. Ultimately, such a framework must account for air transportation-related activity at all origins and destinations within the national system of interconnected major airports. We address this issue from a theoretical perspective by focusing on the generation of interregional air travel and traffic. The findings, in turn, address the formulation of an economic model for assessing the requirements of major airport facilities.

Demand for Air Transportation

A gravity model is the most commonly used analytical framework for estimating the demand for air transportation in the context of the air traffic represented by enplanements, that is, aircraft takeoffs and landings. It takes the form,

$$T_{ij} = T(D_i, D_j, S_i, S_j) \qquad (11\text{-}1)$$

where T_{ij} is the total enplanements between air nodes i and j, D_i and D_j are air node-specific vectors representing the socioeconomic characteristics of the passengers and their air node areas, and S_i and S_j are air node-specific vectors representing transport supply variables.

The gravity model framework reappears in different studies with important variations from the basic model in predicting air travel. The Jönköping study (Hugoson and Karlsson, 1996), for example, incorporates industry- and area-specific explanatory variables in its accounting of the determinants of interregional business travel. The conceptual model is partitioned into four parts based on two sets of differentiating criteria: (1) within-firm and out-of-firm contacts and (2) lower and higher levels of uncertainty and competence. Unlike most other gravity models, this one captures the uniqueness of each industry and area with which each traveler, also differentiated by occupation, is associated. The Jönköping study thus deals directly with the task of estimating the demand for transportation by all long-distance travelers; business travelers are one segment of the long-distance traveler population.

Information transfers characterized by a high level of uncertainty and competence involve a high frequency of business meetings. Conversely, those with a low level of uncertainty and competence require a low frequency of business meetings. The high-order firms engaged in these activities locate in the downtown districts of large metropolitan areas, but their face-to-face contacts may extend over a much larger geographic area. The distant contacts generate the long-distance business travel and the demand for cost-effective air access to the distant contacts and markets.

Each of the studies suffers the common difficulty of a limited database. The Jönköping database, which is the most complete of any for the study of long-distance travel, is built on a new survey each year based on monthly interviews with about 2,000 respondents. The surveys provide data on long-distance trips between 1,940 regions. The survey regions are aggregated into functional regions or LMAs. For the 1991–92 period, the database has over 84,000 respondents, with 5,552 reporting business trips within Sweden between 70 LMAs, generating 7,892 two-way main trips of more than 100 kilometers. Expanding the survey numbers to the entire country of 9.5 million population yields an estimated 110 million long-distance trips yearly, of which 24 million are business trips.

An alternative model of the demand for air transportation starts with the basic gravity model. It includes, however, a series of relationships showing changes in a given set of origin-destination enplanements associated with changes in the composition of intermediate and final demand sectors in destination, as well as originating, market areas. Thus, the total enplanements between air nodes i and j is given by the form,

$$T_{ij} = T(D_{ik}, D_{jk}, P_{ik}, P_{jk}) \qquad (11\text{-}2)$$

and

$$P_{ij} = P(N_{ik}, N_{jk}) \qquad (11\text{-}3)$$

where T_{ij} is the total enplanements between air nodes i and j, D is an air node-specific vector representing the socioeconomic characteristics of the passengers and their air node areas for the k-th commodity-product-passenger category, P is an air node-specific price vector representing average air fares and shipping rates at the two air nodes, and N is a vector of air node-specific market and organizational attributes affecting the cost of air transportation variables between air nodes i and j for each passenger-freight category.

Supply of Air Transportation

Economic impacts of a given airport and its activities are best studied in the context of the larger air transportation system of which each airport is an integral part. An essential part of such a study would be a comparison of the detailed industry structures at competing airports and their air transportation purchases, along with the air transportation purchases of the final demand sectors. This may include simulations of the economic impact of given changes in business travel and exports associated with more congested airport conditions and other restrictions on local access to interregional air transportation.

The representation of air transport supply relationships takes into account the large number of capacity variables, both physical and organizational, affecting total enplanements between air nodes i and j in the form,

$$T_{ij} = T(S_{ik}, S_{jk}) \qquad (11\text{-}4)$$

where S is the air node-specific vector representing the characteristics of the passenger and freight facilities and equipment at their air node areas for the k-th commodity-product-passenger category. These may include the number of direct daily flight capacities between air node pairs during peak periods, the number of direct daily off-peak flight capacities between air node pairs, weighted average aircraft size during the peak periods between air node pairs, weighted average aircraft during off-peak periods between city pairs, and average travel time between air nodes i and j.

Study of the airport and related airline activities, including investments in equipment and delivery systems, would be an additional part of the study. This would involve in-depth examinations of airline business activity at both competing and complementary nodal airports.

Economic Sector Sensitivity to Air Access

Economic sector sensitivity to air access is a real and immediate concern of a wide variety of residents of an economic region. The Hugoson-Karlsson study (1996), as noted in Chapter 2, provides an analytical framework for determining the sensitivity of individual sectors of a region's economy to air transportation. It is highly instructive in the design and construction of a comprehensive database and model for understanding and predicting the demand for long-distance business travel. To this we add the demand for leisure travel and cargo shipments.

Measuring the Value of Air Transportation

Measuring the value of air transportation calls for both concept and tool applications in preparing decision information for its various users in an extended metropolitan region. These economic measures have different implications for the different urban regional decision centers—the dominant airline, the metropolitan airport management, the local corporate community, and the extended civic community—and their clients, constituents, and customers. The full array of economic measures also will vary with the decision center.

One approach to an assessment of the regional implications of air node dominance and profits in the airline industry, for example, is to measure the importance of air transportation and to design, conduct, and process a survey of airline passengers and customers—the destination and purpose of the trip or shipment, the occupation of the passenger, the value of the shipment, the competitive importance of market access to the customer, and so on. Some measures, on the other hand, may relate regional change solely to the facility rather than to the importance to the region's economy of the businesses dependent on air travel; others may equate importance with the profitability of the airline operations. Each approach offers insights into the activities and relationships associated with airport and airline activities. In their totality, they suggest a need to include information about the role and importance of the users of air transportation in the region's economy, the sensitivity of these users to limitations in access to airports and air transportation, the profitability of the air transportation services provided by the airline, and the management goals as they relate to the actual performance of the airport facility.

While a large share of knowledge transfer depends on media-based communication through computers and video conferences, face-to-face communication is increasingly important as a means of transmitting nonstandardized information that must be discussed, interpreted, and analyzed before it can be used. Media-based communication cannot be used to transmit nonstandardized information because "it does not have the capability to perceive, recognize, translate, and transfer information in a way that can reduce the uncertainty connected with nonstandardized information" (Hugoson and Karlsson, 1996). Moreover, nonstandardized information involves much uncertainty that is inherent in processes that involve business research and development, renewal activities, and negotiations of various kinds. Business trips are an increasingly important means of establishing new economic links and rejuvenating existing links.

Of particular importance are the changes in the modern economy that trigger increasing dependence on interregional business communications. According to Hugoson and Karlsson, these are (1) the higher degree of regional specialization that has increased the dependence on interregional links in facilitating transactions and trade, (2) the more rapid technological change that has increased the number of products and the degree of competition through "flexible specialization," (3) the traditional location dependence on sources of raw materials and access to heavy transport modes that has decreased rapidly, and (4) the formal organizations that have increased in total number, with each becoming more specialized as part of a group of companies.

Industry exports and imports, along with total industry outputs and inputs, are measures of this interdependence and specialization. The internal linkages are to local capital and entrepreneurial resources and intermediate input supply sources. Even more focused measures are the specific transactions of one industry with another—sales of the individual businesses in one industry that are the purchases of individual businesses in the same or other industries, either within the labor market area or as exports to destinations outside the area. In technologically advanced economies, air transportation, along with high-order communication infrastructure and related transfer systems, link these businesses and their industries to one another and to their respective markets on a global scale.

Decision Applications of Value Measures

Economic measures of the value of an airport have different implications for the different urban regional decision centers—the dominant airline, the metropolitan airport management, the local corporate community, and the extended civic community, both local and regional. Because the decision information may have tremendous value to its various users, its access may become a source of contention for its various providers. The information itself may incorporate the biases of its particular provider. In addition, it may suffer from the inevitable errors of estimation and forecasting in its basic economic measures.

One goal of a value measures study is to provide the key demographic information on the air node, or economic region, including its quality of life and economic opportunities—past, present, and projected. This could become an important local database and a baseline series of estimates and forecasts for use by various decision centers in the air node region. It would also facilitate periodic updates of the value measures. Estimates and forecasts of area demographics from past studies would be presented. These studies, along with the most recent forecasts of future industry growth, could be used in constructing alternate scenarios of economic activity in the air node area and region, with and without readily available access to regional and global markets. An important part of this goal is to provide specific measures of the value of foreign trade to the local, state, and regional economies, with special emphasis on international air cargo as the fastest growing element of international cargo.

Another goal of a value measures study is to provide comparisons of the strengths and weaknesses of competing air nodes. These would include airports with similar regional settings, dominant airlines, and market organization. This study component would provide the key demographic and economic measures from the reference areas for comparison with the given market area. It would also provide estimates and forecasts of commodity exports, by industry and clusters of supporting and market-linked industries and industry-specific inputs and outputs of the air transportation industry cluster in each of the reference areas.

The tracking of air node markets and market area performance—both past and projected—is only one contribution of a value measures study of air transportation. Also important are the linkages the study uncovers between market area performance and air access to regional and global markets. For a technology-intensive, high value-added economy, like the ones centered on a majority of the 29 air nodes, readily available air access to regional and global markets is an essential precondition for its continued growth.

Notes

1. Saxenian notes further that recent research reveals the interrelations of these three dimensions (p. 174). For example, economists now recognize that "innovation is a product of interactions among customers and suppliers, a firm's internal operating units, and the wider social and institutional environment."

2. Declining rates of growth characterize the industrialized economies of both the United States and Europe. Economists attribute the productivity slowdown to various supply-related factors, like the first oil hike (environmental regulations had the same effect), slowdown in spending for research and development, many inexperienced workers entering the workforce, and financial instability following the breakdown of the Bretton Woods system (Shigehara, 1992, p. 21). Additional explanations come from studies on the effects of declining savings and investment rates, including spending on education. More fundamental causes proposed by Shigehara (pp. 21–32) include the interaction of inflation and productivity performance, the increasing structural rigidities and ossification of economies, increases in rent-seeking activities (for example, growth of nontariff barriers and impediments to trade), problems of some financial markets in channeling investment funds toward long-term productive uses, losses in the efficiency of investment (for example, as insurance against higher trade barriers in the future), and the fragility of the financial system and its institutions.

3. The seven strategies presented were business recruitment, business loans, tax-increment financing, university-based research parks, enterprise zones, self-development, and rural telecommunications. Sears and Reid, the symposium organizers, devised a set of thirteen criteria for scoring the individual strategies. The criteria were breadth of studies, relevance of existing research to the strategy, number of success dimensions, appropriateness of success measures, quality of experimental design, data quality, timeliness of research, objectivity of research, representativeness, sample size, rural applicability, recognition of rural diversity, and policy relevance (pp. 304–307).

4. The U.S. Department of Transportation amended its rules in 1985 to allow airlines to buy and sell slots to one another as a way of minimizing the government's role in the allocation of slots. By the early 1990s, the U.S. General Accounting Office (1996) found that "a few carriers had increased their control of slots to such an extent that could limit access to routes beginning or ending at any of the slot-controlled airports—airports that are crucial to establishing new service in the heavily traveled eastern and midwestern markets" (p. 4).

5. See also: Testimony of Severin Borenstein before the Minnesota State Legislature, June 14, 1996. Department of Economics, University of California, Davis. According to this testimony, the Northwest Airlines monopoly profit premium was more than $540 million in 1995.

6. The GAO (1996) calculated the percentage difference in fares at ten constrained airports compared to other large airports in 1995 as follows: Charlotte, +87.81; Cincinnati, +84.87; Pittsburgh, +72.23; Washington National, +48.39; Minneapolis, +45.32; New York La Guardia, +34.64; Detroit, +26.56; Newark, +24.26; Chicago O'Hare, +23.76; New York Kennedy, −4.08; overall, +31.06 (p. 330).

7. *Wall Street Journal,* November 13, 1996, page A2. An earlier GAO study found a 33 percent premium at these airports.

8. The GAO study (1996) found even higher percentage differences in fares than for other large airports: ±45.32 for Minneapolis-St. Paul and +26.56 for Detroit (p. 33).

References

Alfsen, Knut H., Torstein Bye, and Lorents Lorentsen. 1987. *Natural Resource Accounting and Analysis: The Norwegian Experience, 1978–1986.* Oslo: Central Statistical Bureau.

Alonso, William. 1965. *Location and Land Use: Toward a General Theory of Land Rent.* Cambridge, Mass.: Harvard University Press.

Alward, G.S., E. Siverts, D. Olson, J. Wagner, D. Senf, and S. Lindall. 1989. *Micro IMPLAN Software Manual.* Minneapolis-St. Paul: University of Minnesota.

American Planning Association. 1980 (June). *Local Capital Improvements and Development Management: Analysis and Case Studies.* Washington, D.C.: Government Printing Office.

Amos, Orley M. Jr. 1990. Growth pole cycles: A synthesis of growth pole and long wave theories. *Review of Economic Studies* 20 (1): 37–48.

Andersen, Arthur. 1996. *Economic Impact of a New Twins Stadium.* Minneapolis: Arthur Andersen Consulting.

Andersson, Ake E., and Folke Snickers. 1996. *A Central-Place Theory of Airports.* Stockholm: Institute of Futures Studies and Department of Infrastructure and Planning, Royal Institute of Technology.

Anding, Thomas L. 1990. *Trade Centers of the Upper Midwest, Changes from 1960 to 1989.* Minneapolis: University of Minnesota, Center for Urban and Regional Affairs.

Antonius, Jim. 1994. *Cities Then and Now.* New York: Macmillan.

Armstrong, Harvey W. 1996. European Union Regional Policy: Sleepwalking to a Crisis. *International Review of Regional Science* 19 (3): 193–210.

Aschauer, David A. 1991. Infrastructure: America's Third Deficit. *Challenge* (March–April): 39–45.

Ayres, D. 1991. Sources of municipal revenue. *Public Management* 73 (8): 19–21.

Bailly, Antoine, and Denis Maillat. 1991. Service activities and regional metropolitan development: A comparative study, pp. 129–145. In *Services and Metropolitan*

Development International Perspectives, ed. P.W. Daniels. London/New York: Routledge.

Barker, Roger G. 1968. *Ecological Psychology: Concepts and Methods for Studying the Environment of Human Behavior.* Stanford, Calif.: Stanford University Press.

Barker, Roger G. 1978. *Habitats, Environments, and Human Behavior.* San Francisco: Jossey-Bass.

Barnard, Jerald. 1967. *Design and Use of Social Accounting Systems in State Development Planning. Bureau of Business and Economic Research.* Iowa City: University of Iowa Press.

Barnard, Jerald R., James A. MacMillan, and Wilbur R. Maki. 1970. Evaluation Models for Regional Development Planning. *Papers of the Regional Science Association* 23: 117–138.

Bartik, Timothy J. 1986. Neighborhood revitalization's effects on tenants and the benefit-cost analysis of government neighborhood programs. *Journal of Urban Economics* 19, 234–248.

Bartik, Timothy J. 1989. Small business start-ups in the United States: Estimates of the effects of characteristics of states. *Southern Economic Journal* 55: 1004–1018.

Baumol, William J. 1967. Macroeconomics of unbalanced growth. *American Economic Review* 57: 415–426.

Baumol, William J. 1985. Productivity policy and the service sector, pp. 301–317. In *Managing the Service Economy Prospects and Problems.* New York: Cambridge University Press.

Becker, Gary. 1998. Economic viewpoint. *Newsweek* (May 18): 26.

Beesley, M.E., and R.T. Hamilton. 1984. Small firms' seedbed role and the concept of turbulence. *Journal of Industrial Economics* 33 (2): 217–231.

Bell, Clive, Shata Deuraragan, Peter Hazell, and Roger Glade. 1976. *A Social Accounts Analysis of the Structure of the Muda Regional Economy.* RPO 671–17, Working Paper No. 4. Washington, D.C.: World Bank Development Research Center.

Bendavid, Avrom. 1974. *Regional Economic Analysis for Practitioners: An Introduction to Common Descriptive Methods.* New York: Praeger.

Berry, Brian, and Alled Pred. 1961. *Central Place Studies.* Philadelphia: Regional Science Research Institute, University of Pennsylvania.

Beyers, W.B. 1991. Trends in the producer services in the USA: The last decade, pp. 146–172. In *Services and Metropolitan Development International Perspectives,* ed. P.W. Daniels. New York: Routledge.

Bingham, Richard D., and Robert Mier. 1993. *Theories of Local Economic Development Perspectives From Across the Disciplines.* Newbury Park, Calif.: Sage Publications.

Birch, David L. 1981. Who creates jobs? *The Public Interest* 65: 3–14.

Bluestone, B., and B., Harrison. 1982. *The Deindustrialization of America.* New York: Basic Books.

Bonnen, James T. 1992. Why is there no coherent U.S. rural policy? *Policy Studies Journal* 20 (2): 190–201.

Borchert, John R., and Donald P. Yeager. 1968. *Atlas of Minnesota Resources and Settlement.* St. Paul: Minnesota State Planning Agency.

Borenstein, Severin. 1992. The evolution of U.S. airline competition. *Journal of Economic Perspectives* 6 (2): 45–73.

Borenstein, Severin. 1996. Testimony of Severin Borenstein before the Minnesota State Legislature, June 14, 1996. Davis, Calif.: Department of Economics, University of California.

Braslau, David. 1998. *Air Service and the Minnesota Economy.* Report prepared for the Metropolitan Airports Commission. Minneapolis: Center for Transportation Studies, University of Minnesota.

Bromley, Daniel. 1989. *Economic Interests and Institutions.* Cambridge, Mass.: Baskil Blackwell.

Browne, Lynne Elaine, and Rebecca Hellerstein. 1997. Are we investing too little? *New England Economic Review* (November/December): 29–50.

Brucker, S., S.E. Hastings, and W.R. Latham III. 1987. Regional input-output analysis: A comparison of five readymade model systems. *Review of Regional Studies* 17: 1–16.

Brucker, S., S.E. Hastings, and W.R. Latham III. 1990. The variation of estimated impacts from five regional input-output models. *International Regional Science Review* 13 (1 and 2): 119–139.

Brusco, Sebastiano. 1982. The Emilian Model: Productive decentralization and social integration. *Cambridge Journal of Economics* 6: 167–184.

Bulmer-Thomas, V. 1982. *Input-Output Analysis in Developing Countries: Sources, Methods, and Applications.* Chichester [Sussex, Eng.]; New York: Wiley.

Chinitz, Benjamin. 1961. Contrasts in agglomeration: New York and Pittsburgh. *American Economic Review* 51 (2): 279–289.

Chinitz, Benjamin. 1974. *Regional Economic Policy in the United States. Regional Economic Policy:* Proceedings of a conference sponsored by the Federal Reserve Bank of Minneapolis.

Christaller, Walter. 1966. *Central Places in Southern Germany.* Englewood Cliffs, N.J.: Prentice-Hall.

Citizens League, Planned Unit Development Committee. 1973. *Growth without Sprawl: How the Twin Cities Area Can Provide Amenity in Housing without an Unnecessary Increase in Costs of Urban Services or Damage to the Environment.* Minneapolis: Citizen's League of the Twin Cities Metropolitan Area.

Commission on Global Governance. 1995. *Our Global Neighborhood.* Oxford: Oxford University Press.

Crihfield, John B., and Harrison S. Campbell Jr. 1991. Evaluating alternative regional planning models. In *Growth and Change* 22(2): 1–16.

Crihfield, John B., and Harrison S. Campbell Jr. 1992. Evaluating alternative regional planning models: Reply. *Growth and Change* Fall: 521–530; Spring: 1991: 1–16.

Daly, M.T. 1991. Transitional economic bases: From the mass production society to the world of finance, pp. 26–43. In *Services and Metropolitan Development: International Perspectives,* ed. P.W. Daniels. New York: Routledge.

Daniels, P.W. 1991. Service sector restructuring and metropolitan development: Processes and prospects, pp. 1–25. In *Services and Metropolitan Development International Perspectives,* ed. P.W. Daniels. New York: Routledge.

Deavers, Kenneth. 1992. What is rural? *Policy Studies Journal* 20 (2): 184–189.

Delamaide, Darrell. 1994. *The New Superregions of Europe.* New York: Plume.

DeLong, J. Bradford, and Lawrence H. Summers. 1992. Policies for long-run economic growth, pp. 93–128. In a symposium sponsored by the Federal Reserve Bank of Kansas City, Mo. Eberly, Don E., ed. 1994. *Building a Community of Citizens: Civil Society in the 21st Century.* Lanham, Md.: University Press of America.

Eberly, Don (ed.). 1994. *Building a Community of Citizens: Civil Society in the 21st Century.* Lanham, MD: University Press of America.

Ehrenhalt, Alan. 1995. *The Lost City: Discovering the Forgotten Virtues of Community in the Chicago of the 1950s.* New York: Basic Books.

Ellison, Anthony P. 1982. The structural change of the airline industry following deregulation. *Transportation Journal* 58–69.

Emerson, R.L., M. Jarvin, and F. Charles Lamphear. 1975. *Urban and Regional Economics: Structure and Change.* Boston: Allyn and Bacon.

Etzioni, Amitai, ed. 1995. *New Communitarian Thinking: Persons, Virtues, Institutions, and Communities.* Charlottesville, Va.: University Press of Virginia.

Fox, Karl A. 1985. *Social Systems Accounts: Linking Social and Economic Indicators through Tangible Behavior Settings.* Dordrecht/Boston/Lancaster: D. Reidel.

Fox, Karl A. 1994. Delimitation of regions for transportation planning, pp. 97–117. In *Selected Writings of Karl A. Fox,* eds. James R. Prescott, Paul Van Moeseke, and Jati K. Sengupta. Ames, Iowa: Iowa State University Press.

Fox, Karl. A., and T. Krishna Kumar. 1966. Delineating functional economic areas. In *Research and Education for Regional and Area Development,* eds. Wilbur Maki and Brian Berry. Ames, Iowa: Iowa State University Press.

Freidenberg, Howard L., and Richard M. Beemiller. 1997. Comprehensive revision of gross state product by industry, 1977–94. *Survey of Current Business* 77 (6): 34.

Friedman, David. 1996. The smoke-filled tomb: Which party would be better for the health of the new economy, Republican or Democratic? *INC* (August): 21–22.

Friedman, Lee S. 1984. Designing governance structures, pp. 561–604. In *Microeconomic Policy Analysis.* New York: McGraw Hill.

Gale, Ian. 1994. Competition for scarce inputs: The case of airport takeoff and landing slots. *Economic Review* 30 (2): 18–25.

Galloway, Lowell, and Gary Anderson. 1992 (November 7). *Derailing the Small Business Express.* Prepared for Representative Dick Armey, Joint Economic Committee of the U.S. Congress.

Garreau, Joel. 1981. *The Nine Nations of North America.* New York: Avon.

Ghobrial, A., and A. Kanafani. 1995. Quality-of-service model of intercity air-travel demand. *Journal of Transportation Engineering* 121: 135–140.

Glasmeier, Amy K. 1993. High-tech manufacturing: Problems and prospects. In *Economic Adaptation: Alternatives for Nonmetropolitan Areas,* ed. David L. Barkley. Boulder, Colo.: Westview Press.

Goetz, Edward G., and Terrance Kayser. 1993. Competition and cooperation in economic development: A study of the Twin Cities Metropolitan Area. *Economic Development Quarterly* 7 (1): 63–78.

Goreham, Gary A. 1997. *Encyclopedia of Rural America: The Land and the People.* Santa Barbara, Calif.: ABC-CLIO.

Gottman, Jean. 1964. Megalopolis: The Urbanized Northeastern Seaboard of the United States. Cambridge, Mass.: M.I.T. Press.

Greenhut, M.L. 1963. *Microeconomics and the Space Economy: The Effectiveness of an Oligopolistic Market Economy.* Chicago: Scott Foresman.

Grimes, Donald R., George A. Fulton, and Marc A. Bonardelli. 1992. Evaluating alternative regional planning models: Comment. *Growth and Change* 24: 516–520.

Hall, Peter. 1966. *The World Cities.* New York: McGraw-Hill.

Hamilton, R.T. 1989. Unemployment and business formation rates: Reconciling time-series and cross-sectional evidence. *Environment and Planning* 21: 249–255.

Hansen, Niles M., ed. 1972. *Growth Centers in Regional Economic Development.* New York: Free Press.

Hansen, Niles M. 1993. Endogenous growth centers: Lessons from rural Denmark. In *Economic Adaptation: Alternatives for Nonmetropolitan Areas,* ed. David L. Barkley. Boulder, Colo.: Westview Press.

Hanson, Kenneth A., and Sherman Robinson. 1991. Data, linkages and models: US national income and product accounts in the framework of a social accounting matrix. *Economic Systems Research* 3 (3): 215–232.

Hartley, David, and Ira Muscovice. 1992. *Rural Health in the Northwest Area.* Minneapolis: Institute for Health Services Research, School of Public Health, University of Minnesota.

Hirsch, Werner. 1964. *Elements of Regional Accounts.* Papers presented at the Conference on Regional Accounts, 1962, sponsored by the Committee on Regional Accounts. Baltimore: Johns Hopkins University Press.

Hirschman, Albert O. 1958/1978. *The Strategy of Economic Development.* New Haven, Conn.: Yale University Press; reprinted, New York: Norton.

Hirschman, Albert O. 1992. Industrialization and its manifold discontents: West, east, and south. *World Development* 20: 1225–1232.

Hoben, James E. 1975. *Costs of Sprawl: HUD Challenge.* Washington, D.C.: Government Printing Office.

Hochwald, Werner, ed. 1961. *Design of Regional Accounts.* Papers presented at the Conference of Regional Accounts, 1990, sponsored by the Committee on Regional Accounts. Baltimore: Johns Hopkins Press.

Holland, David, and Peter Wyeth. 1993. *SAM Multipliers: Their Decomposition, Interpretation and Relationship to Input-Output Multipliers.* Research Bulletin XB1027. Pullman, Wash.: Washington State University, College of Agriculture and Home Economics Research Center.

Hoover, Edgar M. 1948. *The Location of Economic Activity.* New York: McGraw-Hill.

Hudson, John. 1989. The birth and death of firms. *Quarterly Review of Economics and Business* 29 (2): 68–86.

Hugoson, Peter, and Charlie Karlsson. 1996. *The Determinants of Interregional Business Trips in Sweden: A Gravity Model Approach.* Paper presented at the 36th European Congress of the Regional Science Association, Zurich, August 26–30.

Hutton, Thomas, and David Ley. 1987. Location, linkages, and labor: The downtown complex of corporate activities in a medium size city, Vancouver, British Columbia. *Economic Geography* 63 (2): 126–141.

Ihlanfeldt, Keith R., and Michael D. Raper. 1990. The intrametropolitan location of new office firms. *Land Economics* 66 (2): 182–198.

Ihlanfeldt, Keith R., and D.L. Sjöquist. 1991. The role of space in determining the occupations of black and white workers. *Regional Science and Urban Economics* 21: 295–316.

Irwin, Michael D., and John D. Kasarda. 1991. Air passenger linkages and employment growth in U.S. metropolitan areas. *American Sociological Review* 56: 524–537.

Isard, Walter. 1956. *Location and the Space-Economy: A General Theory Relating to Industrial Location, Market Access, Land Use, Trade, and Urban Structure.* Cambridge, Mass.: M.I.T. Press.

Isard, Walter. 1972. *Ecologic-Economic Analysis for Regional Development.* New York: Free Press.

Isard, Walter. 1975. *Introduction to Regional Science.* New York: Prentice Hall.

Isard, Walter, and Thomas W. Langford. 1971. *Regional Input-Output Study: Reflections, Reflections, and Diverse Notes on the Philadelphia Experience.* Cambridge, Mass.: M.I.T. Press.

Isard, Walter, Eugene W. Schooler, and Thomas Vietoriz. 1959. *Industrial Complex Development and Regional Analysis.* Cambridge, Mass: M.I.T. Press.

Jacobs, Jane. 1984. *Cities and the Wealth of Nations.* New York: Random House.

Johansen, Leif. 1960. *A Multi-Sectoral Study of Economic Growth.* Amsterdam: North-Holland.

Johansson, B. 1987. Information technology and the viability of spatial networks. *Papers of the Regional Science Association* 61.

Johnson, Peter. 1986. *New Firms: An Economic Perspective.* London: Allen & Unwin.

Johnson, Thomas J., Brady J. Deaton, and Eduardo Segarra. 1988. *Local Infrastructure Investment in Rural America.* Boulder, Colo.: Westview Press.

Judge, George C., and Takashi Takayama, eds. 1973. *Studies in Economic Planning Over Space and Time.* Amsterdam: North-Holland.

Kanter, Rosabeth Moss. 1994. Collaborative advantage: The art of alliances. *Harvard Business Review* (July–August): 96–112.

Karlsson, C. 1994. *From Knowledge and Technology Networks to Network Technology; Patterns of a Network Economy,* ed. B. Johansson. Berlin.

Keeble, David, Sheila Walker, and Martin Robson. 1993. *New Firm Formation and Small Business Growth in the United Kingdom.* Employment Department Research Series 15.

Koh, Young-Kon, Dean F. Schreiner, and Huijune Shin. 1993. Comparisons of regional fixed price and general equilibrium models. *Regional Science Perspectives* 23 (1): 18–32.

Kolderie, Ted. 1972. Governance in the Twin Cities area of Minnesota: Regional councils of government. In *Substate Regionalism and the Federal Government.* Washington, D.C.: Advisory Commission on Intergovernmental Relations.

Kozlowski, Paul J., and Timothy S. Hansen. 1998 . *Do Surveys of Purchasing Managers Predict Business Conditions in Regional Markets?* Paper presented at the 37th Annual Southern Regional Science Association Meeting, Savannah, Ga., April 4.

Krugman, P. 1991. *Geography and Trade.* Cambridge, Mass.: M.I.T. Press.

Kuhn, Thomas. 1970. The Structure of Scientific Revolutions.

Kutcher, Ronald E. 1991. New BLS projections: findings and implications. Monthly Labor Review. November, pp. 3–12.

Levine, Marc V. 1987. Downtown redevelopment as an urban growth strategy: A critical appraisal of the Baltimore renaissance. *Journal of Urban Affairs* 9 (2): 103–123.

Lewis, W. Arthur. 1954. Economic development with unlimited supplies of labour. *Manchester School* (May): 139–191.

Ley, David, and Thomas Hutton. 1987. Vancouver's corporate complex and producer services sector: Linkages and divergence within a provincial staple economy. *Regional Studies* 21: 413–424.

Lipke, Bruce. 1993. Regulatory Costs and Competitiveness of U.S. manufacturing. Presentation to the 35th Annual Meeting of the National Association of Business Economists, Chicago, September 19–22.

Lösch, August. 1964. *The Economics of Location.* New Haven, Conn.: Yale University Press.

Mack, Richard S., and David S. Jacobson. 1996. Core periphery analysis of the European Union: A location quotient approach. *Regional Analysis & Policy* 26 (1): 3–22.

Maki, Wilbur R. 1980. Regional input-output and social accounting systems for agricultural and rural development planning. Staff Paper Series P80–21. St. Paul: Department of Agricultural and Applied Economics, University of Minnesota.

Maki, Wilbur, Scott Loveridge, and Richard Lichty. 1994. Using IMPLAN in regional classes. *Regional Science Perspectives* 24(1): 83–93.

Maki, Wilbur R., Douglas Olson, and Con H. Schallau. 1985. A dynamic simulation model for analyzing the importance of forest resources in Alaska. Research Note PNW-432. Portland, Ore.: Pacific Northwest Forest and Range Experiment Station.

Maki, Wilbur R., and Paul D. Reynolds. 1994. Stability versus volatility, growth versus decline in peripheral regions: A preliminary U.S. application, pp. 27–42. In *Northern Perspectives on European Integration,* eds. L. Lundqvist and L. Persson. Stockholm: Nordiska Institutet for Regionalpolitisk Forskning.

Maki, Wilbur R., and Knut Ingar Westeren. 1992. *Revitalizing a Rural Region's Economic Base*. St. Paul: Department of Agricultural and Applied Economics, University of Minnesota.

Marshall, Ray, and Marc Tucker. 1987. *Thinking for a Living: Education and the Wealth of Nations*. New York: Basic Books.

Mason, Colin M. 1991. Spacial variations in enterprise: The geography of new firm formation. In *Deciphering the Enterprise Culture*, ed. R. Burrows. New York: Routledge.

Mattoon, Richard, and William Testa. 1993 (July/August). *Shaping the Great Lakes Economy: Economic Perspectives*. Chicago: Federal Reserve Bank of Chicago.

McKinley, Charles. 1952. *Uncle Sam in the Pacific Northwest*. Berkeley: University of California Press.

McMahon, C.W. 1965. *Techniques of Economic Forecasting: An Account of the Methods of Short-Term Economic Forecasting Used by the Governments of Canada, France, the Netherlands, Sweden, the United Kingdom, and the United States*. Paris: Organization for Economic Cooperation and Development.

Metropolitan Council. 1994. *Regional Blueprint: Twin Cities Metropolitan Area. A Metropolitan Council Report*. St. Paul: Metropolitan Council.

Metropolitan Council. 1997. *Regional Blueprint Forecast Procedures. Regional Research Note*. St. Paul: Metropolitan Council.

Miernyk, William H. 1982. *Regional Analysis and Regional Policy*. Cambridge, Mass.: Oelgeschlager, Gunn, & Hain.

Miernyk, William H., Kenneth L. Shellhammer, Douglas M. Brown, Ronald L. Coccari, Charles J. Gallagher, and Wesley H. Wineman. 1970. *Simulating Regional Economic Development: An Interindustry Analysis of the West Virginia Economy*. Lexington, Mass.: Heath Lexington Books, D.C. Heath.

Miller, James P. 1993. Small and Midsize Enterprise Development: Prospects for Nonmetropolitan Areas. In *Economic Adaptation: Alternatives for Nonmetropolitan Areas*, ed. David L. Barkley. Boulder, Colo.: Westview Press.

Miller, Ronald E., and Peter D. Blair. 1985. *Input-Output Analysis: Foundations and Extensions*. Englewood Cliffs, NJ: Prentice-Hall.

Mills, Edwin S., and Bruce W. Hamilton. 1988. *Urban Economics*, 4th Ed. Glenview, Ill.: Scott Foresman.

Minneapolis Community Development Agency. 1991. *Group Health Service Area*. Minneapolis: City of Minneapolis.

Minnesota IMPLAN Group, Inc. 1997. *IMPLAN Professional: User's Guide, Analysis Guide, Data Guide*. Stillwater: Minnesota IMPLAN Group.

Morky, Benjamin W. 1988. *Entrepreneurship and Public Policy*. New York: Quorum Books.

Morrill, Richard L. 1970. *The Spatial Organization of Society*. Belmont, Calif.: Wadsworth.

Morrison, Steven A., and Clifford Winston. 1993. *The Evolution of the Airline Industry*. Washington, D.C.: Brookings Institution.

Moss, M.L., and J.G. Brion. 1991. Foreign banks, telecommunications, and the central city, pp. 265–284. In *Services and Metropolitan Development International Perspectives,* ed. P.W. Daniels. London/New York: Routledge.

Moyes, A., and P. Westhead. 1990. Environment for new firm foundation in Great Britain. *Regional Studies* 24 (2): 123–126.

Naftalin, Arthur, and John Brandl. 1980. *The Twin Cities Regional Strategy.* St. Paul: Metropolitan Council.

Normann, Richard, and Rafeal Ramirez. 1993. From value chain to value constellation: Designing interactive strategy. *Harvard Business Review* (July–August): 65–77.

Nourse, Hugh O. *Regional Economics: A Study in the Economic Structure, Stability, and Growth of Regions.* New York: McGraw-Hill.

Noyelle, T. J., and P. Peace. 1991. Information industries: New York's new export base, pp. 285–304. In *Services and Metropolitan Development International Perspectives,* ed. P.W. Daniels. London/New York: Routledge.

Noyelle, Thierry J., and Thomas M. Stanback Jr. 1984. *The Economic Transformation of American Cities.* Totowa, N.J.: Rowman and Allanheld.

Oakey, Ray. 1984. *High Technology Small Firms: Regional Development in Britain and the United States.* New York: St. Martin's Press.

Odum, Howard W., and Harry Estill Moore. 1938. *American Regionalism: A Cultural-Historical Approach to National Integration.* New York: H. Holt.

Ohmae, Kenichi. 1995. *The End of the Nation State: The Rise of Regional Economies.* New York: Free Press.

Okamura, S., F. Sakauchi, and I. Nonaka. 1993. New perspectives on global science and technology policy. In *Proceedings of NISTEP, the Third International Conference on Science and Technology Policy Research.* Tokyo: MITA Press.

Orfield, Byron. 1997. *Micropolitics: A Regional Agenda for Community and Stability.* Washington, D.C.: Brookings Institution and the Lincoln Land Institute.

Osborne, David. 1990. Refining state technology programs. *Issues in Science and Technology* (Summer): 55–61.

Osborne, David, and Ted Gaebler. 1993. *Reinventing Government: How the Entrepreneurial Spirit Is Transforming the Public Sector.* New York: Penguin Books.

Ostrum, Elinor. 1990. *Governing the Commons: The Evolution of Institutions for Collective Action.* Cambridge, Mass.: Cambridge University Press.

O'Sullivan, Arthur. 1993. *Urban Economics.* Chicago: Irwin.

Otto, Daniel M., and Thomas G. Johnson. 1993. *Microcomputer-Based Input-Output Modeling: Applications to Economic Development.* Boulder, Colo.: Westview Press.

Perloff, Harvey S., Edgar S. Dunn Jr., Eric E. Lampard, and Richard F. Muth. 1960. *Regions, Resources, and Economic Growth.* Lincoln, Neb.: University of Nebraska Press.

Peterson, Willis. 1994. Overinvestment in public sector capital. *Cato Journal* 14 (1): 65–73.

Pfouts, Ralph Jr., ed. 1960. *The Techniques of Urban Economic Analysis.* West Trenton, N.J.: Chandler-Davis.

Pierce, Neil R., and Carol F. Steinbach. 1989. *Enterprising Communities: Community-Based Development in America. 1990.* Washington, D.C.: Council for Community-Based Development.

Plaut, Thomas R., and Joseph E. Pluta. 1983. Business climate, taxes and expenditures, and state industrial growth in the United States. *Southern Economic Journal* 50 (1): 99–119.

Plosser, Charles I. 1992. Policies for long-run economic growth, pp. 57–86. In a symposium sponsored by the Federal Reserve Bank of Kansas City, Mo.

Polenske, Karen R. 1993. A property rights perspective on economic development strategies: Venturing beyond Hirschman and Porter. Paper presented at the Concepts in Regional Development session of the 40th meeting of the North American Regional Science Association, Houston, Texas, November 11–14.

Porter, Michael E. 1990. *The Competitive Advantage of Nations.* New York: Free Press.

Porter, Michael E. 1996. Competitive advantage, agglomeration economies, and regional policy. *International Regional Science Review* 19 (1 and 2): 85–94.

Prahalad, C.K., and Gary Hamel. 1990. The core competence of the corporation. *Harvard Business Review* (May–June): 79–91.

Price, Kent A., ed. 1992. *Regional Conflict and Regional Policy.* Washington, D.C.: Resources for the Future.

Puerto Rico Planning Board. 1990. *Insumo Producto 1981–82.* San Juan, P.R.: Puerto Rico Planning Board.

Puerto Rico Study Team. 1994. *Puerto Rico IMPLAN System: Model and Database Construction and Application.* Atlanta, Ga.: Puerto Rico IMPLAN Study Team, U.S. Forest Services.

Putnam, Robert D. 1993a. The prosperous community. *TAP* (Spring).

Putnam, Robert D. 1993b. *Making Democracy Work: Civic Traditions in Modern Italy.* Princeton, N.J.: Princeton University Press.

Putnam, Robert D. 1995. Bowling alone: America's declining social capital. *Journal of Democracy.* (January).

Putnam, Robert D. 1996. The strange disappearance of civic America. *The American Prospect* 24 (Winter): 34–48.

Pyatt, Graham. 1991. Economic systems research. *Journal of the International Input-Output Association* 3 (3): 315–441.

Pyatt, Graham, and Alan R. Rose. 1977. *Social Accounting for Development Planning with Special Reference to Sri Lanka.* Cambridge, Mass.: Cambridge University Press.

Rabinovitch, Jonas. 1992. Curitaba: Towards sustainable urban development. *Environment and Urbanization* 4 (2): 62–73.

Rabinovitch, Jonas, and Josef Leitman. 1996. Urban planning curitaba. *Scientific American* 274 (3): 46–53.

Rapheal, David E., and Claire Starry. 1995. The future of air travel. *Transportation Research Record* 1506.

Rasmussen, N. 1956. *Studies in Intersectoral Relations.* Amsterdam: North-Holland.

Ratajczak, Donald, 1980. A Generalized specification of regional labor markets, pp. 114–139. In *Economic Impact Analysis: Methodology and Applications,* ed. Saul Pleeter. Boston: M. Nijhoff.

Real Estate Research Corporation. 1974. *The Costs of Sprawl: Detailed Cost Analysis.* Washington, D.C.: Government Printing Office.

Regional economic policy. Proceedings of a conference sponsored by the Minnesota Economics Association and the Federal Reserve Bank of Minneapolis.

Reynolds, Paul D. 1991. Predicting new firm births: Interactions of organizational and human populations. In *Entrepreneurship in the 1990s,* eds. D. Saxon and J. Kasarda. Boston: PWS-Kent.

Reynolds, Paul D. 1994. The entrepreneurial process: Preliminary explorations in the United States. Presentation to the First Eurostat International Workshop in Techniques of Enterprise Panels, Luxembourg, February 21–23.

Reynolds, Paul D., and Wilbur R. Maki. 1990a. *Business Volatility and Economic Growth.* Project Report prepared for the U.S. Small Business Administration (Contract SBA 3067-OA-88).

Reynolds, Paul D., and Wilbur Maki. 1990b. *Business Volatility and Regional Economic Change.* Paper prepared for Cross-National Workshop on the Role of Small and Medium Enterprise in Regional Economic Growth, University of Warwick, Coventry, United Kingdom.

Reynolds, Paul D., and Wilbur Maki. 1991. Regional characteristics affecting new firm births. Prepared for presentation to the Second Annual Workshop on Small Business Economics and Entrepreneurship, Harvard Business School, Cambridge, Mass., and 23rd Annual Meeting of the Mid-Continent Regional Science Association, Chicago.

Reynolds, Paul D., Brenda Miller, and Wilbur Maki. 1994a. Regional characteristics affecting business volatility in the United States, 1980–84. In *Small Business Dynamics: International, National, and Regional Perspectives,* eds. C. Karlsson et al. London: Routledge.

Reynolds, Paul D., Brenda Miller, and Wilbur R. Maki. 1994b. Explaining regional variation in business births and deaths: U.S. 1976–88. *Small Business Economics* 6: 1–19.

Ricardo, David. 1817. *On the Principles of Political Economy and Taxation.* London: J. Murray.

Rickman, Dan S., and R. Keigh Schwer. 1993. A systematic comparison of the REMI and IMPLAN models: The case of Southern Nevada. *Review of Regional Studies* 23 (2).

Rodwin, Lloyd. 1970. *Nations and Cities.* New York: Houghton Mifflin.

Rosegrant, Susan, and David R. Lampe. 1992. *Route 128: Lessons from Boston's High-Tech Community.* New York: Basic Books.

Rostow, W. W. 1971. *The Stages of Economic Growth: A Non-Communist Manifesto,* 2nd Ed. Cambridge, London: Cambridge University Press.

Round, Jeffery I. 1991. A SAM for Europe: Problems and perspectives. *Economic Systems Research* 3 (3): 249–268.

Rusk, David. 1993. *Cities Without Suburbs.* Washington, D.C.: Woodrow Wilson Center Press.

Ruttan, Vernon W. 1977. Induced innovation and agricultural development. *Food Policy* 2 (3): 196–216.

Ruttan, Vernon W. 1978. Institutional innovation. In *Distortions of Agricultural Incentives,* ed. T.W. Scults. Bloomington, Ind.: Indiana University Press.

Sanders, Heywood J. 1993. What intrastructure crises? *Public Interest* (Winter): 3–18.

Saxenian, AnnaLee. 1989. The Cheshire cat's grin: Innovation, regional development, and the Cambridge Case. *Economy and Society* 18 (4): 448–477.

Saxenian, AnnaLee. 1994. *Regional Advantage: Culture and Competition in Silicon Valley and Route 128.* Cambridge, Mass: Harvard University Press.

Schaeffer, Peter V., and Richard S. Mack. 1996. *The New International Division of Labor: An Appraisal of the Literature.* Paper prepared for presentation at the Regional Science Association 36th European Congress ETH Zurich, August 26–30.

Schaffer, William A. 1980. The Role of Input-Output Models in Regional Impact Analysis, pp. 156–167. In *Economic Impact Analysis: Methodology and Applications,* ed. Saul Pleeter. Boston: M. Nijhoff.

Schallau, Con H., and Wilbur R. Maki. 1983, Interregional model for analyzing the regional impacts of forest resource and related supply constraints. *Forest Science* 29 (2): 384–394.

Scott, A.J. 1983. Industrial organization and the logic of intrametropolitan location: I. Theoretical considerations. *Economic Geography* 59 (3): 233–250.

Scott, A.J. 1988. Flexible production systems and regional development: The rise of new industrial spaces in North America and western Europe. *International Journal of Urban and Regional Studies* 12: 171–186.

Sears, David W., and J. Norman Reid. 1992. Rural strategies and rural development research: An assessment. *Policy Studies Journal* 20 (2): 301–310.

Selznick, Philip. 1966. *TVA and the Grass Roots: A Study in the Sociology of Formal Organization.* New York: Harper & Row.

Senge, Peter M. 1990. *The Fifth Discipline: The Art and Practice of the Learning Organization.* New York: Doubleday.

Senior, Derek. 1966. The Regional City: *An Anglo-American Discussion of Metropolitan Planning.* Chicago: Aldine.

Shaffer, Ron. 1989. *Community Economics: Economic Structure and Change in Smaller Communities.* Ames, Iowa: Iowa State University Press.

Shigehara, Kumiharu. 1992. Policies for long-run economic growth, pp. 15–40. In a symposium sponsored by the Federal Reserve Bank of Kansas City, Mo.

Smith, Hedrick. 1995. *Across the River.* A Public Broadcasting System documentary produced by Hedrick Smith Productions, Inc., and adapted from Across the River press release, November 24.

Stanback, Thomas M. Jr., and Thierry J. Noyelle. 1982. *Cities in Transition.* Totowa, N.J.: Allanheld, Osmun.

Stenberg, Peter, and Wilbur Maki. 1996. *The Growth of Technology-Intensive Industry Clusters: The Case of Two Rural-Urban Economic Regions.* Paper prepared for presentation at the Regional Science Association 36th European Congress ETH, Zurich, August 26–30.

Stevens, Benjamin H., and Glynnis A. Trainer. 1980. Error generation in regional input-output analysis and its implications for nonsurvey models, pp. 68–84. In *Economic Impact Analysis: Methodology and Applications,* ed. Saul Pleeter. Boston: M. Nijhoff.

Stevens, Benjamin H., George I. Treyz, and Michael L. Lahr. 1989. On the comparative accuracy of rpc estimating techniques, pp. 245–257. In *Frontiers of Input-Output Analysis,* eds. Ronald E. Miller, Karen R. Polenske, and Adam Z. Rose. New York: Oxford University Press.

Stone, C.N. 1987. The study of the politics of urban development, pp. 3–22. In *The Politics of Urban Development,* C.N. Stone and H.T. Sanders, eds. Lawrence, Kansas: University of Kansas Press.

Stone, Kenneth E. 1997. Trade areas, pp. 711–714. In *Encyclopedia of Rural America,* ed. Gary A. Goreham. Santa Barbara, Calif.: ABC-CLIO.

Stone, Richard. 1961. *Input-Output and National Accounts.* Paris: Organization for European Economic Cooperation.

Storey, David J. 1982/1988. *Entrepreneurship and the New Firm.* London: Croom Helm; reprinted by Routledge.

Storey, David J. 1988. *Small and Medium-Size Enterprises and Regional Development.* London: Routledge.

Storey, David J. 1993. *Managerial Labour Markets in Small and Medium-Sized Enterprises.* London/New York: Routledge.

Storper, Michael. 1993. Regional worlds of production: Learning and innovation in the technology districts of France, Italy, and the U.S. *Regional Studies* 27 (5): 433–456.

Tiebout, Charles. 1962. *Community Economic Base Study. Supplementary Paper* No. 16. Committee for Economic Development. New York.

Tolbert II, Charles M., and Molly Sizer Killian. 1987. *Labor Market Areas for the United States.* Washington, D.C.: U.S. Department of Agriculture, Economic Research Service, Agriculture and Rural Economy Division.

Tolbert II, Charles M., and Molly Sizer Killian. 1996. *U.S. Commuting Zones and Labor Market Areas,* ERS Staff Paper Number 9614. Washington, D.C: U.S. Department of Agriculture, Economic Research Service, Rural Economy Division.

Treyz, George I. 1993. *Regional Economic Modeling: A Systematic Approach to Economic Forecasting and Policy Analysis.* Norwell, Mass.: Kluwer Academic Publishers.

Treyz, George, Dan S. Rickman, and Gang Shao. 1992. The REMI economic-demographic forecasting and simulation model. *International Regional Science Review* 14 (3): 221–253.

Turner, Frederick Jackson. 1920. *The Frontier in American History.* New York: H. Holt.

U.S. General Accounting Office. 1996. *Airline Deregulation: Changes in Airfares, Service, and Safety at Small, Medium-Sized, and Large Communities* (GAO/RCED-96-79). April 19.

Vance, Rupert. 1951. The regional concept as a tool for social research, pp. 119–140. In *Regionalism in America,* ed. Merrill Jensen. Madison, Wis.: University of Wisconsin Press.

von Hippel, Eric. 1988. *The Sources of Innovation.* New York: Oxford University Press.

von Thünen, Johann H. 1826. *Der Isolierte Staat ub Bieziehung auf Landwirtschaft and Nationalekonomic.*

Waters, Edward C., David W. Holland, and Richard W. Haynes. 1997. *The Economic Impact of Public Resources Supply Constraints in Northeast Oregon.* General Technical Report PNW-GTR-398. Portland, Ore.: U.S. Forest Service, Pacific Northwest Research Station.

Weber, Alfred. 1909/1929. *Ueber den Standort der Industries.* (Translated by C. J. Freidrick as *Alfred Weber's Theory of the Location of Industries.*) Chicago: University of Chicago Press 1929.

Weisskoff, Richard. 1985. *Factories and Food Stamps: The Puerto Rico Model of Development.* Baltimore: Johns Hopkins University Press.

Westeren, Knut Ingar. 1992. *A Comparison of Regional Policies in Some EC Countries with National Regional Policies in Norway.* Paper presented at 32nd European Congress of the Regional Science Association International, Louvain-la-Neuve, Belgium, August 25–28.

Winters, Jonathan. 1990. After the festival is over. *Governing* (August): 22–34.

Index

Index

Index

ISBN 0-8138-2679-9
90000
9 780813 826790

No Longer Property of
EWU Libraries